The Maryland
Master Naturalist's
Handbook

The *Maryland* Master Naturalist's Handbook

Edited by

McKAY JENKINS and JOY SHINDLER RAFEY

JOHNS HOPKINS UNIVERSITY PRESS BALTIMORE

© 2025 Johns Hopkins University Press
All rights reserved. Published 2025
Printed in the United States of America on acid-free paper
9 8 7 6 5 4 3 2 1

Johns Hopkins University Press
2715 North Charles Street
Baltimore, Maryland 21218
www.press.jhu.edu

Library of Congress Cataloging-in-Publication Data
Names: Jenkins, McKay, 1963– editor. | Rafey, Joy Shindler, 1964– editor.
Title: The Maryland master naturalist's handbook / edited by McKay
    Jenkins and Joy Shindler Rafey.
Description: Baltimore : Johns Hopkins University Press, 2025. | Includes index.
Identifiers: LCCN 2024035790 | ISBN 9781421451589 (paperback) |
    ISBN 9781421451596 (ebook)
Subjects: LCSH: Natural history—Maryland. | Maryland—Guidebooks.
Classification: LCC QH21.U5 M37 2025 | DDC 508.752—dc23/eng/20250206
LC record available at https://lccn.loc.gov/2024035790

A catalog record for this book is available from the British Library.

*Special discounts are available for bulk purchases of this book. For more
information, please contact Special Sales at specialsales@jh.edu.*

EU GPSR Authorized Representative
LOGOS EUROPE, 9 rue Nicolas Poussin, 17000, La Rochelle, France
E-mail: Contact@logoseurope.eu

C O N T E N T S

Becoming a Maryland Master Naturalist

McKAY JENKINS

Over the course of the last 40 years, I have formed a deep affection for the Susquehanna River, the majestic, primordial freshwater source of the Chesapeake Bay, itself one of the most fertile estuaries in the world. Like the Susquehanna, I began my life in New York, but ended up flowing south to Maryland—first to Annapolis, then to Baltimore, where my family and I have spent decades exploring the state's mountains, rivers, forests, and coastline. Over the course of these years and a career spent as an environmental journalist and professor, I have found myself moving deeper and deeper into our adopted place, not just ecologically but socially as well. Like all places, Maryland is a rich tapestry of ecology and history, its creatures and landforms and human communities interacting at levels of complexity it can take a lifetime to understand.

Maryland's Master Naturalist Program and this companion volume are designed to help us all see more deeply, to become more fully (and subtly) aware of the beauty and interconnectedness of our human and ecological communities. We live at a time when the effects of climate change—along with rising seas, droughts and floods, deforestation, species collapse—are becoming more ominous, and when our human communities often seem fragmented, isolated, and in conflict. It is thus a very good time to invest ourselves in the stewardship of the place we call home. "Being naturalized to place means to live as if this is the land that feeds you, as if these are the streams from which you drink, that build your body and fill your spirit," the Indigenous ecologist Robin Wall Kimmerer writes in *Braiding Sweetgrass*. "To become naturalized is to know that your ancestors lie in this ground. Here you will give your gifts and meet your responsibilities. To become naturalized is to live as if your children's future matters, to take care of the land as if our lives and the lives of all our relatives depend on it. Because they do."[1]

In this volume, you will learn from experts in a wide variety of fields— scientists, journalists, and environmental educators—who have made the observation, study, and rehabilitation of the natural world their life's work and passion. Here you will find wisdom on Maryland's geological foundations and its changing meteorology; the complex ecological systems formed by the Chesapeake Bay region's forests, rivers, and mountains (and the creatures who live in them); and the challenges raised by our state's complicated racial, cultural, and economic history.

As you read, you may notice some overlap in these chapters: it's impossible for an urban ecologist not to talk about both forests *and* poverty, for a journalist interested in clean water not to mention chicken farming, or for an insect expert not to talk about the ecological disaster of the suburban lawn. It is our hope that by reading this book—and (if you are able) taking the Master Naturalist training course for which this book serves as the primary text—you will deepen your understanding of historical and ecological systems that have made the Maryland landscape what it is today.

The training course itself has its roots in conversations that began as early as 2005, when University of Maryland Extension administrators saw the value and impact that similar programs were having in other states. With the support of the Maryland Department of Natural Resources, local Parks and Recreation departments, the US Fish and Wildlife Service, the Howard County Conservancy, the Natural History Society of Maryland, the Maryland

Association for Environment and Outdoor Education, and University of Maryland Extension, the Master Naturalist Program was developed and in 2010 trained the inaugural class of 14 Maryland Master Naturalists at the program's first host site, the Howard County Conservancy. By 2024, the program has grown to nearly 50 facilities and organizations serving as host sites throughout the state, with more than 200 people trained each year to serve as volunteer environmental stewards.

We hope that in reading this book, you will join the growing number of people exploring our state's many wonders; committing themselves to deepening their understanding of our ecological systems; and adding their hearts, minds, and bodies to our state's environmental stewardship.

Or, as I often find myself writing on my classroom chalkboards: First you learn. Then you love. Then you heal.

University of Maryland Extension programs, activities, and facilities are available to all without regard to race, color, sex, gender identity or expression, sexual orientation, marital status, age, national origin, political affiliation, physical or mental disability, religion, protected veteran status, genetic information, personal appearance, or any other legally protected class.

* * *

University of Maryland Land Acknowledgment

Every community owes its existence and strength to the generations before them, around the world, who contributed their hopes, dreams, and energy into making the history that led to this moment.

Truth and acknowledgment are critical in building mutual respect and connections across all barriers of heritage and difference.

So, we acknowledge the truth that is often buried: We are on the ancestral lands of the Piscataway People, who are the ancestral stewards of this sacred land. It is their historical responsibility to advocate for the four-legged, the winged, those that crawl and those that swim. They remind us that clean air and pristine waterways are essential to all life.

This Land Acknowledgment is a vocal reminder for each of us as two-leggeds to ensure our physical environment is in better condition than what we inherited, for the health and prosperity of future generations.

Note

1 Robin Wall Kimmerer, *Braiding Sweetgrass: Indigenous Wisdom, Scientific Knowledge, and the Teachings of Plants* (Minneapolis: Milkweed Editions, 2013), 214–15.

ACKNOWLEDGMENTS

The Maryland Master Naturalist Program is made possible by dedicated experts who generously devote time and talent as host site program facilitators and volunteer service coordinators; training instructors; field study leaders; and resource contributors—including for the chapters in this handbook. Without their support and engagement, this program would not exist.

Thanks to Bob Tjaden, professor emeritus and former University of Maryland Extension (UME) assistant director for environmental, natural resources, and sea grant programs, for having the vision to explore the possibility of Maryland establishing a Master Naturalist Program. Thanks also to Bill Hubbard, current UME assistant director and state program leader for environmental, natural resources, and sea grant programs, for providing continuing, robust support.

Even with their vision and support, this program would not exist without the tireless efforts of Wanda MacLachlan, Maryland Master Naturalist program coordinator from 2005 to 2022, and resources from the Alliance of Natural Resource Outreach and Service Programs.

Joy Shindler Rafey wishes to thank Lisa Marini for her comprehensive review of the Maryland Master Naturalist Program's resource manuals, which set us on the course toward this new handbook, and McKay Jenkins for making the handbook a reality. Both Lisa and McKay are Maryland Master Naturalists and represent the breadth of talent and depth of commitment that thousands of Marylanders contribute through this program.

Letha Grimes would like to acknowledge Dr. Joseph W. Love, statewide operations manager, Fishing and Boating Services, Maryland Department of Natural Resources.

Morgan Grove and Steward Pickett acknowledge support from the Baltimore Ecosystem Study, NSF DEB 1855277. The findings and conclusions in this article are those of the authors and should not be construed to represent any official USDA or US government determination or policy.

McKay Jenkins would like to thank the University of Delaware for its support of a sabbatical leave to complete work on this project. Thanks also to the University of Delaware Press for permission to reprint Douglas Tallamy's chapters from *The Delaware Naturalist Handbook* (edited by McKay Jenkins and Susan Barton, 2000). Thanks to Tiffany Gasbarrini and Ezra Rodriguez at Hopkins Press for all their guidance; to Lia Purpura and Tim Wheeler for their editorial eyes; and to Joy Rafey for being such a pro. And thanks, as always and for everything, to Katherine, Steedman, and Annalisa.

David Ruppert would like to thank the faculty of the University of Maryland for discussions and his Maryland-based soils education, in particular, Martin Rabenhorst, Raymond Weil, and his thesis advisor, Brian Needelman. Dr. Weil's book, now in its 15th edition, remains the definitive general soils textbook. Thanks also to the faculty and students at Texas A&M University–Kingsville for the opportunity learn about soils in the southern Plains; and thanks to Natural Resources Conservation Service employees in Texas (Gary Harris, Jim Akin) and Maryland (Phil King, Annie Rossi Gill, Ben Marshall). NRCS staff are generous public servants, providing the United States with invaluable information about the all-important soils resource.

Angela Yau dedicates her chapter to the memory of Nat Frazer, her graduate advisor, who embodied service and dedication to resource conservation through communication.

The Maryland
Master Naturalist's
Handbook

CHAPTER 1

Maryland Land Use History

McKAY JENKINS

Once a semester for the past 30 years, I have led my college students on ca-
noe trips down the Susquehanna River, the main trunk of the Chesapeake
Bay and the longest river east of the Mississippi. The Susquehanna offers
stunning displays of beauty: one afternoon, our little group had just rafted
together in a narrow stretch of the river when a bald eagle flew immediately overhead, a fish dangling from its talons. Before we could reach for
our binoculars, a *second* bald eagle dive-bombed the first, and the two—
leaning backward in midair, their talons extended—battled over the fish.
In what must have been some weird sort of karma, the squabbling birds
dropped the fish back into the water, and away it swam.

The Susquehanna also provides a powerful opportunity to think in
multidimensional ways about the evolving relationships between people
and their place in the world. First, there is the mind-bending sense of geological time. The Susquehanna is one of the oldest rivers on the planet; my
students and I often break for lunch on outcroppings of gneiss that are
hundreds of millions of years old. The ridges rising above the banks evoke
the height of a river that, roughly 18,000 years ago, drained a mile-thick

sheet of ice that covered eastern North America from Pennsylvania all the way to Quebec and Ontario. The melting ice carved the Susquehanna, which then carved the Chesapeake itself. Although the Bay itself averages a depth of just 21 feet (and 40% of it is less than 6 feet deep), parts of the Susquehanna gorge running down the middle of the Chesapeake still reach down 174 feet. The river delivers some 9 trillion gallons of fresh water to the Bay every year—about half of all the water in the entire Chesapeake.

And then, inevitably, there are the indelible exhibits of human impact. A river that was once jammed bank-to-bank by trees clear-cut from Pennsylvania's forests has—for many decades—been choked by four hydroelectric dams, which have created carbon-free electricity but also destroyed a once legendary shad run. Perhaps more ominous: backed up behind these dams are hundreds of millions of tons of sediment—more than 185 million tons behind the Conowingo alone—the result of soil washing off Pennsylvania farms. Much of this soil is contaminated with agricultural herbicides and pesticides.[1]

The landing where we put our boats in the river is just downhill from a natural gas pipeline and not far from an intake that provides cooling water for the Peach Bottom nuclear reactor. The launch is downriver from the Holtwood Dam and upriver from the Conowingo Dam, which, when it was built in 1928, was the largest power plant in the world (see figure 1.1). Sediment along the riverbanks still contains visible coal dust left over from Pennsylvania's anthracite mines, and the water itself is glutted with nitrogen-saturated runoff from Lancaster County farms. A quarry on the south side of the river—viewable from the Interstate 95 bridge—still mines the rock that was used to build Fort McHenry, Goucher and Haverford Colleges, and the US Naval Academy.

Also below Holtwood—and hidden from view by Lower Bear Island—is yet another industrial plant, this one pumping water straight up a steep ridge to a 100-acre lake nestled inside the Muddy Run Recreation Park. The facility (and the park) is operated by Constellation Energy, one of the largest energy companies in the country. The man-made "lake" is actually a "pumped storage facility": at night, water is pumped uphill from the

FIG. 1.1. The Conowingo Dam, once the largest energy-producing facility in the world, remains a source of environmental debate. It produces renewable electricity but also impedes migrating fish. Photo courtesy of David Harp.

Susquehanna into the reservoir, which holds some 60,000 acre-feet of water. During the day, when energy demand and prices are higher, the company releases the water into its hydroelectric turbines. Every 24 hours the reservoir rises and falls by 80 feet—a dramatic boom-and-bust cycle that (had we been able to examine it more closely) would no doubt have revealed an aquatic habitat in crisis.[2]

The environmental history of Maryland and the broader Chesapeake Bay region mirror the history of the Susquehanna. They all bear the marks of forces both cataclysmic and subtle, both destructive and regenerative, both "natural" and man-made. Tectonic collisions can thrust entire mountain ranges into the air, Martin Schmidt reminds us in his chapter on geology, and millions of years of rainfall can wash these same mountains into the sea. These cycles have also left us with deeply fertile farmland, robust aquatic and bird life, and dense forests that support a great diversity of life.

At the human scale, as we see in this volume's environmental justice chapter, we find similar forces: cataclysms like Indigenous genocide and human enslavement, followed by centuries of fits and starts of damage and repair: industrial pollution, suburban deforestation, and global-scale agriculture, but also burgeoning commitments to reforestation, toughening pollution standards, and a resurgence of local food production.

Some of the changes we are now struggling to address have evolved over decades (or centuries) and have left us with highly complex and intertwined social and ecological challenges.

Consider the most recent period, during which most of us have grown up: the Era of Suburban Sprawl. In 2011, state planners released a major report on the state of Maryland's land use and drew some daunting conclusions. While it took three centuries to develop the first 650,000 acres in Maryland, more than a million acres had been paved over in the previous 50 years. In the previous half century, Maryland had lost 873,000 acres of farmland and nearly 500,000 acres of forest. Across the Chesapeake region today, an area the size of Washington, DC, is covered in pavement and buildings every five years.[3]

The consequences of this change have been legion. The average Marylander now has one of the longest commutes in the country. "Bay cleanup goals pay lip service to population growth, but essentially accept it," Tom Horton, the veteran Chesapeake journalist and essayist, writes in the *Bay Journal*: "Maryland and Virginia absolutely embrace it. Between 2010 and 2020, Maryland added 385,000 people to the Bay watershed and Virginia a whopping 630,000. And a lot of Virginia's people boom occurred on forested lands, with 60,000 acres lost in the most recent four years of record. Population growth is listed almost nowhere officially as a source of pollution or as a threat to Bay health. Almost everywhere, governments treat only its symptoms: sewage, paving, the loss of wetlands and forest, and air pollution."[4]

Close to another million people are expected to move to the region in the next couple of decades, and these new residents—and the 500,000 homes they are likely to demand—will add enormous pressure on natural and man-made systems. By 2035, the state is projected to have lost another 226,000 acres of farmland and 176,000 acres of forest.[5]

And all those deer munching their way through our forests? Populations that roamed the dense woods of pre-colonial Maryland were virtually wiped out by the late 18th century and remained close to local extinction for 200 years. It was only when the state's farms stopped growing crops and started growing houses that the deer population exploded. Today there are roughly 250,000 deer in the state, devouring tree seedlings wherever they roam. Although hunters killed nearly 82,000 deer in the 2020–2021 season, there are still between 30,000 and 35,000 car-deer accidents a year, a result of too many deer, too many roads, and too many cars. All of which have come from too much development.[6]

And our beloved lawns? Since the mid-20th century, Americans have become obsessed with grass (see figure 1.2). If you add up the country's 80 million home lawns and nearly 16,000 golf courses, you get close to 50 million acres of cultivated turf in the United States. This space is growing by 600 square miles a year. By 1999, more than two-thirds of America's home lawns had been treated with chemical fertilizers or pesticides—14 million of them by professional lawn-care companies, I note in my book *ContamiNation* (later excerpted in *Reader's Digest*). Americans are spraying 67 million pounds of synthetic chemicals on their grass every year, and annual sales of lawn-care pesticides have grown to $700 million. "All those trucks rolling around our suburban neighborhoods seem to represent something more than a communal desire for soft carpets of monoculture grass. They seem to represent a relief from anxiety. (Why else call a company 'Lawn Doctor'?) But anxiety from what, exactly?"[7]

The double insult of cutting down forests and replacing them with suburban sprawl—with all the pavement and pesticides that such development implies—has had a devastating impact on bird habitat, for example. Cornell University's David Pimentel has estimated that, nationwide, some 72 million birds are killed each year by direct exposure to pesticides. And logging, industrial agriculture, and other landform changes have led to a "staggering" loss of some 3 billion birds since 1970.[8]

FIG. 1.2. Expansive lawns—a common feature of suburban landscaping—are ecologically bereft. It is far better to emphasize native plants and trees, which are superior for supporting food webs and absorbing stormwater. Photo by Douglas Tallamy.

Another issue: lawn chemicals are engineered to preserve grass but kill "weeds," like clover. Clover does not deserve such disrespect. Unlike grass, clover pulls nitrogen out of the air and fixes it in the soil, so without plants like clover, soil—and the plants embedded in soil—can become starved for nitrogen. Chemical companies know this and are quite happy to sell you *synthetic* fertilizer to replace the depleted nitrogen, which homeowners used to get (for free) from clover. Unfortunately, synthetic fertilizers are water soluble, so they tend to run off our lawns after a rain and continue to do what they were designed to do: feed plants. But now they feed *aquatic* plants (especially algae) in our rivers and bays, where algae "blooms" explode, quickly die, and—during decomposition—suck so much oxygen from the water they can create "dead zones" big enough to see from outer space.

And climate change? With a long coastline and split by the enormous Chesapeake Bay, Maryland is especially vulnerable to sea level rise. The Blackwater National Wildlife Refuge is already at risk, and half of Dorchester County may be underwater by 2100, the *Washington Post*'s Scott Dance reports.[9] With increasing rainfall, worsening storms, and rising seas, civil infrastructure is under increasing risk of failing. The *New York Times* recently reported that one in nine residents of the lower 48 states, "largely in populous regions including the Mid-Atlantic and the Texas Gulf Coast, is at significant risk of downpours that deliver at least 50 percent more rain per hour than local pipes, channels and culverts might be designed to drain."[10]

Indeed, much of this is worrisome news for a state that is already the fifth-most densely populated in the country. And while the last few years have moved the needle on things like wind and solar power and electric cars, Maryland Master Naturalists are still left with fundamental questions that writers (me included) have wrestled with for years. How do you get Marylanders to understand that "virtually every person in the state lives within a single living, complex watershed, and that every decision we make—from the houses we buy and the places we shop, to the length of our commutes and the way we eat—has an effect not only on the quality and cost of our lives but on the fundamental systems on which our lives depend?"[11]

Tom Horton considers these questions not just ecological, but existential. "Carl Jung, the great psychoanalyst, once said he had never been able to cure a patient who did not have a firm belief that he or she was part of something larger," Horton writes in his book *Bay Country*. "Thus we need our religions, our cosmologies, and equally, I think, a greatly expanded appreciation of all the ways in which we and nature fit together."[12]

In this chapter we'll take a look at the ways people have lived on, used, damaged, and begun to repair the place we call home.

Indigenous History

As we will explore more fully in the chapter on environmental justice, the moment that Europeans arrived in North America, they ushered in not just a new culture but an entirely new vision of the relationship people have with their land. Whatever their reasons for exploring the mysterious world beyond their own shores, the Europeans who first hit North America realized this continent had—in incredible abundance—what Europe had lost, or exhausted, or destroyed: fantastically fertile soil, infinite forests, marine life so abundant that early settlers said they could walk across the Bay on islands of oyster beds. This fabulous plenitude made the newcomers see not just a place to settle down, but a series of "resources" to extract and sell.

Here was an entirely new way of thinking about the relationship between human beings and the land: where Indigenous communities—100 million strong in the Americas—lived within their systems, Europeans lived outside them. The newcomers looked in, liked what they saw, and began looking for ways to pull things out and sell them. All they needed to do was clear the land of its ancestral inhabitants, then find the cheapest labor possible to work the land. Timber, for example, was considered "the oil of the seventeenth and eighteenth centuries, and any shortages created similar anxieties about fuel, manufacturing and transport," writes the journalist and historian Andrea Wulf. "Without trees there would be no iron and glass industries, no blazing fires warming homes during cold winter nights, nor a navy to protect the shores of England."[13]

Conveniently, Europeans also had centuries of philosophy to support their extractive forays into the wider world. From Aristotle forward through the Enlightenment, European thinkers (taking their cues from a decidedly extractive interpretation of the Bible) had converged on the idea that everything in nature was (in the words of Thomas Aquinas) "made for the sake of man."[14]

Starting in the 18th century, the idea that people could "perfect" nature by engineering it to their own purposes dominated Western thinking. Geometrically ordered farms, clear-cut forests, dammed rivers—the control of nature became both a source of profit and a powerful (and resilient) aesthetic standard. The primeval forests of the New World, by contrast, represented "a 'howling wilderness' that had to be conquered," Wulf writes. "Chaos had to be ordered, and evil to be transformed into good."[15]

These standards—nature engineered for profit and refashioned to satisfy a European sense of beauty—hit the North American continent like a meteor, and we have been negotiating the impact ever since. The European economic system, based on extractive economics and privatized resources held by a very small ruling class, was already in place before the first ships sailed for the Americas. Denied access to land and prosperity in their home countries, Europe's indentured poor viewed the New World as an "escape valve" from their own economic oppression, notes the historian Roxanne Dunbar-Ortiz.[16]

While indentured laborers were looking for promises of work and land in exchange for safe passage across the Atlantic, Europe's already wealthy landowners arrived looking for expanded sources of profit. Europe itself had long since become ecologically wrung out—soils exhausted, forests chopped down—so when colonial profiteers laid eyes on North America, they were stunned. They did not see the continent's primordial forests as redolent with ecological and spiritual vitality; they may not even have seen the forest as a collection of "trees." What they saw was "timber," and they moved in quickly to cut it down. Likewise, over the years, for the exploitation of mineral deposits, fertile soil, fish and deer and ducks and geese. As a haunting proverb (variously attributed to the Cree and the Abenaki, both Indigenous peoples in what is now Canada) reminds us: "Only when the last tree is cut down, the last fish eaten, and the last stream poisoned, will you realize that you cannot eat money."[17]

Of course, there had been a lot going on here before the Europeans arrived. It may be hard to fathom, but human beings and their ancestors have only occupied the Earth itself for about 300,000 years—or about 0.007% of the Earth's 4.6 billion years.[18] Many millennia before Captain John Smith sailed up the Chesapeake in 1607–1609, the upper reaches of what is now the Chesapeake Bay watershed were covered in ice that was thousands of feet thick. There was so much ice tied up in the world's glaciers that the world's oceans were 300–400 feet lower than they are today. Locally, the state's ocean coastline reached nearly 200 miles farther into the Atlantic.

Before there was a Chesapeake Bay, there was just the ancient Susquehanna, winding down through what is now Pennsylvania before taking a hard right by what is now Havre de Grace, then cutting a long channel to the sea. As the glaciers melted and drained into the deepening river, fertile lands emerged and became covered in mixed hardwood forests dense with everything from bears, bison, and beavers to prehistoric mammoths and mastodons.

Early Indigenous communities lived in villages along the banks of the region's rivers, often traveling hundreds of miles in the frigid temperatures to hunt, fish, and gather food. Coastal geologist Darrin Lowery has found 13,000-year-old stone tools buried in soil at Paw Paw Cove on Tilghman Island, including spear points that "might very well have been stuck into the side of a mastodon."[19]

By 1000 BC, Maryland was already home to some 8,000 people in 40 different tribes, according to the Chesapeake Bay Program. With glaciers long since melted and temperatures far more temperate, people found bountiful food supplies in emergent oyster beds, massive shad and striped bass runs, and expansive marshlands. The word *Chesapeake* had long been thought to derive from the Algonquian word for Great Shellfish Bay, though Algonquian linguist Blair Rudes says it is closer to "great water"—and may have originally referred to a village at the Bay's mouth.[20]

The region's Native communities harvested an abundance of fish and shellfish, cultivated agricultural food crops and mast from the region's nut trees, and hunted the region's diverse forests; by periodically using fire to clear the forest understory, deer and turkey became staples. The "three sisters"—corn, beans, and squash—that had long been staples of Indigenous cultures on the continent's desert southwest didn't arrive in the region's piedmont and coastal plain until perhaps 1100 AD, in part because there were so many other local sources of food.

The region's people mostly spoke dialects of Algonquian. Among the historical tribes, the largest in what is now Maryland were the Accohannock, the Assateaque, the Choptank, the Lenape, the Matapeake, the Nanticoke, the Piscataway, the Pocomoke, and the Shawnee. Tribes often created allegiances—referred to as *confederacies* or *nations* by Europeans—for protection and trade; the largest, organized under Chief Powhatan, stretched from Maryland south to the Carolinas.[21]

Contrary to the European approach to "natural resources," a highly reverent and subtle understanding of the relationships between people and their ancestral lands was (and remains) fundamental to Indigenous spiritual life, the historian Vine Deloria writes in *God Is Red*.

> The task of the tribal religion, if such a religion can be said to have a task, is to determine the proper relationship that the people of the tribe must have with other living things and to develop the self-discipline within the tribal community so that man acts harmoniously with other creatures. The world that he experiences is dominated by the presence of a power, the manifestation of life energies, the whole life-flow of creation. Recognition that the human beings hold an important place in such a creation is tempered by the thought that they are dependent on everything in creation for their existence.[22]

Indeed, there was so much bounty in the land that tribes rarely came into conflict over resources, wrote Chief Quiet Thunder, a 21st-century leader of New Jersey's Lenni-Lenape tribe. Especially during times of scarcity, communities were flexible enough to follow game, seasons, or weather. "The European concept of 'home' was that you lived where you built your cabin, your barn, your outbuildings, and where you had your garden, livestock, and pastures," Quiet Thunder wrote.

> Everything was fenced in, and they called it "real estate." That was their concept, where the Lenape concept was so different. We believed that land was a gift from the Creator for all to share, just as the air, sunshine, and water. We did not stay in one spot forever. We followed the harvest through the seasons in several locations. In the spring, we traveled from the winter villages near the Delaware River and Bay area for the fish migration. Later on, in the summer, we walked to the seashore to harvest fish, crabs, clams and oysters. In the winter, we came back to our winter villages and camps in the forest. In that way, we gave Nature a chance to rejuvenate itself. The Europeans' concept was that life must be stationary, but we moved around.[23]

It was only when Europeans arrived and began forcing people off their ancestral lands—when sacred land was reconceived of as a source of profit—that extraction, displacement, and war became an existential threat to Indigenous life. As elsewhere, the region's Indigenous people had little immunity to European illnesses, especially smallpox; in some places, Europeans knowingly traded blankets infected with this scourge as a kind of germ warfare. Remaining populations either were absorbed into other tribes, assimilated with Europeans, or fled for their lives. The trauma and grief this caused to communities and their ancestral lands left an indelible imprint on the region's (and the continent's) landscape that remains to this day.

Yet in places around the Chesapeake, Indigenous communities remain vibrant. As of the 2020 Census, more than 128,000 people with American Indian heritage reported living in Maryland, though—as my friend Dennis Coker, who has served as chief of the Lenape Indian Tribe of Delaware for 25 years, likes to say—many have had to become expert at "hiding in plain sight."[24]

In 2012, the state gave formal recognition to the Piscataway-Conoy Tribe (which includes the Cedarville Band of Piscataways) and the Piscataway Indian Nation; later, state recognition was also give to the Accohannock.[25]

Elsewhere in the state, smaller communities, like the Nause-Waiwash in Dorchester County, continue to maintain tribal traditions. Other communities, particularly those with roots in the Deep South, fled the violence of Jim Crow segregation and moved into cities like Baltimore, swapping sharecropping for jobs in factories, construction, and the service industry. Members of North Carolina's Lumbee tribe—the largest tribe east of the Mississippi and the ninth largest in the country—have long lived in Baltimore's Upper Fells Point and Washington Hill neighborhoods; by the 1960s, the *Smithsonian* magazine reports, "there were so many Native Americans living in the area that many Lumbee affectionately referred to it as 'The Reservation.'" By the early 1970s, however, these neighborhoods had been largely razed in the name of urban renewal, and "almost every Lumbee-occupied space was turned into a vacant lot or a green space," Ashley Minner, a Baltimore community artist, educator, and enrolled member of the Lumbee Tribe of North Carolina, told the *Smithsonian*. The population of The Reservation dropped off, and today—like members of other tribes living in urban areas across the country—Lumbee often suffer from cases of "mistaken identity," Minner said.

"I've been called Asian, Puerto Rican, Hawaiian—everything but what I am," she told the *Smithsonian*. "Then you tell people that you're Indian, and they say, 'No, you're not.' We run the gamut of skin colors, eye colors and hair textures. When the Lumbee came to Baltimore, Westerns were all the rage. But we didn't look like the Indians on TV."[26]

Slavery

Once early colonists had forced Indigenous people off their land, they set about changing the land itself. Just 20 years after Europeans touched down in Jamestown, Virginia, England's Calvert family secured a charter from King Charles I granting them 10 million acres north of the Potomac River, the National Park Service's Ethnography Program reports; the Maryland colony would be named from the king's consort, Henrietta Maria.[27]

Over the next 200 years, settlers proceeded to make their new land resemble their old land. They chopped down forests, dammed rivers, filled wetlands, built forges and canals and railroads, and established a plantation economy that—given the previously unimaginable scale of the land itself—required a tremendous amount of human labor. Like all employers everywhere, wealthy colonial landowners went looking for the cheapest workers possible. At first, these landowners were able to expand both their wealth and their political power by exploiting the labor of indentured (that is to say, poor) white Europeans. But as their land holdings grew, Maryland's landowners began forcing enslaved Africans to work the land.

The Calverts owned several plantations, enslaved hundreds of people, and controlled Maryland politics for decades (the state flag reflects some of this troubled history: yellow and black are the colors of the paternal family of George Calvert, the first Lord Baltimore; the red and white represent Calvert's mother's side, the Crosslands). In 1663, under Governor Charles Calvert, the Maryland Assembly ruled that enslaved people and their children were to be enslaved for life.[28]

Meanwhile, in 1642, under George's son Leonard, who became Maryland's first colonial governor, the Native Susquehannock "were officially declared enemies of the Maryland province and were to be shot on sight if seen in colonial territory."[29]

When it came to enslaved labor, what started as a trickle quickly became a flood. While fewer than 1,000 Africans arrived in Maryland between 1619 and 1697, nearly 100,000 were forced ashore during the 75 years prior to the American Revolution. By 1755, fully one-third of Maryland's population—in some places as much as one-half—had African roots. By this time (and, one might say, ever since), Maryland became—in the words of the state archives—"as much an extension of Africa as of Europe."[30]

In Baltimore County, some 400 slaves worked 25,000 acres at the Ridgely estate, not more than three miles from where my wife Katherine and I raised our family. Our son Steedman devoted his 133-page college honors thesis to bringing the story to light. He discovered that the plantation was one of the larger slave operations in the region, with enslaved workers (along with around a dozen white indentured servants) also running the Northampton iron furnace, "the primary source of the Ridgely's fortune and a main supplier of iron to the Continental Army during the American Revolution."[31]

Yet, he found, more than a hundred years after the end of the Civil War, a culture of nostalgia about the plantation system remained intact—not just in the Deep South but in our own mid-Atlantic suburban backyard. In the mid-1970s, a *New York Times* story was still waxing romantic about the plantation's sense of "cosmic order that it still dominates the top of its mild Maryland hill and gives the viewer a powerful sense of another time and another truth. . . . When you get to the edge of the lawn, stand still and look back to the beautiful house behind you. Give it a minute, and you'll begin to feel the Georgian assumptions of order and stability soothe away your twentieth century frustrations."[32]

Steedman's reaction? The plantation's "cosmic order" was in fact "an order of master and slave, [and] inherently *un*stable. These assumptions are the original source of many of our twentieth and twenty-first century frustrations, not the antidotes to them."[33]

It wasn't until 1948, with Baltimore County beginning to succumb to suburban development, that John Ridgely Jr. turned the mansion property over to the National Park Service.[34]

Especially when compared to the small-scale, regenerative agriculture practiced by Native Americans, the extractive plantation economy—and especially tobacco—was devastating to the land itself. Wealthy plantation owners may have lounged in luxurious homes, but enslaved workers—the people who actually worked the land—were forced to live in small units called "quarters" and had to keep moving because the repeated planting of monoculture crops burned out soil health. In the state's northern and western counties, smaller farms planted with more diverse crops required fewer enslaved workers, but in the state's eastern and southern counties, tobacco (and slavery) remained the root of the region's economy for years.

As profitable as it became for the few white landowners who controlled it, the country's plantation economy was also ruinous for the laborers, especially enslaved Africans and their offspring. By the eve of the American Revolution, 90% of the colony's enslaved population were native born, the state archives report. Plantation work isolated and destroyed families, and enslaved workers were exposed to deadly diseases to which—like the Indigenous people who had been devastated before them—they had no immunity.

Then, in the early 19th century, the economics of slavery in Maryland began to change, as cotton production—which required even more enslaved labor—began to take hold farther South. By 1850, slave dealers in the Mississippi and other Deep South states were offering between $1,200 and $1,600 for male field workers, and Maryland's slave owners saw a new source of profits in the sale of their enslaved workforce. Between 1830 and 1860, they sold some 20,000 people to southern cotton planters, devastating Black families on an even wider scale. "Sales south shattered approximately one slave marriage in three and separated one fifth of the children under fourteen from one or both of their parents," the state archives report.[35]

By 1860, Maryland had become home to large numbers of free African Americans; Baltimore's Black population alone had grown to 27,000, over 90% of whom were legally free. Abraham Lincoln delivered his Emancipation Proclamation in 1863, and the following year, Maryland ratified a new constitution prohibiting slavery. But it took until the year 2000 for Maryland's state legislature to create a Commission to Coordinate the Study, Commemoration, and Impact of Slavery's History and Legacy, and seven years beyond that for the legislature to pass a resolution expressing "profound regret for the role that Maryland played in instituting and maintaining slavery and for the discrimination that was slavery's legacy."[36]

Industrial History

When Captain John Smith first laid eyes on the "great red bank of clay" flanking Baltimore's Federal Hill in 1608, he could hardly have imagined the extent to which this material would come to define the city. By 1793, a visitor to Baltimore noted that "this town is built chiefly of Brick," and as early as the 1830s, the city was already producing (for construction and export) some 32 million bricks a year.[37]

Sure enough, as Ned Tillman relates in his excellent book *The Chesapeake Watershed*, Baltimore became architecturally defined by brick, and fortunes were made shipping brick to Boston, New York, Philadelphia, and England. Mining companies in Maryland and Virginia made fortunes pulling sedimentary deposits from the coastal plain and turning them into iron ore in furnaces in Baltimore, Cecil, and Anne Arundel Counties. And since forges required waterpower—to run their bellows, to drive the hammers that turned pig iron to bar iron, and to form iron into nails or sheet metal—furnace owners dammed rivers and built riverside water wheels along the state's fall line.[38]

Eighteenth- and 19th-century miners used the region's abundant quartzite, serpentine, gneiss, granite, and gabbro to build roads, bridges, and buildings. Marble mines—some of which are still in operation in Cockeysville and Marriotsville—provided the stone for both Baltimore rowhomes and the columns of both the US Capitol and Baltimore City Hall (see figure 1.3). Cockeysville marble—pulled from a deep mine still visible near the state fairgrounds—was used to build virtually all of the 555-foot-tall Washington Monument in DC.[39]

Another source of mining wealth with a more complicated history, chromite—pulled from some of the country's richest deposits of serpentine in Baltimore County's Bare Hills and Soldier's Delight—was used from the 1840s to the 1980s to make chemicals, paints, and dyes. For many decades, Allied Signal's 27-acre Chrome Works was the world's largest processor of chrome in the world. It was only later that chromium processing was

FIG. 1.3. The Beaver Dam Quarry in Cockeysville was mined during the 19th and early 20th centuries for its dolomitic marble, known to geologists as "Cockeysville Marble." The marble was used in constructing both the Washington Monument in Baltimore and the one on the national mall in Washington, DC. Beaver Dam Quarry was flooded in 1934 and became a swimming hole. Artist unknown. Courtesy of the Maryland Center for History and Culture [CC1003].

discovered to have resulted in "widespread chromium contamination" in Baltimore's Inner Harbor and that hexavalent chromium compounds "are considered carcinogenic to workers" and can lead to "increased rates of lung cancer mortality," according to the federal Occupational Safety and Health Administration. Cleanup of the site required a decade and some $110 million.[40]

Major industries like Bethlehem Steel dominated waterfront properties around the region for many decades (see figure 1.4). Ned Tillman's own father spent his entire working life at the American Smelting and Refining Company on Clinton Street in East Baltimore. "From this and other steel, copper, and chromium plants scattered around the harbor came the emissions, the waste, and the by-products of manufacturing," Tillman writes.

FIG. 1.4. The Bethlehem Shipbuilding Corporation's shipyard at Sparrows Point, Maryland, southeast of Baltimore, 1940. Photo by Robert Kniesche. Courtesy of the Maryland Center for History and Culture [PP79.324].

Plants were built right on the harbor to more easily receive raw materials brought in by ship and rail. Their products were shipped out from these same docks. Gaseous by-products were emitted into the air—no smokestack scrubbers in those days. Acids, stormwater runoff, and gray water were discharged into settling ponds or holding tanks or just down storm drains or sewers. Solid waste was used to fill in low spots that were once vibrant wetlands, or shipped off-site to landfills. Much of the residue ended up in the bay.[41]

Since the 1960s, thanks largely to advances in environmental science, public activism, and signature laws like the federal Clean Water Act and the Clean Air Act—both the state and the country have moved a significant distance in mitigating industrial pollution. And thanks to probing environmental journalism, especially in newspapers like the *Baltimore Sun* and the *Washington Post* and nonprofit outlets like the excellent *Bay Journal* (and, more recently, the *Baltimore Banner*), the Maryland public has become far more aware of the damage caused by these and other legacy industries. The Chesapeake Bay Foundation, the region's most visible environmental organization, makes a public splash every year with its release of an environmental "report card" on the health of the Bay. In 2023, despite decades of efforts, CBF still gave the Bay a grade of D+—a reflection of just how difficult it has been to repair the ecological damage done by centuries of the magnifying pressures of extractive industries, an ongoing reliance on fossil fuels, and explosive population growth.

Of course, not all of Maryland's environmental challenges are homegrown. As beautiful (and comparatively unburdened by urban development) as it remains today, the Susquehanna River itself remains fundamentally industrial, both downstream in Maryland and upstream in Pennsylvania. Consider, for example: 90% of Pennsylvania was fully forested when William Penn arrived in 1692. Settlers soon began mowing down the forests for fuel and building materials, but they also removed trees to clear land for farming. By the mid-1800s, the timber industry grew exponentially, and the Susquehanna (as well as the Delaware and Allegheny Rivers) were used as pipelines from the forests to the sawmills. For a time, a sawmill in Williamsport, Pennsylvania, on the west branch of the Susquehanna, created "more millionaires per capita than anywhere in the country," the *Bay Journal* reports; in 1875 alone it produced 190 million board feet of white pine lumber.

Slash left by the clear-cutting of forests was also vulnerable to wildfires; soils washed downstream to the Chesapeake Bay, and ecosystems became so ecologically denuded they were unable to regenerate themselves. By the 1920s, Pennsylvania's forests had dropped from covering 90% of the state to just 30%. Soon afterward, a "tree army" hired by the Civilian Conservation Corps began repairing the damage by planting 50 million trees, and Pennsylvania's forests are now back up to covering 60% of the state. But as in Maryland, these forests face an explosion in invasive species, deer populations, and other impediments to natural regeneration. The state still annually exports some $1.2 billion of hardwood lumber and wood products, the *Bay Journal* reports.[42]

In the 20th century, the Susquehanna became an "energy river." Where tens of millions of shad once migrated up the river all the way from the Atlantic, their runs were essentially destroyed by the construction of four hydroelectric dams, most famously the Conowingo, the only one of the four dams in Maryland. At 4,648 feet long and rising 108 feet above the riverbed, the Conowingo captures enough power to drive seven turbines, second in power only to Niagara Falls. My students and I have toured the inside of the Conowingo; it is somehow even more stunning from the inside than it is from the outside. In the 1990s, the power company Exelon spent millions to insert fish elevators in three of the Susquehanna dams and a fish ladder on the fourth; when it was completed, the lift on the Conowingo was the largest of its kind in the country. Still, no one considers these projects a success. Only a vanishing few fish make it past all four dams each year. And only a tiny handful make it back to the sea. In 2015, for example, more than 8,300 shad were counted at the Conowingo dam. How many were counted at the York Haven dam, farther upstream? Forty-three. The year before, the number was eight.[43]

Today, the Conowingo stands as a monument to environmental complexity: it generates carbon-free electricity, but it has compromised virtually every element of the river's ecological integrity. It prevents sediment from cascading into the Chesapeake from Lancaster County, Pennsylvania, but fears remain about what might happen if a major storm causes that sediment to release downstream. River enthusiasts always push for dam removal, but what happens to a hundred years (and, reportedly, more than 185 million tons) of agricultural sediment behind the Conowingo is a legal, environmental, and engineering problem that no one seems to know how to solve. Blowing the dam would—just for starters—bury the downstream Susquehanna Flats, a broad stretch of open water where decades of work has been done to reestablish aquatic grasses vital to waterfowl and fish alike.

At a smaller scale, on Baltimore's Patapsco River, ecologists believe the removal of the century-old, obsolete Bloede Dam led to a sharp rebound in the numbers of American eels making their way upstream. Before the Bloede Dam came down in 2018, scientists would capture just over two dozen eels a year. Two years later, they captured 361. A year later, they caught more than 3,000, and a year after that, more than 36,000.[44]

Suburbia

Even by the late 1930s, state land-use planners were already complaining that the arrival of the automobile—and the thousands of miles of roads they required—had begun to turn Maryland into a series of "miserable stringtowns" that gobbled up the state's farms and forests. This trend was massively exacerbated in the middle of the century by the completion of the Interstate Highway System and its related feeder roads: with feverish cheerleading from real estate developers, county officials spent decades pushing people to abandon cities and towns where infrastructure (like roads, sewage systems, schools, and hospitals) already existed, and move into new homes built "out in the country." Almost overnight, it seemed, the state's former forests and farms were turned into subdivisions. As the saying goes, farms no longer grew food, they grew houses.[45]

For years, large-scale deforestation has had scientists worried about everything from climate change to the land's diminished ability to filter drinking water. Trees now cover roughly 3 million acres in the state, which is roughly 40% of the state's land. "It is curious that the news media have drawn our attention to the loss of tropical forests yet have been silent when it comes to how we have devastated our own forests here in the temperate zone," writes Doug Tallamy, an entomologist at the University of Delaware whose chapters appear in this book. "Only 15 percent of the Amazonian basin has been logged, whereas over 70 percent of the forests along our eastern seaboard are gone."[46]

Over the last 20 years, the state's forests have continued to see a slight decline but, especially compared to earlier periods of far wider deforestation, seem to be stabilizing. This is remarkable given that from 2000 to 2020, the state added more than 880,000 people, a population increase of 17%.

Yet inside cities, according to a 2022 study of the state's forests, Maryland still saw a net loss of more than 13,000 forested acres from 2013 to 2018, and only 58% of the state's streamside "riparian" areas are covered in trees. Under the 2014 Chesapeake Bay Watershed Agreement, federal, state, and regional environmental planners want to see at least 70% of riparian buffers covered in forest by 2025. And since forests in state, county, and local parks are already well protected, and 60% of forests are on privately held land, it's become increasingly clear that planting and protecting trees on "personal property" is critical to regenerating forest health. The state's Tree Solutions Now Act of 2021 set a statewide goal of planting 5 million trees (about 12,500 acres) by 2031; some 500,000 of these trees are slated to be planted in urban areas. The study's researchers tabulated that there are more than 373,000 acres of nonagricultural lands in the state that could accommodate substantial tree planting; reforesting just 3.3% of these lands would achieve the state's goal.[47]

Suburban development, of course, has caused problems beyond the loss of trees. Countless acres of shopping mall parking lots send contaminated runoff into local creeks. Car culture drives fossil fuel consumption (Maryland public transit ridership has grown by only 1% in 20 years, despite nearly 80%

of the state's population living within a 10-minute drive of public transit). In recent years, with the advent of online shopping, abandoned "ghost malls" have begun to be replaced by warehouses, and distribution centers have gobbled up land along interstate corridors like I-95. Subdivisions using septic tanks release 10 times more nitrogen (the most damaging pollutant in the Chesapeake) per household than homes on sewer lines. Every year, across the state, more than 430,000 septic systems release 4 million pounds of nitrogen in Maryland.[48]

And what of all the garbage generated by these expanding populations? In 2021, Baltimore County residents recycled a third *less* than they had a decade before, and fully one-third of the county's waste was shipped elsewhere, including (as you will read about in the environmental justice chapter) to an incinerator in Baltimore City, the *Baltimore Banner* reports. The county's Eastern Sanitary Landfill is thought to have just six years of capacity left.[49]

Despite the passage of the Clean Water Act in 1972, pollution still impairs 30% of river and stream miles in Pennsylvania, 73% in Virginia, and 97% in Delaware, writes the veteran environmental journalist Timothy Wheeler, whose chapter on Bay ecology appears in this volume. In Maryland, some 80% of the state's 5,300 river and stream miles contain so much fecal bacteria they are unsafe for recreation. "Forever chemicals"— per- and polyfluoroalkyl substances (PFOA and PFOS)—have been found in 83% of waterways sampled in 29 states, including more than a dozen in the Chesapeake watershed, including the Anacostia River and tributaries of the Susquehanna. These chemicals, Wheeler writes, affect human fertility, development, and "an increased risk of some cancers."[50]

And then there is industrial agriculture. Especially on the Eastern Shore, agriculture—particularly the monoculture planting of corn and soybeans, and the tremendous explosion of chicken farming—has become one of the state's most alarming environmental challenges. During hot summer months, nitrogen- and phosphorous-rich runoff from industrial-scale farms can create hypoxic "dead zones" that can cover 40% of the Chesapeake, the *Bay Journal*'s Karl Blankenship reports. Fully 25% of the Chesapeake watershed is covered in farms, and nearly half the nitrogen reaching the Bay comes from the region's 83,000 farms (and the 600 pounds of farm animals for every person in the watershed).[51]

As of 2019, Maryland had nearly 2,200 chicken houses—each longer than a football field—raising about 300 million chickens a year, according to the Environmental Integrity Project's report "Blind Eye to Big Chicken." These birds produce "more than 600 million pounds of manure and tons of airborne ammonia, which create a host of downstream problems, including runoff of nitrogen and phosphorous." Despite the industry's large footprint on the Eastern Shore, state oversight is "an empty paperwork exercise" and ineffective at protecting water quality, the EIP concluded.[52]

When it comes to drinking water supplies, Marylanders might take comfort that it isn't (yet) like the Southwest and Southern California, where housing developments, golf courses, and industrial agriculture are well on their way to draining not just the region's rivers but the colossal underground aquifers as well. Maryland has no natural lakes, but it does have a great wealth of rivers. In 1881, engineers constructed a dam on the Gunpowder River to create the Loch Raven Reservoir to provide drinking water to Baltimore City. The dam was enlarged twice over the years, and now—at 240 feet high—holds back 23 billion gallons of water, "enough to sustain the state of Maryland for approximately 50 days."[53]

But with crumbling water and sewer pipes in cities, hundreds of thousands of leaky septic tanks, petrochemicals running off roads and parking lots, and fertilizers and pesticides pouring off industrial farms, the safety of Maryland's drinking water remains threatened.

On the plus side: five Maryland counties rank in the country's top dozen for preserving land, and plans are underway to continue this work. In Baltimore County, for example, limiting water and sewer lines to urban areas has helped place more than 70,000 acres of land in permanent protection. Baltimore City is one of the few cities in the country with an expanding tree cover; the city wants its canopy to grow from 29% to 40% in the next 25 years.[54]

The Susquehanna and the Potomac Rivers have begun to show the benefit of major forest restoration efforts on farms, including the planting of 5 million trees (so far) in Pennsylvania. In 2022, Pennsylvania state lawmakers approved using $700 million to preserve and improve the state's water quality, including $220 million to create a Clean Streams Fund to reduce polluted runoff. And—after failing for 12 years—the state also passed a law to reduce fertilizer use on home lawns, golf courses, parks, and athletic fields.[55]

Other promising news: Maryland lawmakers recently passed a law seeking to produce 8.5 gigawatts of electricity off the state's coastline by 2030. They also voted to support "community solar" projects that help provide carbon-free power to residents who can't afford solar panels, and to require at least 40% of community solar development be reserved for low- and moderate-income households. The state has also revised its Forest Conservation Act for the first time in 30 years, moving from a policy of "no net loss" to efforts to increase forest cover. The Maryland the Beautiful Act has set a goal of bumping up its preserved state land from 23% to 30% by 2030, and 40% by 2040. And the state wants to see 5 million trees planted on public and private land by 2031—with 500,000 of them slated to be planted in urban areas.[56]

All of which is a good beginning. Master Naturalists can join in this work, and much more. For though our generations did not cause all our environmental challenges, we have surely inherited them.

Notes

1 Susan Q. Stranahan, "The River That Could Choke the Chesapeake," *Washington Post*, April 7, 2001, https://www.washingtonpost.com/archive/opinions/2001/04/08/the-river-that-could-choke-the-chesapeake/c033d28d-a9d9-46f4-840e-8285be497a2c/. See also "The Susquehanna River Drains Directly into the Chesapeake Bay. It Carries Pollutants from Pennsylvania with It," Stormwater PA, https://www.stormwaterpa.org/cumberland-susquehanna-river-health.html.

2 McKay Jenkins, "Canoes but No Kayaks: Thoughts on Environmental Studies," *Huffington Post*, November 18, 2011. See also "Muddy Run Pumped Storage

Facility," Constellation, https://www.constellationenergy.com/our-company /locations/location-sites/muddy-run-pumped-storage-facility.html.

3 Maryland State Department of Planning, "Plan Maryland: A Sustainable Growth Plan for the 21st Century," December 2011. See also Tim Wheeler, "Tree Cover Declines as Pavement Spreads across Chesapeake Watershed," *Bay Journal*, October 14, 2023.

4 Tom Horton, "If You Point the Finger at Pennsylvania, the Other Three Are Pointing at You," *Bay Journal*, September 26, 2023.

5 Maryland State Department of Planning, "Plan Maryland." See also McKay Jenkins, "The Era of Suburban Sprawl Has To End. Now What?" *Urbanite*, March 30, 2012.

6 Maryland State Department of Natural Resources, *Maryland Annual Deer Report, 2020–2021* (Annapolis, MD: Wildlife and Heritage Service), https://dnr.maryland .gov/wildlife/Documents/Maryland-Annual-Deer-Report-2020-2021.pdf.

7 McKay Jenkins, *ContamiNation* (New York: Avery, 2011), 163–73. See also McKay Jenkins, "The Dark Side of the American Lawn: Is It Giving You Cancer?" *Reader's Digest*, updated May 9, 2023, https://www.rd.com/article/lawn -fertilizer-dangers/.

8 David Pimentel, "Silent Spring Revisited," *Journal of the Royal Society of Chemistry*, October 2002. See also Kenneth Rosenberg, et al., "Decline of the North American Avifauna," *Science* 366, no. 6461 (September 2019): 120–24, https://www.science.org/doi/10.1126/science.aaw1313.

9 Scott Dance, "A County in Maryland's Lower Eastern Shore Is Washing Away, Leaving Its Residents with Hard Choices," *Washington Post*, August 24, 2020, https://www.washingtonpost.com/local/a-county-in-marylands-lower-eastern -shore-is-washing-away-leaving-its-residents-with-hard-choices/2020/08/24 /0724bdf8-e628-11ea-bc79-834454439a44_story.html.

10 Raymond Zhong, "Intensifying Rains Pose Hidden Flood Risks across the U.S.," *New York Times*, June 26, 2023.

11 Jenkins, "The Era of Suburban Sprawl Has To End."

12 Tom Horton, *Bay Country* (Baltimore, MD: Johns Hopkins University Press, 1987), xiii.

13 Andrea Wulf, *The Invention of Nature: Alexander von Humboldt's New World* (New York: Vintage Press, 2015), 65–68.

14 Marie I. George, "Christian Environmentalism from St. Augustine to Recent Popes (Part 1)," *Society of Catholic Scientists*, August 31, 2022, https://

catholicscientists.org/articles/christian-environmentalism-from-st-augustine-to
-pope-francis-part-1/.

15 Wulf, *The Invention of Nature*, 65–68.

16 Roxanne Dunbar-Ortiz, *An Indigenous People's History of the United States* (Boston: Beacon Press, 2014), 35–36.

17 Oxford Reference, http://www.oxfordreference.com/display/10.1093/acref
/9780199539536.001.0001/acref-9780199539536-e-1221.

18 Gil Oliveria, "Earth History in Your Hand," Carnegie Museum of Natural History, https://carnegiemnh.org/earth-history-in-your-hand.

19 Chris Guy, "Local Treasure Unearthed; History: A Tilghman Island Native Returns as a Graduate Student to Uncover Relics Buried 13,000 Years Ago," *Baltimore Sun*, March 10, 2000.

20 David A. Fahrenthold, "A Dead Indian Language Is Brought Back to Life," *Washington Post*, December 12, 2006.

21 Chesapeake Bay Program, "Indigenous Peoples of the Chesapeake," https://
www.chesapeakebay.net/discover/history/indigenous-peoples-of-the
-chesapeake.

22 Vine Deloria Jr., *God Is Red: A Native View of Religion* (Wheat Ridge, CO: Fulcrum Publishing, 2003), 88.

23 Chief Quiet Thunder and Greg Vizzi, *The Original People: The Ancient Culture and Wisdom of the Lenni-Lenape People* (N.p.: Nature's Wisdom Press, 2020), 56.

24 Maryland State Archives, "Maryland at a Glance: Indigenous Peoples," February 8, 2024, https://msa.maryland.gov/msa/mdmanual/01glance/native
/html/00list.html.

25 Martin O'Malley, "Recognition of the Maryland Indian Status of the Piscataway Indian Nation," Executive Order 01.01.2012.02, https://msa.maryland.gov
/megafile/msa/speccol/sc5300/sc5339/000113/016000/016239/unrestricted
/20130091e-002.pdf.

26 Isabel Spiegel, "A Native American Community in Baltimore Reclaims Its History," *Smithsonian*, October 5, 2020.

27 National Park Service, "The Peopling of Maryland Colony," African-American Heritage and Ethnography, https://www.nps.gov/ethnography/aah/aaheritage
/chesapeakeb.htm. Used with permission of the National Park Service, Park Ethnography Program.

28 National Park Service, "The Peopling of Maryland Colony."

29 Celeste Marie Gagnon and Sara K. Becker, "Native Lives in Colonial Times: Insights from the Skeletal Remains of Susquehannocks, A.D. 1575–1675," *Historical Archaeology* 54 (2020): 262–85, https://link.springer.com/article/10 .1007/s41636-019-00221-8.

30 Maryland State Archives and the University of Maryland, *A Guide to the History of Slavery in Maryland* (Annapolis: Maryland State Archives, 2007), https://msa .maryland.gov/msa/intromsa/pdf/slavery_pamphlet.pdf.

31 Steedman Jenkins, "Unlearning Terra Nullius: History and Memory in a Baltimore Suburb," unpublished senior thesis, Amherst College, 2023. See also Nancy Goldring, Carol Allen, and Nancy Horst, "The History and Development of East Towson," posted October 15, 2021, by AIA Baltimore and Baltimore Architecture Foundation, YouTube, www.youtube.com/watch?v=9ulRcJ8z5KI; John Whitfield, "Out of the Shadows of History: The Batty and Spencer Families," in *Tracing Lives in Slavery: Reclaiming Families in Freedom*, Ethnographic Overview and Assessment Report, Hampton National Historic Site, 2020, 113.

32 Chaplin and Chaplin, "Back to an Elegant Past," *New York Times*, October 22, 1976, Somerset County Library, *Digital Maryland*.

33 Jenkins, "Unlearning Terra Nullius"; Goldring, Allen, and Horst, "The History and Development of East Towson"; Whitfield, "Out of the Shadows of History."

34 Lynne Dakin Hastings, *A Guidebook to Hampton National Historic Site*, Historic Hampton, Inc., 1986, 21.

35 Maryland State Archives and the University of Maryland, *A Guide to the History of Slavery in Maryland*.

36 "House of Delegates Passes Resolution Acknowledging State's Part in Slavery," *Baltimore Sun*, March 27, 2007, https://www.baltimoresun.com/2007/03/27 /house-of-delegates-passes-resolution-acknowledging-states-part-in-slavery-2/.

37 James Kent, "A New Yorker in Maryland: 1793 and 1831," *Maryland Historical Magazine* 47, no. 2 (June 1952): 139, https://archive.org/details/maryland historic47brow. See also J. Thomas Scharf, *History of Baltimore City and County* (Philadelphia: Louis H. Everts, 1881 [1874]), 418, https://www.loc.gov/item/ rc01003473; Federal Hill Online, "History," http://www.federalhillonline.com /history.htm; and Emmanuel Mehr, "City of Brick: Bricks and Early Baltimore," Baltimore Histories Weekly, November 18, 2023, https://baltimorehistories .substack.com/p/city-of-bricks-bricks-and-early-baltimore.

38 Ned Tillman, *The Chesapeake Watershed: A Sense of Place and a Call to Action* (Baltimore, MD: Chesapeake Book Company, 2009), 28.

39 Karen Huff and James R. Brooks, "Building Stones of Maryland," *Maryland Geological Survey*, 1985, http://www.mgs.md.gov/geology/building_stones_of _maryland.html.

40 Andrew M. Graham, Amar R. Wadhawan, and Edward J. Bouwer, "Chromium Occurrence and Speciation in Baltimore Harbor Sediments and Porewater, Baltimore, Maryland, USA," *Environmental Toxicology and Chemistry* 28, no. 3 (March 2009): 471–80, https://doi.org/10.1897/08-149.1. See also Occupational Health and Safety Administration, "Hexavalent Chromium: Health Effect," https://www.osha.gov/hexavalent-chromium/health-effects, and Eric Cox and Timothy Wheeler, "Harbor Point projects stirs environmental concerns," *Baltimore Sun*, August 31, 2013, https://www.baltimoresun.com/2013/08/31 /harbor-point-project-stirs-environmental-concerns/.

41 Tillman, *The Chesapeake Watershed*, 195.

42 Karl Blankenship, "Timber! Museum Chronicles the Decline and Return of Forests," *Bay Journal*, November 2022.

43 US Fish and Wildlife Service, "Susquehanna River Fish Passage," https://www .fws.gov/project/susquehanna-river-fish-passage.

44 Christine Condon, "With Dam Demolished, Eel Numbers Explode," *Baltimore Sun*, December 28, 2022.

45 Maryland State Department of Planning, "Plan Maryland." See also Jenkins, "The Era of Suburban Sprawl Has To End."

46 Doug Tallamy, *Bringing Nature Home: How Native Plants Sustain Wildlife in Our Gardens* (Portland: Timber Press, 2007), 25.

47 Susan Minnemeyer et al., "Technical Study on Changes in Forest Cover and Tree Canopy in Maryland," a joint report by the Harry R. Hughes Center for Agro-Ecology; the University of Maryland College of Agriculture and Natural Resources; the Chesapeake Conservancy; and the University of Vermont Spatial Analysis Lab, November 2022, https://www.chesapeakeconservancy.org/wp -content/uploads/2022/12/MarylandForestStudy2022.pdf.

48 Jessica Grannis et al., "A Model Sea-Level Rise Overlay Zone for Maryland Local Governments," Georgetown Climate Center, November 2011, https://dnr .maryland.gov/ccs/Publication/GCC_MD-SLROrdRpt_FINALv3_11-2011.pdf.

49 Taylor Deville, "Baltimore County Adding Recycling Options to Extend Life of Only Landfill," *Baltimore Banner*, October 10, 2023.

50 Timothy B. Wheeler, "'Forever Chemicals' Found in More than a Dozen Bay Waterways," *Bay Journal*, November 2022.

51 Karl Blankenship, "Bay Cleanup Faces Difficult Trade-Offs with Agriculture," *Bay Journal*, May 2023.

52 Environmental Integrity Project, *Blind Eye to Big Chicken: Frequent Violations but Few Penalties for Maryland's Poultry Industry* (Washington, DC: Environmental Integrity Project, October 2021), https://htv-prod-media.s3.amazonaws.com/files /md-cafo-enforcement-report-embargoed-for-10-28-21-1635354222.pdf.

53 Baltimore City Department of Public Works, "Operations," https://publicworks .baltimorecity.gov/pw-bureaus/water-wastewater/water/operations. See also Loch Raven Trails, "About Loch Raven," https://lochraventrails.com/about-loch -raven-reservoir.

54 Baltimore County Government, "Land Preservation," https://www.baltimore countymd.gov/departments/planning/land-preservation.

55 Coalition for the Delaware River Watershed, "Pennsylvania State Budget Includes Historic Investment in Clean Water," July 8, 2022, https://www .delriverwatershed.org/news/2022/7/11/pennsylvania-state-budget-includes -historic-investments-in-clean-water. See also "Healthy Water, Healthy Communities: Pennsylvania's Investments in Water Quality," https://files.dep .state.pa.us/Water/Drinking%20Water%20and%20Facility%20Regulation /WaterQualityPortalFiles/IntegratedWatersReport/2024/CleanStreamsFund.pdf, and Penn State Extension, "New Pennsylvania Law Regulates Lawn Fertilizers," https://extension.psu.edu/programs/master-gardener/counties/allegheny /additional-resources/soil-management/new-pennsylvania-law-regulates-lawn -fertilizers.

56 Aman Azhar, "The Biden Administration's Scaled-Back Lease Proposal for Atlantic Offshore Wind Projects Prompts Questions, Criticism," *Inside Climate News*, December 16, 2023, https://insideclimatenews.org/news/16122023 /maryland-wind. See also Caroline Petterson, "Maryland's Community Solar Program," Institute for Local Self-Reliance, August 9, 2023, https://ilsr.org /marylands-community-solar-program/.

CHAPTER 2

Chesapeake Bay Ecology

TIMOTHY WHEELER

The Chesapeake Bay is in my blood. I was born hundreds of miles from it, but I grew up eating its oysters and shad. It got into me via my stomach. The rest came later, when our young family moved to Maryland, and I spent four decades exploring and writing about the Bay. There's still something invigorating and soul-refreshing about paddling out into the brackish creek by our home on the Eastern Shore.

Nature gets up close and personal around the Bay. In warm weather, you see dark little dots on the cove's surface: turtles poking their heads above water for air, then disappearing as my kayak approaches. Audible splashes signal fish feeding on insects—or each other. A muskrat cruising the reed-lined shore senses me and ducks under. A great blue heron, irritated by my intrusion, erupts from the trees with a guttural squawk and storms off to seek solitude elsewhere.

If my identity is inextricably linked to the Chesapeake, so is Maryland's. Look at a map and it's easy to see why: the Bay juts into the heart of the state, nearly severing it in two, and its reach extends well beyond its immediate shores. About 95% of the state's rivers, streams, and branches ultimately flow into the Bay. So vast is the Chesapeake's scale, it's hard to fully grasp it as an ecosystem. Its waters, and the cornucopia of plants and creatures that inhabit it, vary widely and are constantly changing—by the season, certainly, but sometimes from one day to the next or even hour to hour.[1]

The Chesapeake is in fact the largest estuary in the United States and the third largest in the world. An *estuary* is a semi-enclosed body of water where fresh water from rivers mixes with salty seawater, ebbing and flowing with the ocean tides.[2]

About half the Bay's water comes from the Atlantic Ocean, while the rest flows in from rivers draining portions of six states—New York, Pennsylvania, Maryland, Virginia, West Virginia, and Delaware—plus the District of Columbia. The Bay's footprint can perhaps best be quantified by the size of its coastline: Though only the 42nd largest state by land area, Maryland ranks 10th for the most coastline. Yet of its 3,190 miles of coast, just 31 are Atlantic beach, while the rest border the Bay and its tidal tributaries. All told, this relatively small state has the highest ratio of coastline to land mass of any state in the nation.

There are more than 100 such estuaries along the nation's coasts and hundreds more globally. They are among the most productive ecosystems in the world, providing habitat for thousands of species of birds, mammals, fish, and other wildlife and serving as the source of more than two-thirds of the fish and shellfish eaten in this country. The Chesapeake is—or was—widely considered the most productive of them all. In 1986, it accounted for 20% of the oysters and 50% of the crabs and soft-shell clams harvested nationwide (see figure 2.1).[3]

Though we tend to think of it as timeless, the Chesapeake Bay (as such) is only about 20,000 years old—the blink of an eye in geologic time. Back then, the Northern Hemisphere was covered by the mile-thick Laurentide ice sheet, and the Atlantic Ocean was more than 300 feet lower than it is today. As warming temperatures marked the end of the last Ice Age, the glaciers that covered everything as far south as Pennsylvania began to melt.

The meltwater runoff carved the underlying terrain with small streams and massive rivers (especially the Susquehanna) as they flowed down to the

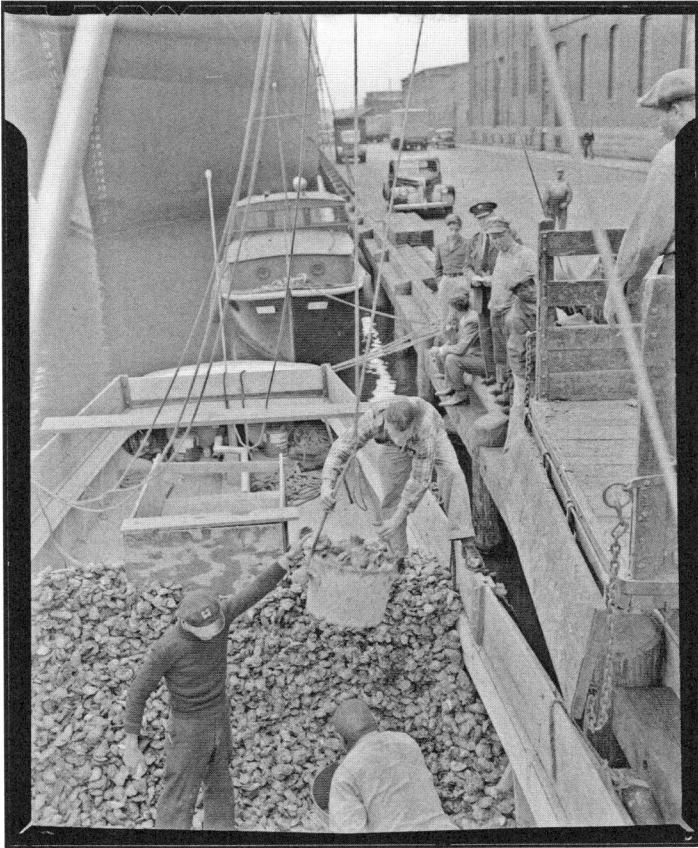

FIG. 2.1. Workers unloading oysters from a boat docked in the harbor of Baltimore.
Photo by Aubrey Bodine. Courtesy of the Maryland Center for History and Culture [B1605–2].

Atlantic coast, which then was about 180 miles east of where it is today. As the glaciers melted and sea levels rose, the coast itself became inundated; by about 3,000 years ago, the lower Susquehanna River valley broadened out into the Bay we know today. Remnants of that drowned river channel remain, in a deep trough that runs down the middle of the Bay.

As Martin Schmidt explains in his chapter on geology, the Chesapeake's location and shape were predetermined by a cataclysmic event 35 million years ago, when a large comet (or meteor) known as a *bolide* struck the

Earth near what is now the town of Exmore, Virginia, on the lower tip of the Delmarva Peninsula. The whole region became inundated by a shallow sea.

This celestial wrecking ball, about two miles in diameter, hit with such explosive force that it carved out a vast depression in the continental shelf below the sea. The crater was twice as large as Rhode Island and as deep as the Grand Canyon. It quickly filled with sediment, but the disruption caused by that cosmic collision continues to affect land subsidence in the region, exacerbating the modern-day rise in sea level.

The bolide impact also broke up the *aquifers*, or water-bearing layers of rock, sand, and gravel, underlying the coastal plain in Virginia and Maryland. In their place, it carved out a huge underground reservoir for super-salty groundwater, unsuitable for use as drinking water.[4]

Since its creation, the Chesapeake estuary has continued changing, from both geologic and human forces. In more recent times, the Bay has played a major role in Maryland's history, economy, and culture. Long before European explorers and settlers arrived on the East Coast, the Bay's oysters and fish helped sustain the region's Indigenous peoples, and its rivers provided a watery highway for moving about the watershed. The Maryland state seal features a waterman as well as a farmer, in recognition of the importance of fishing after Europeans settled here beginning in the 17th century. The Chesapeake's most-prized finfish, Atlantic striped bass (or *rockfish*), is the official state fish; the blue crab is the state crustacean; and the Bay-dwelling diamondback terrapin is the state reptile.

Although the Chesapeake's primary freshwater source starts in the mountains of central New York State, the Bay itself is generally said to stretch 195 miles from Havre de Grace in Maryland to Norfolk, Virginia. Its width varies from 4 miles near Aberdeen to 30 miles just north of where it joins the ocean in Virginia. Its deepest spot, southeast of Annapolis, drops down 174 feet. But overall, the Bay is relatively shallow, averaging just 21 feet in depth. Across much of its 2,500 square-mile surface, in fact, a person who is 6 feet tall could wade around without getting their hair wet.[5]

The surface area of the Bay, including the tidal portions of its river tributaries, spans approximately 4,400 square miles. But the watershed—all the land that drains into the Chesapeake—encompasses 64,000 square miles, extending as far north as the source of the Susquehanna in Cooperstown, New York.[6]

Besides the Susquehanna, which furnishes half of the fresh water entering the Bay, major rivers—including the Patapsco, Patuxent, and Potomac—flow into the Chesapeake from the region's western mountains. From the east come the Chester, Choptank, Nanticoke, Wicomico, and Pocomoke.

The Chesapeake holds more than 18 *trillion* gallons of water—enough to fill 27 million Olympic-sized swimming pools.[7] Most of this comes from the Atlantic Ocean, which is why the vast majority of the estuary's water is a *brackish* mixture of fresh and salt water. Yet each day, about 51 *billion* gallons of fresh water flow into the Bay from 150 rivers and streams, though the average daily volume in any given year can vary widely, ranging (since 1990) from 29 billion gallons to more than 80 billion gallons.[8] This vast network of tributaries means that the Bay's health depends tremendously on the quality of the water flowing into it.

The Bay's physical condition varies both geographically and seasonally. *Salinity*—the amount of salt dissolved in the water—is a critical variable. It is highest (averaging 25 to 30 parts per thousand) from the influx of salty Atlantic Ocean water at the estuary's mouth in Virginia. Salinity gradually declines northward, with little (or none) usually detectable at the head of the Bay and at the head of tide in river tributaries.[9]

Salt water is heavier than fresh water, so salinity tends to vary both by depth (from lower at the water's surface to higher at the bottom) and by geography: saltier ocean water tends to move north along the Bay's bottom, while lighter river water flows down along the surface. This water stratification, as we will see later, is a vital component of the Bay's ecology. Salinity also fluctuates by time of year. Spring rains and snowmelt increase flows of fresh water from river tributaries, decreasing the Bay's overall salinity levels. Salinity usually rises in summer and autumn, when rainfall typically declines, though hurricanes and other major rainstorms across the watershed can temporarily depress salinity again.

As if that's not complicated enough, salinity also increases from west to east in the Bay because of what's called the *Coriolis effect*. As the Earth rotates, fresh water flowing into the Bay from its largest rivers tracks along the Western Shore, while salt water moving up from the ocean hews to the Eastern Shore. As a result, salinity levels are generally higher along the Eastern Shore. Throughout the Bay and its tributaries, water salinity determines what plants and animals are found in any given location. Fish (such as sharks and red drum) that thrive in salty water frequently turn up near the Bay's mouth. Freshwater species, such as largemouth bass and catfish, often roam the upper Bay. Many fish, such as white perch, striped bass, blue catfish, and snakehead, can do quite well in waters of varying salinity.

Temperature is another variable. Bay water temperatures fluctuate widely throughout the year, ranging from 34 degrees in winter to 84 degrees in summer.[10] The Bay's overall shallowness means its waters heat up rapidly as air temperatures climb in spring and summer; in winter, they cool off more quickly than ocean water. These temperature fluctuations substantially influence where underwater grasses can grow, and thus where (and when) fish and crabs feed, reproduce, and migrate.

Tides are yet another constantly changing force in the Bay. Ocean levels rise and fall twice daily, generally by a foot or two, in response to the gravitational pull of the moon and the centrifugal force of the Earth's rotation. With each high tide, salty ocean water enters the Bay and moves northward. It reaches the head of the Bay and begins to recede about the same time another high tide brings in a new pulse of ocean water. That rhythmic tidal fluctuation creates currents, with water ceaselessly flowing up the Bay as the tide rises, then reversing as the tide ebbs or falls. These currents are almost imperceptible, averaging just 0.3 miles per hour, running faster in narrow water passages.

Water levels in the Bay and its tributaries rise and fall in sync with the tides. The vertical difference between high and low tides is greatest—about

3 feet—at the Bay's mouth in Virginia. It declines as you go northward, to about 1 foot around Annapolis, then increases again to about 2 feet as the Chesapeake narrows toward its head.[11]

The up-down cycle of the tides is influenced by the most unpredictable of variables—the wind. A persistent northwest wind can push water out of the Bay, dampening high tide and exaggerating its ebb. Extreme "blowout" low tides can lay bare vast areas of bottom, which become potentially life-threatening for oysters, clams, and other shellfish left exposed for long periods of time to freezing winter air.

Strong northeast winds, on the other hand, can push water higher onto the shore, causing tidal flooding. Such brackish inundations affect the types of vegetation that can survive along the waterfront.

The shallowness of the Bay magnifies the impact of all those variables. Although the Chesapeake hosts about 2,700 species of plants and animals, salinity shifts and temperature swings (plus wind-driven waves) can make it tough for many of them to establish and maintain a permanent presence. During the summer, for example, more than 265 fish species visit the Chesapeake to find food and reproduce, but only about 29 stick around through the winter.[12]

Challenging as those ever-changing conditions may be, estuaries like the Chesapeake are still an all-you-can-eat buffet for many migratory species. The Bay provides a variety of habitats for plants and animals, from *sandy beaches* to *wetlands*, *intertidal flats*, *shallow water*, and *deeper open water*. Within those broad categories are also *oyster bars*, *seagrass beds*, and even *piers*, *rocks*, or *jetties* that attract, protect, and provide for various aquatic or marine life.[13]

The Bay's ecological table is set with a soup of nutrients, minerals, and organic matter that flows in from rivers. This mix provides essential ingredients for *phytoplankton*, the microscopic plants that form the base of the food web. These algae are in turn fed upon by zooplankton, their tiny animal counterparts, as well as by shellfish, small finfish, snails, and jellyfish. Especially in the shallows, sunlight penetrating the water fuels the growth of *underwater grasses*, or submerged aquatic vegetation, which provides vital habitat for a great variety of other aquatic life. Finfish hide and hunt amid the grasses; molting blue crabs also seek cover from predators there as their new shells harden.

Such *seagrass beds*, or *underwater meadows*, also absorb carbon and nutrients and settle suspended sediments to the bottom. Grass beds are highly sensitive to changes in water clarity and nutrient concentrations, so their relative abundance is tracked as a barometer of the Chesapeake's vitality.

Wetlands—and especially *salt marshes*—are another major contributor to the estuary's productivity, feeding the ecosystem by producing an estimated 10 tons of organic matter a year. A critical transition zone between land and water, wetlands are one of the most important habitats in the Chesapeake, corralling stormwater runoff from uplands and absorbing coastal storm surges. They also filter nutrients, sediment, and other pollutants and provide food, shelter, and nursery grounds for many different animals.

Maryland currently has about 757,000 acres of wetlands, by official state estimates. About 57% of these wetlands are *palustrine* or *freshwater wetlands*, with about 240,000 acres in estuarine or tidal wetlands. Wetlands are most abundant on the Eastern Shore; they make up fully one-quarter (or more) of low-lying Dorchester County.[14]

As critical as wetlands and aquatic vegetation are to the Chesapeake's health and vitality, those vital habitats have been under enormous pressure for many, many decades. The state had nearly 1.7 million acres of wetlands in the 1780s and lost nearly 75% over the next two centuries.[15] Since the 1780s, experts estimate, 45–65% have been filled or drained—mostly for use as farmland but in more recent decades for housing development as well.[16]

As for aquatic vegetation, scientists say that—based on old aerial photos going back to the 1930s—plants once covered anywhere from 200,000 to 600,000 acres of Bay and river bottom. By 1984, aerial surveys estimated only 38,000 acres remained.[17]

One of the most daunting problems faced by these plants is the flow of *nutrients* and *sediment* into the Bay. Though nutrients are essential to a productive estuary, the Chesapeake has become over-enriched with them. Excessive nitrogen and phosphorus gush into the Bay from wastewater treatment plants. When it rains or when snow melts, unabsorbed fertilizers containing the same nutrients wash off farm fields and suburban lawns. Stormwater runoff from urban and suburban streets, parking lots, and buildings adds to this nutrient overdose. Pollution from fossil-fueled power plants and motor vehicles spews nitrogen into the atmosphere, only to have it precipitate back to Earth—and into our water.

Sediments—the fine-grained soil particles running off from farms, development, and eroding shorelines—do triple damage. First, they cloud the water, preventing submerged grasses from getting the sunlight they need to grow. When the sediment settles to the bottom, it can smother fish eggs and bottom-dwelling organisms, including oysters and clams. And it carries additional phosphorus into the water, adding to the nutrient overload.

Here's how nutrients degrade water quality and render it inhospitable (if not downright toxic) to fish. Nitrogen and phosphorus from wastewater and runoff pour into the Bay and its tributaries, where they fuel the growth of algae. When those floating masses (or *blooms*) of microscopic plants die, they sink to the bottom, where the chemical process of algal decay pulls dissolved oxygen out of the water. Because of the stratification of fresh and saltier water, oxygen near the surface does not readily mix into the depths, so fish and crabs struggle to breathe—and in extreme cases even suffocate.

The Chesapeake's resulting *hypoxic dead zone* that forms every summer (and which is so big it can be seen from outer space) has been the principal target of the multistate effort to restore the Bay's water quality and ecological abundance. This effort started with a federal study of the Bay's water quality that began in the 1970s. Its findings led to the first formal Bay restoration agreement in 1983, a four-paragraph pledge to work together to "fully address the extent, complexity and sources of pollutants entering the Bay" signed by governors of Maryland, Virginia, and Pennsylvania; the mayor of the District of Columbia; the administrator of the US Environmental Protection Agency; and the chair of the Chesapeake Bay Commission, a tri-state legislative advisory body.

That simple (and voluntary) compact has been supplanted three times since then—in 1987, 2000, and 2014—with more detailed vows to address the range of ills afflicting the Bay's ecological vitality, including toxic pollutants, sprawling development, loss of habitat, and climate change. Meanwhile, in 2010, the federal Environmental Protection Agency (EPA), working with the states, established a kind of "pollution diet" for the Bay known as the *total maximum daily load (TMDL)*. The agency set nutrient and sediment reduction targets for each state; states were then expected to lean on local governments to hit their own targets.

With the signing of the *Chesapeake Bay Watershed Agreement* in 2014, states pledged to work toward 31 varied restoration outcomes, with the deadline for achieving many of them set for 2025 (in 2022, state and federal leaders acknowledged their efforts were falling short in many key respects, including on water quality and on key pledges to expand streamside tree buffers and wetlands).

Progress has been frustrating. The federal-state effort has been touted as a model of ecosystem restoration and a leader in estuarine science, but it also has struggled to mitigate the negative impacts of a rapidly growing population on the Bay's 64,000-square-mile watershed. Over the decades, the federal government and Bay watershed states—including not just Maryland and Virginia but also Delaware, New York, and West Virginia—have collectively spent billions of dollars (including about $16 billion just in the past decade) to upgrade wastewater treatment plants and to help pay for installation of runoff-controlling "best management practices" on farms and in communities. Yet so far, the results have been mixed.[18]

Since the mid-1980s, the US Geological Survey has tracked improvements in nitrogen, phosphorus, and sediment levels in some rivers but worsening water quality in others. Indeed, like almost everything else about the Bay, its ecological health seems to vary from year to year. Weather plays a big part: nutrient and sediment pollution tends to improve in dry years, when there is less precipitation to cause runoff, and worsen in wet ones. In 2021, in the aftermath of record-breaking rainfall in 2018–2019, officials reported that just 28% of the Bay and its tributaries met water quality standards, down from 42% a few years earlier.[19]

But there are political reasons as well. Among the states, Pennsylvania has been—by far—the biggest laggard when it comes to meeting TMDL standards, in part because of the state's powerful farm lobby. But Maryland has fallen short as well. In a 2022 assessment, the EPA determined that—while Maryland hit its overall mark for reducing sediment getting to the Bay—the state had not met its 2021 targets for reducing nitrogen and phosphorus. Although state officials worked to get more runoff-limiting "best management practices" installed on farms, EPA officials found Maryland's efforts inadequate. The state also lagged in enforcing standards for curbing urban and suburban stormwater runoff. Efforts to promote more compact "smart growth" to curb habitat loss and runoff pollution from sprawl have been resisted—often successfully—by real estate interests and local officials whose election campaigns benefit from developer donations.

It remains unclear what, if anything, will replace, update, or improve the 2014 Chesapeake Bay agreement. The federal-state Chesapeake Bay Program's Executive Council—which includes leaders of the six states, the District, and the EPA—has called for recommendations for how to proceed beyond 2025.

But a 2023 report from Bay scientists cautioned that it may be impossible to eliminate the dead zone and suggested that might be the wrong goal anyway. It urged shifting the focus to improving habitat in shallow waters, where most of the living resources are, while warning that rising temperatures from climate change threaten to complicate matters in ways difficult to predict.

Perhaps most significantly, the scientists' report suggests it is time to stop talking about "restoring" the Bay. Although working to make the Bay healthier remains a worthy and urgent ecological goal, they cautioned, it may be that human impact has so changed the Bay that it simply cannot be restored to what it was like decades (if not centuries) ago.[20]

The challenges remain complex and (as always) are directly linked to our collective social, economic, and industrial systems. Beyond suffering from the nutrients and sediment generated by industrial farms, urban streets, and suburban development, the Bay is impaired by *toxic contaminants*. Crabs and some finfish caught in parts of the Bay (and many of its tributaries) harbor potentially harmful levels of polychlorinated biphenyls, or PCBs. Some fish caught in freshwater lakes—including Maryland's state fish, the striped bass—are unsafe to eat because of mercury contamination, deposited in Maryland waters by prevailing winds blowing in from coal-burning power plants in other states.

PCBs, on the other hand, are a homegrown problem. Once widely used as insulators and lubricants, they have been banned since 1979 because of their link to cancer and other health effects. Yet the compounds do not readily break down and can linger in soil and sediments, where they leach into the water and are picked up by passing or foraging fish. The process of *bioaccumulation*—in which small things (and their contaminated bodies) are eaten by bigger things—means that toxic loads tend to be greater the higher you go on the food chain. And that includes us.

The Bay's waters, wildlife, and human inhabitants also face uncertain consequences from an "emerging" group of contaminants even more widely used than PCBs: per- and polyfluoroalkyl substances (PFAS for short)—commonly called "forever chemicals" because (like PCBs) they don't readily break down.

PFAS comprise a group of about 9,000 fluorinated chemicals widely used for decades in firefighting foam and a variety of consumer products, such as nonstick cookware, stain-resistant or water-repellant fabrics, and food packaging. They dissolve easily and spread throughout groundwater, which is why PFAS contamination has been discovered in private and community wells across the state (and the nation).

In December 2023, because of PFAS contamination, the Maryland Department of the Environment (MDE) issued fish consumption advisories, warning recreational anglers and subsistence fishers to limit their consumption of 15 different species of locally caught finfish—again in some instances including striped bass—in dozens of different locations around the state.[21] Levels measured in oysters and crabs were low enough not to pose health risks, according to MDE, but some independent sampling by nonprofit groups has not been as reassuring.[22]

The Bay's overall ecological condition will always reflect the health of its broader watershed, the intricate web of brooks, streams, wetlands, and rivers that feed into the Chesapeake. Remember: 95% of Maryland's land drains into the Bay. The state has 19,127 miles of rivers and streams, and 61% of assessed waterways are considered *impaired* in one way or another. A water body is deemed impaired when it does not support a designated use, such as swimming or wading, shellfish harvesting, or fishing for trout (the most sensitive of all freshwater fish to changes in water temperature, clarity, and overall quality). Nearly 94% of the state's nearly 22,000 acres of publicly owned lakes and reservoirs are also impaired, as are 100% of the state's 2,451 square miles of estuarine water in the Chesapeake Bay.[23]

Oysters

Historically, oysters have been among the Bay's most significant creatures. Early European settlers marveled at their abundance. The bottom-hugging Eastern oyster sustained Native Americans and European settlers alike, and the intensively managed fishery grew exponentially in the latter half of the 19th century. By the late 1800s, the Chesapeake was the greatest oyster-producing region of the world, employing more than 32,000 Marylanders in harvesting or processing a peak of 15 million bushels in 1885.[24] The haul plummeted in the decades that followed, leveling off at around 2 million bushels a year in the 1930s. Then, in the 1980s (and again in the early 2000s), a pair of parasitic diseases, MSX and Dermo, ravaged the Chesapeake oyster population. Maryland's harvest fell to just 30,000 bushels in 2004, then began a slow recovery that hit 600,000 bushels in 2023.[25]

Along the way, scientists have successfully impressed on policymakers that oysters play an important ecological role in the Bay. They are filter feeders, helping remove algae and sediment from the water. They also are marine engineers, building underwater habitat for worms, barnacles, and other marine creatures. Attaching themselves layer by layer to other oysters (living or dead), they create reefs, which attract blue crabs and a variety of finfish preying on the marine life lurking there or seeking their own places to hide.

With the oyster fishery at a low ebb, the state moved in 2010 to eliminate historic limits on leasing the bottom for private oyster cultivation and set aside large areas as *sanctuaries* where no harvesting would be allowed. Oyster aquaculture has blossomed in Maryland since, though its output continues to be outpaced by the public harvest.[26]

In 2014, recognizing the ecological value of oysters, Maryland joined with Virginia in pledging to undertake large-scale restoration of bivalve populations in five of its Bay tributaries: Harris Creek, the Tred Avon River, the Little Choptank River, the St. Mary's River, and the Manokin River. Because salinities are so low in all but the southernmost section of the Maryland portion of the Bay, oysters there have had trouble reproducing.

The restoration effort has thus been driven by hatcheries, particularly a large operation overseen by the University of Maryland Center for Environmental Science at Horn Point in Cambridge. Scientists there nurture billions of oyster larvae a year and either set the baby oysters on shells themselves or sell them to others who do so. Since 2014, the state has planted almost 6 billion hatchery-reared juvenile oysters. By 2022, with Harris Creek fully restored and work continuing on the other four waterways, nearly $71 million had been spent on the effort—a further indication of how costly ecosystem restoration can be.[27]

Crabs

Though once the king of Maryland's seafood industry, oysters have long since been supplanted by the blue crab. In the 1950s and '60s, the Bay's commercial harvest averaged 70 million pounds, with Maryland and Virginia periodically trading places in catching the larger share. The overall harvest varied but climbed significantly in the 1990s to top 100 million pounds before sinking back to around 50 million pounds by the early 2000s.[28] That swoon prompted the two states to agree to coordinate their fishery management by preserving enough females to rebuild the population.

Here again, Bay-wide (as opposed to state-by-state) management makes sense because crabs use all of the Bay in their life cycle. The young hatch from fertilized eggs as free-floating *zoea* carried by currents and winds out into the Atlantic Ocean and back again into the Bay. As they grow and mature, crabs move northward and into the brackish stretches of the Bay's tributaries. There, they mate, and the impregnated females migrate back down the Bay to near its mouth to release their eggs.

Winter conditions can affect crabs' abundance, with intense cold causing increased mortality. But the relative health and abundance of underwater grasses is also a factor, as crabs hide in submerged vegetation while molting (when they must survive for a brief period without their hard shells for protection). Even though the numbers of females currently remain at what is believed to be sustainable levels, in recent years scientists have grown concerned over surveys finding relatively few juvenile crabs.

Rockfish

Atlantic striped bass, known to many Marylanders as *rockfish*, are the state's most important commercial and recreational fish species. A silvery fish that gets its name from several dark, continuous stripes along the side of its body, stripers are migratory and *anadromous*, meaning they spend most of their lives roaming the Atlantic coast or brackish Bay waters but swim every spring up to freshwater rivers to spawn. The Chesapeake is the primary spawning and nursery ground for 70–90% of the entire Atlantic coastal striped bass stock.

Once abundant, striped bass declined dramatically during the 1970s and early '80s, prompting Maryland to impose a catch moratorium as other states also took conservation measures. In just five years, the fish population rebounded dramatically, prompting officials to lift the moratorium in 1990. In recent years the fish has lost ground again, amid renewed concerns about overfishing and multiple years of poor recruitment of juvenile fish. Since 2020, in a bid to halt their slide and rebuild the coastwide stock, regulators have again imposed restrictions on commercial and recreational fishing for striped bass.

Shad

American shad offer an object lesson in the risk of delayed response to evidence of a fishery's decline. Prized for their roe as well as their bony flesh, shad once outclassed striped bass as the Bay's premier finfish. Shad are also anadromous, spending most of their life at sea and only swimming up Bay rivers to fresh water in the spring to spawn. So plentiful in the 18th and 19th centuries that they were spread on farm fields as fertilizer, shad populations began to decline as dams were built across rivers, preventing the fish from reaching their spawning areas. Completed in 1928, the Conowingo Dam completely blocked the Susquehanna River, the Bay's best shad spawning run; three more major dams were constructed farther upstream in Pennsylvania. Although Maryland moved to preserve its dwindling shad stocks by imposing a catch moratorium in 1980 (and shad numbers have recovered some in the decades since), stocks remain far too low to merit lifting the catch ban.

Non-Native Invasive Fish

While native fish have struggled to hold on in the Bay, new species have found a home there, raising concerns about their impacts. *Northern snakeheads*, imported from China to supply restaurants and fish markets, were first discovered in the wild in a pond in Crofton in 2002. Though the pond was drained and those fish eradicated, more snakeheads turned up in the Potomac River in 2004 and have since spread throughout the Bay.

Another import of even greater concern is the *blue catfish*, a native of midwestern rivers first introduced in the 1970s in Virginia to give sports anglers a bigger fish to catch. Unlike other catfish, blue catfish are omnivorous and a top predator, feeding voraciously on underwater vegetation when young—and on other fish and crabs as they mature. They can grow quite large: as of 2023, the biggest caught in Maryland weighed fully 84 pounds. Blue catfish have become the dominant fish in some Maryland rivers, notably the Patuxent, and scientists fear they may have thrown longstanding food webs into disarray.

Both snakeheads and blue catfish are most at home in fresh water, but scientists have found they can tolerate moderate salinity, a fact that is likely key to their spread throughout the Bay. Scientists are also worried about the impacts of both species on native fish and crab populations; a Virginia study estimated that blue catfish in a single stretch of the lower James River consumed *5 million* crabs. Maryland and Virginia have encouraged development of commercial fisheries for snakeheads, blue catfish, and other invasive fish, and in 2023 Maryland sought federal disaster relief for the impact of all invasive fish species on its commercial and recreational fisheries.[29]

Changed as the Chesapeake's ecology is—especially compared to what it was before European contact—its systems seem certain to face additional profound challenges from climate change. In the last 30 years, the average water temperature in the Bay has increased by a full 1° Celsius, or 1.8° Fahrenheit. Over the last century, waters have risen about 1 foot (half from rising global sea level, half from land subsidence related to that ancient bolide hit eons ago). Water levels are predicted to rise another 1.3 to 5.2 feet over the next 100 years.[30] Precipitation patterns are changing as well: between 1958 and 2012, according to the EPA, the amount of rain falling during heavy storms increased more than 70% in the northeastern United States—more than any other region in the nation.[31]

How exactly climate change will affect water quality and living resources remains uncertain. One study by Morgan State University predicted oysters may fare better as a result of rising temperatures and altered salinity. But researchers foresaw the same scenarios would lead to declines in two other Bay staples, crabs and rockfish.[32]

From my little cove on the Eastern Shore, I see ample evidence of the changing Bay. The water now rises up and covers the street in front of our house several times a year, bringing little fish and crabs with it. The biggest fish I've caught close to home so far has been a snakehead. It's clear the Chesapeake of my grandchildren will be different from the one that captivated me in the late 1970s when I read James Michener's epic saga of life on and beside the Bay.

Yet as I celebrate the annual spring return of ospreys, striped bass, and crabs, there's much there still to enchant, appreciate, and fight for. At my dock, I raise tiny oyster "spat" spawned in a hatchery, to be planted the following year on a sanctuary reef. And as owls murmur in the dark, I drift off to sleep reliving another day's exploration on the water.

Notes

1 Chesapeake Bay Commission, *The Chesapeake Bay and Its Watershed: General Facts* (Annapolis: Chesapeake Bay Commission, 2020), https://www.chesbay.us /library/public/documents/Fact-Sheets/Bay-Factoids-FINAL.pdf.

2 Catherine Krikstan, "Eight Reasons the Chesapeake Bay Is an Exceptional Estuary," 2012, Chesapeake Bay Program, https://www.chesapeakebay.net/news /blog/eight-reasons-the-chesapeake-bay-is-an-exceptional-estuary.

3 Christopher P. White, *Chesapeake Bay: Nature of the Estuary, A Field Guide* (Centreville, MD: Tidewater Publishers, 1989), 3.

4 US Geologic Survey, "The Chesapeake Bay Bolide Impact: A New View of Coastal Plain Evolution," USGS Fact Sheet 049–98, pubs.usgs.gov/fs/fs49-98.

5 "Chesapeake Bay Facts and Figures," Maryland Sea Grant, accessed December 13, 2023, www.mdsg.umd.edu/topics/ecosystems-restoration /chesapeake-bay-facts-and-figures.

6 White, *Chesapeake Bay*, 8.

7 "Chesapeake Bay Facts and Figures," Maryland Sea Grant, accessed December 13, 2023, www.mdsg.umd.edu/topics/ecosystems-restoration/chesapeake-bay-facts -and-figures.

8 "Pollution Loads and River Flow to the Chesapeake Bay (1990–2021)," Chesapeake Progress, accessed December 13, 2023, https://www.chesapeakeprogress.com /charts/pollution-loads-and-river-flow-to-the-chesapeake-bay-1990-2021.

9 "Physical Characteristics," Chesapeake Bay Program, accessed December 14, 2023, chesapeakebay.net/discover/ecosystem/physical-characteristics.

10 "Physical Characteristics."

11 Alice Jane Lippson and Robert L. Lippson, *Life in the Chesapeake Bay*, 3rd ed. (Baltimore, MD: Johns Hopkins University Press, 2006), 5.

12 White, *Chesapeake Bay*, 5.

13 Lippson and Lippson, *Life in the Chesapeake Bay*, 7–13.

14 Denise Clearwater, Paryse Turgeon, Christi Noble, and Julie LaBranche, *An Overview of Wetlands and Water Resources of Maryland*, report prepared for the Maryland Wetland Conservation Plan Workgroup of the Maryland Department of the Environment, January 2000, 7, https://mde.maryland.gov /programs/Water/WetlandsandWaterways/DocumentsandInformation/Pages /maps.aspx.

15 Thomas E. Dahl, *Wetlands Losses in the United States, 1780's to 1980's* (Washington, DC: US Department of the Interior, Fish and Wildlife Service, 1990), 5–6, https://www.fws.gov/sites/default/files/documents/Wetlands-Losses -in-the-United-States-1780s-to-1980s.pdf.

16 Dahl, *Wetlands Losses in the United States.* See also Clearwater et al., *An Overview of Wetlands and Water Resources of Maryland.*

17 Chesapeake Bay Program, Chesapeake Progress, "Submerged Aquatic Vegetation," https://www.chesapeakeprogress.com/abundant-life/sav.

18 Karl Blankenship and Timothy B. Wheeler, "After 40 Years, Chesapeake Bay Program Yields Mixed Results," *Bay Journal*, December 4, 2023, https://www .bayjournal.com/news/pollution/after-40-years-chesapeake-bay-program-yields -mixed-results/article_4af88180-92b0-11ee-9d06-ab0f3bb0d72f.html.

19 "Water Quality Standards Attainment and Monitoring," Chesapeake Progress, accessed December 2023, https://www.chesapeakeprogress.com/clean-water /water-quality.

20 Scientific and Technical Advisory Committee, *Achieving Water Quality Goals in the Chesapeake Bay: A Comprehensive Evaluation of System Response* (Edgewater, MD: Chesapeake Bay Program, 2023), https://www.chesapeake.org/stac/cesr.

21 Maryland Department of the Environment, "Maryland Department of the Environment Issues New Fish Consumption Advisory and Guidelines," press release, December 8, 2023, https://news.maryland.gov/mde/2023/12/08 /maryland-department-of-the-environment-issues-new-fish-consumption -advisory-and-guidelines/.

22 Timothy B. Wheeler, "'Forever Chemicals' Found in Chesapeake Seafood and Maryland Drinking Water," *Bay Journal*, November 17, 2020, https://www .bayjournal.com/news/fisheries/forever-chemicals-found-in-chesapeake-seafood -and-maryland-drinking-water/article_2aa7a82a-28fa-11eb-ac61-9f14273a6e14. html.

23 Environmental Integrity Project, "Clean Water Act at 50: Promises Half-Kept at Half-Century Mark," Appendix A, March 17, 2022.

24 Victor S. Kennedy, *Shifting Baselines in the Chesapeake Bay, An Environmental History* (Baltimore, MD: Johns Hopkins University Press, 2018), 42; Maryland Department of Natural Resources, Oyster landings, 1870–2021.

25 Timothy B. Wheeler, "Amid Oyster Bounty, Maryland Worries about Overfishing, Eyes Harvest Limits," *Bay Journal*, June 15, 2023.

26 Alicia Pimentel, "Maryland Proposes New Regulations for Oyster Sanctuaries, Aquaculture," Chesapeake Bay Program, 2010, https://www.chesapeakebay.net/news/blog/maryland-proposes-new-regulations-for-oyster-sanctuaries-aquaculture.

27 Maryland Department of Natural Resources, *2022 Chesapeake Bay Oyster Restoration Update* (Annapolis: Maryland Department of Natural Resources, 2023), https://dnr.maryland.gov/fisheries/Documents/2022_Maryland_Oyster_Restoration_Update.pdf.

28 "Blue Crab Landings, 1950 to 2022," NOAA Fisheries, https://www.fisheries.noaa.gov/foss/f?p=215.

29 Timothy B. Wheeler, "Can Chesapeake's Blue Catfish Shift from Disaster to Dinner Plate?," *Bay Journal*, May 3, 2023, https://www.bayjournal.com/news/fisheries/can-chesapeakes-blue-catfish-shift-from-disaster-to-dinner-plate/article_746c3c88-e602-11ed-81e0-c7d020fde338.html.

30 "Chesapeake Bay: Climate Change," NOAA Fisheries, accessed December 2023, https://www.fisheries.noaa.gov/topic/chesapeake-bay/climate-change.

31 "Climate Change," Chesapeake Bay Program, accessed December 2023, https://www.chesapeakebay.net/issues/threats-to-the-bay/climate-change.

32 Kira L. Allen, Thomas Ihde, Scott Knoche, Howard Townsend, and Kristy A. Lewis, "Simulated Climate Change Impacts on Striped Bass, Blue Crab and Eastern Oyster in Oyster Sanctuary Habitats of Chesapeake Bay," *Estuarine, Coastal and Shelf Science* 292 (October 5, 2023): 108465.

CHAPTER 3

Urban Ecology

J. MORGAN GROVE
AND
STEWARD T. A. PICKETT

On Memorial Day, 2018, the day Pastor Michael Martin began his ecological journey at Baltimore's Stillmeadow Community Fellowship, it took nearly an hour for him to drive less than a mile to his church. Stillmeadow sits on Frederick Avenue, close to the boundary line with Baltimore County, and just the day before, the church's Westgate neighborhood had been flooded by a "thousand-year" storm and massive flash flood. The flooding was so intense that the Baltimore City Fire Department had to use its boats in seven feet of water to rescue more than 20 people from car rooftops and a stranded bus. Over the next year, it would become clear that more than 200 homes had experienced substantial damage within those few hours.

This was not the first time the area had flooded. A similar deluge struck two years earlier in 2016, and only two months after the 2018 Memorial Day weekend flood, another storm would flood the neighborhood, challenging ongoing recovery efforts. These severe storms and floods are likely to continue and intensify. Pastor Michael finds himself and his church, congregation, and community exposed and vulnerable to the perils of climate change in the city. This situation is exacerbated by aging stormwater and transportation infrastructure: Westgate, a community of older residents and working families, sits in a floodplain within a city that is inexperienced at handling the interacting social, economic, physical, and environmental responses needed to address these types of events.

Pastor Michael has a diverse professional background, ranging from human resources to music composition, but his pastoral training did not include disaster response. Along with other faith leaders in the community, he has nonetheless been forced to become an expert not only in disaster response but in disaster preparedness as well. And as Pastor Michael's attention has turned to preparedness, his focus has turned to the forested stream and 10-acre forest area that are part of his church's property.

Pastor Michael's focus on the forest is paired with his concern for the community. The recent floods have exposed and exacerbated stresses in the community, including physical, emotional, and mental trauma, in addition to the economic costs of recovery and people deciding to move away from the neighborhood over concerns about crime and schools. Beyond its usefulness in flood control, Pastor Michael also imagined that the forest could be a place for people to walk on paths and sit on benches for solace, stress reduction, and peaceful contemplation. He envisioned a time when the stream could be clean enough that the church could perform baptisms in its waters.

But the church's forest is stressed as well. Members of the Pastor's community with "forest experience" began clearing invasive plants and vines from the ground layer and trees. Pastor Michael and the church used the organizational skills they had developed in recovering from the flood and repurposed them for the forest. They began to build networks of individuals and organizations to help restore the forest and stream, and they mobilized volunteers to join in by removing invasive plants and vines, building trails, and planting an orchard along the forest edge.

This process revealed that the forest was even more ecologically compromised than previously understood. Many of its trees were ash trees, which had died quickly and catastrophically due to a recent emerald ash borer infestation. Further, the growing herd of deer in the forest foraged for seedlings, preventing a new succession of trees. Because of the challenges the forest faces and an unknown future with climate change, Stillmeadow's narrative and approach has shifted from the management of the forest for the community to the stewardship of community and forest together. Through this process, the spirit of individuals and community, and streams and forests, have become interconnected.

Stillmeadow Church and its forest project are representative of many of the conservation and restoration issues that challenge Maryland and its lands. Pastor Michael's journey with his community and forest also represent fundamental ideas in ecology and its subdiscipline, urban ecology.

What Is Urban Ecology?

Ecology is the scientific study of the processes that influence the distribution and abundance of organisms, the interactions among organisms, and the interactions between organisms and flows and transformations of energy, matter, and information. This definition of ecology emphasizes several features:

1. A focus on organisms, aggregations of organisms, or systems incorporating organisms and their byproducts;

2. The consideration of both biotic and abiotic aspects of nature;

3. The relationships between organisms and the physical world can be bidirectional;

4. The boundary between the abiotic and the biotic aspects of ecology can be blurry;

5. A key focus is on "processes," "interactions," and "relations" in addition to physical entities. These processes can be complex, with many interdependent parts that include positive and neg-

ative feedbacks, thresholds, multi-scalar dynamics, and path dependencies or historical legacies.

Several features of this definition benefit from further explanation. *Flows* refers to how energy, matter, and information move through a system, and *transformations* refers to how energy, matter, and information change as they flow through the system. For instance, solar radiation is transformed by plants through photosynthesis to make sugars for growth. These sugars are another form of energy. Plants also transform matter in the form of nutrients such as nitrogen, phosphorous, and potassium for plant growth. Plants can be eaten by omnivores for their own growth. And when organisms die, they are consumed by other organisms with energy and matter transformations and fluxes. Essentially, the fluxes and transformations of energy, matter, and information underly the "web of life."

Systems can be distinguished by whether they are *complicated* or *complex*. Both complicated and complex systems can include many parts. While a spaceship may have many parts, it is a *mechanical system*, and the *interdependent feedbacks* in the system are *predictable*. By contrast, a forest is an *organic system*, and interdependent feedbacks in the system—including evolution itself—are often *unpredictable*. Organic systems have other characteristics not found in mechanical systems, including *unique properties* at *different scales* as well as *historical legacies* or *path dependencies*. Past conditions of an organic system can affect the current and future state of the system. For instance, forests that have been farmed in the past may have very different soil structure, soil nutrients, and seed sources than a forest that has not experienced other previous land uses.

Urban ecology is a subdiscipline of ecology, and humans are one of the organisms included in this definition. As we know, a key feature of humans is that, like numerous other species, we exhibit *social behaviors*. When we include humans in understanding ecological systems, it is important to embrace our own social roles and responses in ecological systems, at multiple scales of social organization. Thus, urban ecology broadens ecology to include understanding the dynamic feedbacks among biophysical and human components of the system, their spatial and temporal contexts, and their effects on ecosystems at various social scales.[1]

The approaches used in urban ecology to include humans in ecology have grown significantly since the mid-1990s. An important leader in this effort is based in Maryland: the Baltimore Ecosystem Study (BES). Originally supported by the National Science Foundation and US Department of Agriculture Forest Service, BES is a long-term ecological research (LTER) project seeking to understand the social-ecological dynamics of the Baltimore region. A significant outcome of BES has been what we call the Baltimore School of Urban Ecology. When we say "school," we are referring to a way of thinking, a set of assumptions about what is important or included, a network of empirical approaches, and a set of goals and approaches. The idea of a school is analogous to an "invisible college," which contains a group of researchers who share goals, communicate regularly, recruit colleagues (including students), and assess the progress of a rapidly changing field of scientific study and decision-making.[2]

FOUR ORGANIZING PRINCIPLES OF THE BALTIMORE SCHOOL OF URBAN ECOLOGY

The Baltimore School of Urban Ecology is based on four propositions. *First, urban ecology addresses the complete mosaic of land uses and management in metropolitan systems.* This comprehensive approach requires that we understand *urban mosaics* as an integrated ecosystem, consisting of biotic, physical, social, and constructed components (see figure 3.1). This proposition follows the definition and "use of the ecosystem" concept in mainstream ecology. Indeed, the original discussion of the ecosystem concept in 1935 emphasized the role of humans.[3] While the basic definition of the ecosystem as a *biotic complex interacting with a physical context in a specified spatial frame* can apply to urban systems, it is useful to explicitly add human social and constructed components to clarify the idea that physical and biotic complexes include social systems and human infrastructure within urban areas.[4]

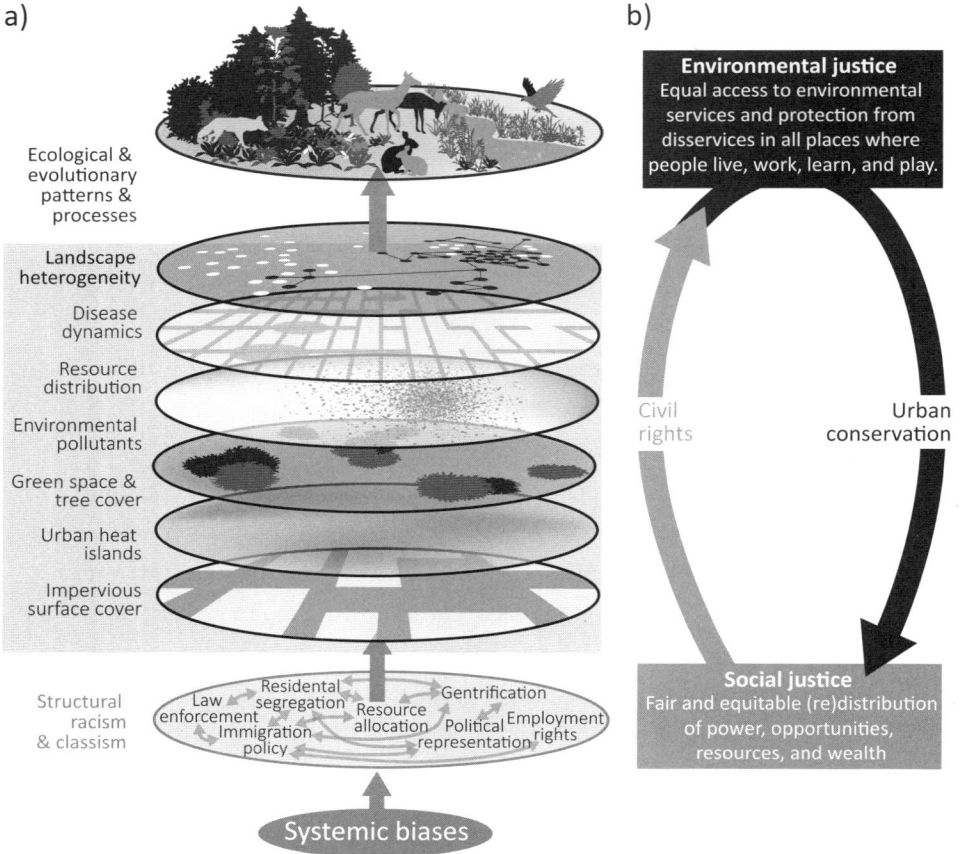

a)

Ecological & evolutionary patterns & processes

Landscape heterogeneity

Disease dynamics

Resource distribution

Environmental pollutants

Green space & tree cover

Urban heat islands

Impervious surface cover

Structural racism & classism

Law enforcement · Residental segregation · Gentrification · Resource allocation · Immigration policy · Political representation · Employment rights

Systemic biases

b)

Environmental justice
Equal access to environmental services and protection from disservices in all places where people live, work, learn, and play.

Civil rights

Urban conservation

Social justice
Fair and equitable (re)distribution of power, opportunities, resources, and wealth

c) Definitions

Inequality: The unequal distribution of wealth and resources across social groups.

Inequity: The unjust allocation of resources driven by power dynamics, discrimination, stereotypes, and systemic biases.

Racism: Stereotypical norms that disadvantage communities of color (typically Black, Asian, Latinx, and Indigenous groups), including the interdependent forces of "prejudice plus power," which dictate how racial inequalities persist even after elimination of racist actors or policies.

Classism: Discriminatory actions based on wealth, income, or social class, usually directed at barring people with working-class backgrounds from accessing benefits and social spaces dominated by middle or upper classes.

Intersectionality: The intersection, interaction, and compounding of marginalized identities, causing individuals and communities at such intersections to experience greater social inequities.

FIG. 3.1. Social and ecological relationships are always highly complex, multidimensional, and interconnected. From Christopher J. Schell, Karen Dyson, Tracy L. Fuentes, Simone Des Roches, Nyeema C. Harris, Danica Sterud Miller, Cleo A. Woelfle-Erskine, and Max R. Lambert, "The Ecological and Evolutionary Consequences of Systemic Racism in Urban Environments," *Science* 369, no. 6510 (2020). Reprinted with permission from AAAS.

The Stillmeadow community and its forest can be understood as a *human ecosystem*. Key characteristics of the *biotic complex* include that the current forest is in an early stage of forest development, with pioneer trees species such as tulip poplar, ash, and slippery elm emerging after agricultural abandonment. Continued forest development has been significantly inhibited by invasive plant competition and deer browse. Further, all of the ash trees have died in the past few years due to the emerald ash borer (see figure 3.2).

FIG. 3.2. In Baltimore's Stillmeadow Peace Park, as in much of the eastern forests, ash trees have been devastated by the invasive emerald ash borer (and in this case also by invasive English ivy). Photo courtesy of McKay Jenkins.

The *physical complex* has been affected by previous agriculture practices, with diminished soil structure and nutrient levels. The forest has numerous sloping topographies, which affect soil moisture and the distribution of tree species from the top to bottom of hill slopes. One part of the forest has a steep, southern facing slope, which increases the amount of sunlight reaching the ground and prevents trees from shading out invasive plants.

The *social complex* includes the church's existing social organization and networks, as well as its adaptability and effectiveness at building new networks to mobilize material and financial resources and volunteers. Many of the participants in the forest restoration activities are motivated by theology and a "call to service," beginning with community and including Earth's care. This involves the care of both communities and forests that are vulnerable and in significant need. The *built complex* includes the church's parking lot, bathrooms, meeting spaces, kitchen, and housing, which support volunteer engagement for forest restoration and education activities. The configuration and condition of the stormwater system in the neighborhood are also part of the built complex and a factor in the flooding that the community experiences.

Our second proposition is that the urban mosaic is complex in terms of space, scale, and time. Since the middle of the 20th century, ecology has become increasingly aware of the need to understand *spatial heterogeneity*.[5] Gone are the days when ecosystems could be considered *homogenous* or *uniformly mixed* and the system to be in *equilibrium* or *balance*.[6] Landscape ecology has emerged as a specialty in ecology to address how ecosystems can be *spatially heterogeneous* or *unevenly mixed*, from the distribution of mushrooms on a log to the mix of residential areas, forests, and agricultural land uses in an urban region.[7] This awareness has prepared ecologists to examine the *fine scale heterogeneity* so often encountered in urban systems, where simply turning a neighborhood corner might reveal a new forest patch defined by differences in biotic, social, or physical components or, more likely, all three. The functional significance of such a *patchwork* has been important in order to understand and identify threats and opportunities for urban ecological conservation and restoration.

Spatial heterogeneity can be examined with increasing levels of analytical complexity. Complexity increases as the analysis moves from patch type and the number of each type, to spatial configuration, and to changes in the mosaic over time.[8] At the simplest level of spatial complexity, systems can be described in terms of a set of spatial patch types. Richness of patch types summarizes the number of patch types making up the set. Analytical complexity is increased when the number of each patch type in an area is quantified. This measurement is expressed as *patch frequency*. How those patches are arranged in space relative to each other increases the complexity of understanding the spatial heterogeneity and structure of the system. Finally, each patch can change over time. Which patches change and how they change is a higher level of spatial complexity. The most complex characterization of system heterogeneity occurs when the system is quantified as a shifting mosaic. The spatial complexity of the area around the Stillmeadow church and forest has changed from 1927 to 2020 in terms of agricultural, forest, residential, and transportation land use "patches"; the size and frequency of these patches; and their spatial arrangement and configuration. These changes in spatial heterogeneity have important implications for the fluxes and transformations of energy, matter, and information in the forest and neighborhood.

Organizational complexity relates to the interactions within and among social and ecological scales of organization. For instance, humans organize and interact at multiple scales of social organization, from individuals to households, neighborhoods, and complex and persistent government jurisdictions. Organizational complexity can also be examined with increasing levels of analysis, reflecting the increasing connectivity of the basic units that control system dynamics within and among scales of social and ecological organization. Within organizational hierarchies, influence can move upward or downward.[9]

Organizational complexity drives *system resilience*, or the capacity to adjust to shifting external conditions or internal feedbacks.[10] Following our structural approach, we can return to a spatial patch as an example of the basic functional unit of a system to explain organizational complexity more fully. The simplest level of organizational complexity is within-patch processes, such as the ecology within a forest patch. When the interactions among patches are incorporated, analytical complexity increases, such as the interactions among forest and residential patches. Understanding how interactions may be regulated by the boundary among patches constitutes a still-higher level of complexity. For instance, how does the structure of forest edge affect the interactions between forest and residential patches? The analytical complexity increases further by examining whether patch interactions are controlled by features of the patches themselves in addition to the boundary. For example, how does the slope of patches affect the flow of water from one patch to another? Finally, the highest level of analytical complexity on the organizational axis is the functional significance of patch connectivity for patch dynamics, both of a single patch and of the entire patch mosaic within and between scales. For instance, how does water flow through the different land use patches of a watershed and what are the effects on water quality?

Temporal complexity addresses the historical contingencies that include legacies, path dependencies, and temporal lags in urban ecological systems.[11] The historical distribution of physical environmental conditions, soils, and biota can influence contemporary and future ecological and social conditions. The built environment is itself a *legacy* in many urban systems. Certainly this is the case in Baltimore, which was established in 1729. The persistent template of the old market roads combined with a series of newer road networks presents a powerful legacy. The clashing street grids, with their alteration of hydrology and demarcation of neighborhoods, and the partially implemented Olmsted Brothers parks and parkways plan is another form of legacies.

As McKay Jenkins describes in his chapter on environmental justice, *social legacies* include the legal segregation of Blacks during the Jim Crow era and other segregation practices epitomized by the federal Home Owners' Loan Corporation "redlining" maps. Persistent social legacies and current

discriminatory processes have important environmental and social consequences. For example, the historic distributions of social groups, economic
classes, and housing characteristics affect the uneven distribution (spatial
heterogeneity) of tree cover and tree species diversity in Baltimore and other
cities in the United States.[12]

Temporal complexity refers to relationships that extend beyond direct,
present-day interactions. Historical contingency includes the influence of
indirect effects, lagged effects, legacies, and the presence of slowly appearing indirect effects. To illustrate the analytical levels of this axis, we start
with simple or contemporary interactions. Contemporary interactions include those interactions when element A influences element B directly. For
instance, removing invasive plants (A) increases forest regeneration (B). Indirect contemporary interactions involve a third component, C, to transmit
the effect of A on B. In this case, removal of invasive plants (A) and the in-
stallation of deer fencing (C) increases forest regeneration (B). An interaction is lagged if the influence of element A on element B is not immediate
but manifested over some period of time. For example, removal of invasive
plants (A) leads to forest regeneration, which leads to changes in tree species diversity (B) as mid- and late successional species grow in the forest. A
higher level of temporal complexity is invoked by legacies. Legacies are
created when element A modifies the environment and that modification,
whether structural or functional, eventually influences element B. Historical agricultural land use and abandonment (A) affects contemporary forest
biodiversity (B) due to changes in social structure, soil nutrients, and seed
sources. At the high end of the temporal complexity axis are slowly emerging indirect effects. Here, climate change (A) affects forest biodiversity
(B) due to changes in temperature and precipitation.

Stillmeadow Community Fellowship and its forest are spatially, temporally, and organizationally complex. From 1927 to 2020, the spatial richness, frequency, and configuration of patches of land use adjacent to (and
upstream of) the area have changed significantly in terms of forest, agricultural, residential, and transportation land uses. Changes in spatial complexity have interacted with organizational complexity, particularly patch

interactions that are adjacent or upstream. Residential areas adjacent to Stillmeadow have been important food sources for deer, sustaining substantial deer populations that use the Stillmeadow forest for both food and refuge. These adjacent areas have also been sources for invasive plant species, including English ivy, porcelain berry, Asian bittersweet, and Norway maple. Changes in patches upstream of Stillmeadow and its Westgate neighborhood—particularly increases in impervious surfaces and stormwater connectivity—have contributed to flooding. Temporal complexity spans contemporary forest restoration, lagged effects of invasives and deer on forest succession, legacies of agricultural land use, and slowly emerging indirect effects of climate change.

Our third proposition is that urban ecology is an integrative pursuit. We have already identified the human ecosystem framework as an integrative frame for urban ecology.[13] This framework requires concepts, theories, methods, and data from both the social and biophysical sciences. We recognize that the social and biophysical sciences can be further categorized in terms of disciplines such as psychology, anthropology, geography, sociology, political science, economics, physics, chemistry, geology, biology, meteorology, and ecology and that these disciplines can be further categorized into subdisciplines. We recognize too that some disciplines include spatial dynamics in their questions and explanations while others do not. Some disciplines tend to focus on one level or scale of organization over another. And some disciplines tend to focus on long-term changes measured in centuries to millennia while others address changes measured in seconds to days. Thus, urban ecology needs to be open to (and capable of) integrating diverse disciplines across the urban mosaic in terms of spatial, organizational, and temporal complexity.

We purposefully use the term *pursuit* in our third proposition to signal a goal. With urban ecology, our goal is to pursue a more general scientific understanding of urban ecological systems and to increase our practical capacity to solve urban ecological problems. Missing from this framework is the role that *race* plays in *social order* and the *inequitable distribution of resources* (figure 3.1). An important contribution of the Baltimore School of

Urban Ecology has been to explore the long-term role of *institutionalized racism* in urban ecological systems.[14]

The work at Stillmeadow and its forest continues to be an integrative pursuit. *Critical socioeconomic resources* include its access to and ability to mobilize information and labor. Its organization (and networks) and beliefs are *cultural resources*. Faith, education, health, and justice are essential *social institutions*, interacting with *social order* factors of social status, knowledge, and territory. Our experience with Stillmeadow and its forest highlights the importance of race. These components of critical resources, social institutions, and social order are essential parts of what makes Stillmeadow church and its forest a complex system with numerous interdependent feedbacks.

Our fourth proposition is that urban ecology can be useful to link and advance both decision-making and research. The typical types of questions that decision-makers ask in order to address a problem are what to do, where, how and how much, by whom, when, and for how long? These types of questions can be seen as components of a decision-maker's solution, and they correspond to our propositions. For instance, the decision-maker's questions "what, how, and how much" are typically questions about the *parts of the system*. "Where" corresponds to *spatial complexity*. "Who" matches to *organizational complexity*. "When" and "how long" correspond to *temporal complexity*. Of course, these types of questions are not exclusive to "environmental problems," and are indeed characteristic of many issues—such as public safety, health, recreation, and community and economic development—routinely faced in urban areas. These problems are in fact often interconnected and may be effectively addressed with an urban ecology perspective.

Urban ecology often engages with what has become popularly known as *wicked problems*. Wicked problems have several characteristics: they occur in situations that are composed of complex systems of interacting and interdependent parts.[15] Stakeholders often hold diverse and conflicting values and perspectives. Solutions are often uncertain and suboptimal.[16]

For example, urban stream restoration projects are wicked problems. How can urban stream restoration projects be designed and implemented to minimize flooding, improve nutrient processing, and enhance aesthetics while—at the same time—minimizing the loss of trees and introduction of invasive species? Are stream restoration projects more effective for minimizing fluxes of water and nutrients than increasing tree canopy cover on upstream areas of the urban watershed mosaic? How do watershed revitalization projects affect property values, risk of gentrification, and existing residents' sense of place? Stakeholders from government agencies, environmental groups, communities, and businesses may disagree, leading to conflicts over preferred solutions, especially when large trees must be removed to reengineer and restore the stream channel and riparian areas. Both the value and scientific validity of the stream restoration may be called into question. Other stakeholders, including local communities, may question whether such expenditures are even a priority given other pressing social or economic concerns.

Wicked problems can also have different temporal dimensions. They can be short term, long term, or immediate.[17] Our stream restoration example may be an illustration of a short-term event, but it is also part of longer-term wicked problems, including the needs to mitigate or adapt to changes in climate and urbanization. Wicked problems are also evident in immediate responses to emergencies and recurrent acute disasters (RADs) such as catastrophic storm events, epidemics, heat waves, droughts, fires, disease outbreaks, and toxic spills.[18] BES has contributed to solutions of wicked problems, for example, in streamside vegetation function,[19] neighborhood restoration,[20] increase of urban tree canopy,[21] environmental justice relative to polluting industry,[22] and downsides of different kinds of stormwater management interventions.[23]

Stillmeadow church and its forest are also an example of *linking research and decision-making* in urban ecology. Ecological conservation and restoration involve assessing ecological conditions, intervening, and assessing what has been successful, what has been unsuccessful, and why. In its most basic form, this process is research. It is also a form of decision-making called *adaptive management*. Based on what is learned, some restoration approaches may be retained, others may be modified or abandoned, and new approaches may be developed. When adaptative management is engaged over time, cycles of research and decision-making develop together. It is important to remember that ecological conservation and restoration are social and cultural processes. For instance, the link between research and decision-making can consider *volunteer engagement* and how volunteer motivations, skills, safety, team size, and organization can affect the project, including the benefits that accrue to participants. At Stillmeadow, volunteers helped plant a tree nursery that has led to the planting of many hundreds of trees in the church forest (see figure 3.3).

A classic 1997 article entitled "Human Domination of Earth's Ecosystems" by the ecologists Peter Vitousek, Harold Mooney, Jane Lubchenco, and Jerry Melillo brought to prominence the need to understand the role of humans in local, regional, and global ecosystems.[24] The paper provided scientific justification for addressing the fact that the world's population was becoming predominantly urban. Since then, with increasing ecological attention focused on urban regions, there have been several progressive phases of how urban regions have been ecologically conceived, understood, and managed.

The *first phase* tended to see urban regions as *complicated* systems, composed of many predictably interconnected parts, and to focus on the negative impacts of urban regions on the environment. A *second phase* consolidated early in the 2000s,[25] recognizing that urban regions are actually *complex* systems, with many interdependent parts that include positive and negative feedbacks, thresholds, multi-scalar dynamics, and path dependencies or historical legacies.[26] Often, this complex urban system was assumed to exist in a *static environment*. The *third phase* recognizes that not only are urban regions complex but they exist in environments that are themselves *spatially and temporally complex*. For example, while urban regions were typically designed for relatively stationary climates, technologies, and economies, today's urban regions need to be

FIG. 3.3. To help repopulate the forest canopy, volunteers at the Stillmeadow Peace Park nurtured thousands of trees in a makeshift nursery. Photo courtesy of McKay Jenkins.

understood in the context of rapid global and regional change and uncertain futures.

These three changing conceptions of urban regions create scientific and practical challenges. Science is progressing from a focus on human impacts, through coupled "natural-human" systems, to the simultaneity of co-produced social-ecological systems.[27] Co-production, in this sense, implies that urban ecology needs to address urban systems that are simultaneously "a biophysical entity, a territory, a commodity, a habitat for nonhuman species, a resource for productive activities, [a place of consumption], and a buffer for absorbing pollutants. These systems are allocated, regulated, and administrated by various laws, norms, and rules and a source of meaning and sense of place, a landscape component, and symbolically loaded."[28] There is also the complex simultaneity of time. To quote William Faulkner's famous line, "The past is never dead. It's not even past."[29] One might also observe that in urban systems the future is never realized, but it is always present.

This progression in how we conceive of urban regions and their environments—beginning from a complicated system in a stable context, to a complex system in a changing world—will require both enhanced and new scientific and practical capacities. There is a need for diverse scientific perspectives on how an urban system works, the nature of its multistranded past, and the shape of its likely futures. There may be incomplete data, uncertain knowledge, or varying levels of confidence in the data and knowledge. Analytical models may not currently exist, or may be insufficient to deal with the complexity of urban ecological problems. Finally, there may not be resources for monitoring and evaluating the long-term impacts of societal interventions for adaptive management. We must continue to reach for scientific and social understanding that is interdisciplinary and synthetic if we are to effectively address these challenges. Such understanding is critical for urban regions and the Master Naturalists who hope to work in them.

Notes

1 J. Morgan Grove, Mary L. Cadenasso, Steward T. Pickett, Gary E. Machlis, and William R. Burch, *The Baltimore School of Urban Ecology: Space, Scale, and Time for the Study of Cities* (New Haven, CT: Yale University Press, 2015).

2 Diana Crane, *Invisible Colleges: Diffusion of Knowledge in Scientific Communities* (Chicago: University of Chicago Press, 1972), https://www.biblio.com/book /invisible-colleges-diffusion-knowledge-scientific-communities/d/1546765946.

3 A. G. Tansley, "The Use and Abuse of Vegetational Concepts and Terms," *Ecology* 16 (1935): 284–307.

4 Mary L. Cadenasso and Steward T. A. Pickett, "Urban Principles for Ecological Landscape Design and Management: Scientific Fundamentals," *Cities and the Environment (CATE)* 1, no. 2 (2008): 1–16; Steward T. A. Pickett and J. Morgan Grove, "Urban Ecosystems: What Would Tansley Do?," *Urban Ecosystems* 12 (2009): 1–8.

5 John A. Wiens, "Ecological Heterogeneity: An Ontogeny of Concepts and Approaches," in *Ecological Consequences of Environmental Heterogeneity*, ed. by Michael J. Hutchings, Elizabeth A. John, and Alan J. A. Stewart (Malden: Blackwell, 2000).

6 Mary L. Cadenasso and Steward T. A. Pickett, "Three Tides: The Development and State of the Art of Urban Ecological Science," in *Resilience in Ecology and Urban Design: Linking Theory and Practice for Sustainable Cities*, ed. by Steward T. A. Pickett, Mary L. Cadenasso, and Brian P. McGrath (New York: Springer, 2013), 29–46; Jianguo Wu and Orie L. Loucks, "From Balance of Nature to Hierarchical Patch Dynamics: A Paradigm Shift in Ecology," *Quarterly Review of Biology* 70 (1995): 439–66.

7 Richard T. T. Forman, "Some General Principles of Landscape and Regional Ecology," *Landscape Ecology* 10, no. 3 (June 1995): 133–42.

8 John A. Wiens, "Landscape Mosaics and Ecological Theory," in *Mosaic Landscapes and Ecological Processes*, ed. by Lennart Hansson, Lenore Fahrig, and Gray Merriman (New York: Chapman and Hall, 1995), 1–26; H. Li and J. F. Reynolds, "On Definition and Quantification of Heterogeneity," *Oikos* 73 (1995): 280–84.

9 Valerie Ahl and Timothy F. H. Allen, *Hierarchy Theory: A Vision, Vocabulary, and Epistemology* (New York: Columbia University Press, 1996); J. Morgan Grove and William R. Burch Jr., "A Social Ecology Approach and Applications of Urban Ecosystem and Landscape Analyses: A Case Study of Baltimore, Maryland," *Urban Ecosystems* 1, no. 4 (1997): 259–75; Rinku Roy Chowdhury, Kelli Larson, J. Morgan Grove, Colin Polsky, and Elizabeth Cook, "A Multi-Scalar Approach to Theorizing Socio-Ecological Dynamics of Urban Residential Landscapes," *Cities and the Environment* 4, no. 1 (2011).

10 C. S. Holling and Lance H. Gunderson, "Resilience and Adaptive Cycles," in *Panarchy: Understanding Transformations in Human and Natural Systems*, ed. by Lance H. Gunderson and C. S. Holling (Washington, DC: Island Press, 2002).

11 Mary L. Cadenasso, Steward T. A. Pickett, and J. Morgan Grove, "Dimensions of Ecosystem Complexity," *Ecological Complexity* 3, no. 1 (March 2006): 1–12.

12 Dexter H. Locke, Billy Hall, J. Morgan Grove, Steward T. A. Pickett, Laura A. Ogden, Carissa Aoki, Christopher G. Boone, and Jarlath P. M. O'Neil-Dunne, "Residential Housing Segregation and Urban Tree Canopy in 37 US Cities," *npj Urban Sustainability* 1, no. 1 (2021): 15; Karin T. Burghardt, Meghan L. Avolio, Dexter H. Locke, J. Morgan Grove, Nancy F. Sonti, and Christopher M. Swan, "Current Street Tree Communities Reflect Race-Based Housing Policy and Modern Attempts to Remedy Environmental Injustice," *Ecology* 104, no. 2 (2023): e3881.

13 Gary E. Machlis, Jo Ellen Force, and William R. Burch Jr., "The Human Ecosystem Part I: The Human Ecosystem as an Organizing Concept in Ecosystem Management," *Society and Natural Resources* 10, no. 4 (1997): 347–67.

14 J. Morgan Grove, Laura Ogden, Steward T. A. Pickett, Chris Boone, Geoff Buckley, Dexter H. Locke, Charlie Lord, and Billy Hall, "The Legacy Effect: Understanding How Segregation and Environmental Injustice Unfold over Time in Baltimore," *Annals of the American Association of Geographers* 108, no. 2 (2018): 524–37; Steward T. A. Pickett, J. Morgan Grove, Christopher G. Boone, and Geoffrey L. Buckley, "Resilience of Racialized Segregation Is an Ecological Factor: Baltimore Case Study," *Buildings and Cities* 4, no. 1 (2023).

15 Jianguo Liu, Thomas Dietz, Stephen R. Carpenter, Marina Alberti, Carl Folke, Emilio Moran, Alice N. Pell, et al., "Complexity of Coupled Human and Natural Systems," *Science* 317, no. 5844 (2007): 1513–16.

16 David Simon and Friedrich Schiemer, "Crossing Boundaries: Complex Systems, Transdisciplinarity and Applied Impact Agendas," *Current Opinion in Environmental Sustainability* 12 (2015): 6–11; Dominic Duckett, Diana Feliciano, Julia Martin-Ortega, and J. Munoz-Rojas, "Tackling Wicked Environmental Problems: The Discourse and Its Influence on Praxis in Scotland," *Landscape and Urban Planning* 154 (2016): 44–56.

17 Scott L. Collins, Stephen R. Carpenter, Scott M. Swinton, Daniel E. Orenstein, Daniel L. Childers, Ted L. Gragson, Nancy B. Grimm, et al., "An Integrated Conceptual Framework for Social-Ecological Research," *Frontiers in Ecology and the Environment* 9, no. 6 (2011): 351–57.

18 Gary E. Machlis, Miguel O. Román, and Steward T. A. Pickett, "A Framework for Research on Recurrent Acute Disasters," *Science Advances* 8 no. 10 (2022): eabk2458, https://doi.org/10.1126/sciadv.abk2458.

19 Peter M. Groffman, Daniel J. Bain, Lawrence E. Band, Kenneth T. Belt, Grace S. Brush, J. Morgan Grove, Richard V. Pouyat, Ian C. Yesilonis, and Wayne C. Zipperer, "Down by the Riverside: Urban Riparian Ecology," *Frontiers in Ecology and the Environment* 1, no. 6 (2003): 315–21.

20 Guy W. Hager, Kenneth T. Belt, William Stack, Kimberly Burgess, J. Morgan Grove, Bess Caplan, Mary Hardcastle, Desiree Shelley, Steward T. A. Pickett, and Peter M. Groffman, "Socioecological Revitalization of an Urban Watershed," *Frontiers in Ecology and the Environment* 11, no. 1 (2013): 28–36.

21 Dexter H. Locke, J. Morgan Grove, Michael Galvin, Jarlath P. M. O'Neil-Dunne, and Charles Murphy, "Applications of Urban Tree Canopy Assessment and Prioritization Tools: Supporting Collaborative Decision Making to Achieve Urban Sustainability Goals," *Cities and the Environment (CATE)* 6, no. 1 (2013): 7.

22 Christopher G. Boone, Geoffrey L. Buckley, J. Morgan Grove, and Chona Sister, "Parks and People: An Environmental Justice Inquiry in Baltimore, Maryland," *Annals of the Association of American Geographers* 99, no. 4 (2009): 767–87.

23 Nicholas B. Irwin, H. Allen Klaiber, and Elena G. Irwin, "Do Stormwater Basins Generate Co-benefits? Evidence from Baltimore County, Maryland," *Ecological Economics* 141 (2017): 202–212; Joanna P. Solins, Amanda K. Phillips de Lucas, Logan E. G. Brissette, J. Morgan Grove, Steward T. A. Pickett, and Mary L. Cadenasso, "Regulatory Requirements and Voluntary Interventions Create Contrasting Distributions of Green Stormwater Infrastructure in Baltimore, Maryland," *Landscape and Urban Planning* 229 (2023): 104607, https://doi.org/10.1016/j.landurbplan.2022.104607.

24 Peter M. Vitousek, Harold A. Mooney, Jane Lubchenco, and Jerry M. Melillo, "Human Domination of Earth's Ecosystems," *Science* 277 (1997): 494–99.

25 Michael Batty, "Cities as Complex Systems: Scaling, Interaction, Networks, Dynamics and Urban Morphologies," in *Encyclopedia of Complexity and Systems Science*, ed. Robert A. Meyers (New York: Springer-Verlag, 2009), 1041–71.

26 Timothy F. H. Allen, Preston Austin, Mario Giampietro, Zora Kovacic, Edmond Ramly, and Joseph Tainter, "Mapping Degrees of Complexity, Complicatedness, and Emergent Complexity," *Ecological Complexity* 35 (2018): 39–44, https://doi.org/10.1016/j.ecocom.2017.05.004.

27 A. Rademacher, Mary L. Cadenasso, and Steward T. A.Pickett, "From Feedbacks to Coproduction: Toward an Integrated Conceptual Framework for Urban Ecosystems," *Urban Ecosystems* (2019), https://doi.org/10.1007/s11252-018-0751-0.

28 Patrick Meyfroidt, Rinku Roy Chowdhury, Ariane de Bremond, Erle C. Ellis, Karl-Heinz Erb, Tatiana Filatova, Rachael D. Garrett, et al., "Middle-Range Theories of Land System Change," *Global Environmental Change* 53 (2018): 63.

29 William Faulkner, *Requiem for a Nun* (New York: Vintage, 2011).

CHAPTER 4

Environmental Justice

McKAY JENKINS

On a recent Monday in June, a pair of Baltimore high school students named Caleb and Tariq spent their morning harvesting carrots and collard greens at the nonprofit Rock Rose Food Justice Project, a small urban farm in the city's Jones Falls watershed. They boxed up the produce, tossed it in the back of my pickup truck, and we drove across town to Love and Cornbread and Soul Kitchen, a pair of nonprofit community kitchens, where the vegetables would be prepared by a team of volunteer cooks and distributed in the form of nutritious meals to about 500 neighbors in two Baltimore communities. This was just the start of another robust growing season at Rock Rose, where we harvest and deliver—every week from March through November—boxes stuffed with eggplants and tomatoes, okra and butternut squash, bush beans and sweet potatoes, all of it grown with volunteer labor and given away free of charge. The farm also functions as an education hub, teaching kindergartners how to plant peas, high school kids how to turn compost into raised beds, and retirees how to harvest tomatillos. From seed to kitchen to table, the farm satisfies many forms of hunger.

The morning's task completed, Caleb and Tariq hopped back in the truck, and we drove across town to a 10-acre urban forest attached to a church in West Baltimore. The Stillmeadow forest, neglected and ecologically degraded for decades, had been suffering mightily from invasive insects that had killed dozens of mature trees and invasive vines that had badly compromised hundreds of others. But that afternoon, Caleb and Tariq were joining a highly energized team of scientists, educators, and teenage volunteers to help repair this damage. Over the past three years, teams organized by the church's visionary Pastor Michael Martin, along with local universities, nonprofits, and the United States Forest Service, have planted thousands of trees, built hiking trails, and created open-air learning spaces where children of all ages have come to work, learn, and play. Terris King, an energetic former kindergarten teacher, leads young people on walks in the park, teaching them about trees but also just letting them stomp around in the woods and splash in the stream. A once-overlooked urban forest patch has now been reimagined as the Stillmeadow Peace Park: a place of respite, rejuvenation, and ecological health. And the church itself has been named a "resiliency hub," not just for its rejuvenated land, but for its deep capacity for delivering food, shelter, and vaccines during times of heat waves, pandemics, or flood.

A few weeks after this, I delivered 140 8-foot white oak planks—milled by Baltimore City's inspirational Camp Small wood recycling center—to the Nepali American Cultural Center, located atop a beautiful hill in rural Baltimore County. There, I joined a group of men and women constructing 40 raised vegetable beds alongside a building complex that serves as a gathering place for some 25,000 Nepali immigrants living in the Baltimore-DC area. The project will serve as a symbol of food sovereignty for hundreds of first- and second-generation families seeking to establish their place in a country far from their ancestral Himalayan homes.

These projects are just a few of countless examples of urban restoration and ecological regeneration work being done across the state under the guiding principle of environmental justice, a growing field in which a Master Naturalist can do some of their most exciting and healing work. These projects go beyond restoring compromised landscapes; they offer the chance to help repair and restore our communities. And they also offer the chance to reconsider, integrate, and redress important parts of our collective history.

This work has become transformative for me personally. For nearly three decades, I have split my time between my home in Baltimore, one of the Chesapeake Bay's largest and most socially and ecologically complex cities, and my work as a professor at the University of Delaware, which sits in the heart of suburbia at the very edge of the Chesapeake watershed. This life has allowed me to spend a great deal of time working with scientists, scholars, activists, and community organizers across many landscapes and social dynamics to restitch some of our frayed social and ecological fabric.

Around 10 years ago, I helped launch a program at my university called the Environmental Humanities. My goal was to introduce aspiring journalists, historians, and literature majors to the ecological sciences, and to introduce young scientists to our country's complex racial, political, and geographic history. As the years have passed, it has become inescapably clear to me that there is no repair work of lasting value that does not engage both our land and the human communities that call it home. Repairing ecological systems and repairing human systems should go hand in hand. As we become more intimate with our land, we can become more sensitive to its health—and the land will return the favor. Robin Wall Kimmerer, the ecologist and Indigenous elder, calls this back-and-forth a form of ecological "reciprocity."

"It's not just the land that is broken, it is our relationship to the land," she writes in her best-selling book *Braiding Sweetgrass*. "For all of us, becoming indigenous to a place means living as if your children's future mattered, to take care of the land as if our lives, both material and spiritual, depended on it."[1]

Master Naturalists understand that ecological systems are profoundly and subtly interconnected. Healthy bird populations require abundant forests and native plants, which serve as hosts to abundant native insects, which serve as bird food. When any of these fundamentals become compromised—

through water contamination, soil degradation, human-caused species extinction, or climate disruption—ecological systems can collapse.

Likewise, nutritious human food requires clean water and sunshine, robust soil microbes, beneficial insects, and a climate in equilibrium. But it also requires intact *social* systems: safe and healthy land on which to grow food and just and equitable economic arrangements around food distribution and access, including places to distribute it, cook it, and eat it. Human communities thrive when people can rely on nutritious food, good jobs, affordable housing and health care, quality schooling, and other dimensions of social equity. When any of these fundamentals become compromised— disinvestment in housing or schools, disregard for public health in the zoning of polluting industries, a willingness to disregard food apartheid—social systems can collapse. And since all functioning human communities are dependent on functioning ecological systems—clean drinking water, healthy tree cover, breathable air—disruption in any of these systems can cause cascading problems for all of us.

So, everything is connected to everything else, and always has been. And when systems become compromised, it's up to all of us to see things clearly and honestly, consider holistic solutions, and do our best to join our neighbors in putting things right. We are talking here not just about ecology and sociology but also about *ethics*, about restoring both ecological and social systems so that they work well for all people.

This work requires an expanded vision of ecological restoration work to include the additional imperative of expanding our vision of who we consider members of "our communities." As Morgan Grove and Steward Pickett point out in this book's chapter on urban ecology, there is simply no way to separate the threads connecting ecology and human culture. Drive anywhere in Baltimore and you will see "racialized segregation, disinvestment, extraction of wealth by landlords and speculators, lack of employment opportunities, and sparse recreational green space in minoritized neighborhoods," they write. Meanwhile, white suburban children learn that city neighborhoods—and their residents—are places to fear and (at all costs) to avoid. In the most literal sense, this is a system designed to perpetuate segregation.[2]

In other words, these outside perceptions are reductive and harmful. As you will see below, Baltimore (and Maryland more broadly) is in fact brimming over with inspiring, creative, and deeply committed citizens, journalists, scientists, and community organizations that are setting national standards for ecological restoration and environmental justice.

Consider the value of restoring the ecological health of an urban neighborhood, protecting the drinking water in a rural agricultural community, or mitigating the damage done to Indigenous lands or communities built in the shadows of industrial sites or coal mines. These places may not fill the pages of tourist brochures, but they are the places where we live, work, and raise our children. In other words, "nature"—and especially the ecological systems we rely on locally to survive—is not a remote place that we visit on vacation. Nor is it something contained within the limits of a county park that offers us respite from the "degraded" places where we live. Nature *is* where we live—whether we live in cities, in the suburbs, or on the farm. And all of it—*all of it*—deserves and requires our deepest and most careful attention. But when we speak of environmental justice, we must also grapple with dynamics not typically studied in ecology, notably politics, race, and class, because environmental damage that affects all of us affects our most vulnerable the most.

Everywhere we look, there are deep, pervasive, and *identifiable* reasons why poor (and typically non-white) communities are most exposed to contaminated air, soil, and water; degraded landscapes; and intensive pressures on community health. Low-income neighborhoods often lack the political or economic clout necessary to prevent industrial development or hazardous waste dumping from being zoned into their neighborhoods. And powerful, often insidiously racist political structures—especially those dictating housing policy, transportation networks, and the zoning of industrial sites—have their own kind of staying power, remaining in place for decades even in the face of public outrage.

Just as ecologists desire to repair *systems* rather than individual *members* of that system, environmental justice work asks that we seek standards of ecological and human health for our communities that have long struggled under the weight of racism and economic disempowerment. Grove, who has overseen the Baltimore office of the United States Forest Service—and has served as co–principal investigator in the Baltimore Ecosystem Study (BES) since its beginning in 1997—views the racism that undergirds so many of our social and political structures as a "complex, adaptive, and resilient system" in its own right. Environmental justice requires that we "understand the dynamic feedbacks that make institutionalized racism so resilient," he writes. "Understanding environmental racism as a system can help identify potential interventions to dismantle the racist system and create new alternatives."[3]

A medical waste incinerator company in South Baltimore's Curtis Bay neighborhood recently agreed to pay $1.75 million—one of the largest criminal environmental fines in state history—for inadequately burning contaminated medical waste it receives from all over the country, the *Baltimore Sun* reported. Now owned by a private equity firm in Los Angeles, the company was also fined $126,000 for air pollution violations between 2012 and 2022. More recently, researchers found coal dust in "100% of samples taken across eight different sites in the neighborhood, ranging from adjacent to the CSX coal terminal to as far as three-quarters of a mile away," the *Baltimore Banner* reports. Because of the high density of industry, including the CSX coal piers, where an explosion occurred in 2021, community groups had asked that Maryland Governor Wes Moore consider issuing an air pollution "state of emergency" for Curtis Bay and the surrounding area, the *Sun* reported.[4]

To Nicole Fabricant, a Towson University professor who supports community activists in South Baltimore fighting for clean air, environmental toxicity is "yet another form of state-sanctioned violence." Such pollution "wreaks havoc upon lands as well as bodies and in so doing affects all aspects of human daily life," Fabricant writes in her powerful book *Fighting to Breathe: Race, Toxicity, and the Rise of Youth Activism in Baltimore*. "Like the criminal justice system that too often provides police with impunity for brutality and even murder, market-driven logics make it difficult to hold industrial polluters, for example, accountable for their toxic impacts and consequent high rates of illness and death."[5]

Take lead contamination, which can cause profound neurological damage in children. For years, the lead paint industry blamed lead poisoning on "decaying cities" and thus on the people living in them, including mothers, "for their neglectful or ignorant childrearing practices, and children for being stupid a priori and thus prone to eat lead paint," the Johns Hopkins scientist Ellen Silbergeld has written. Nine decades ago, the secretary of the Lead Industries Association referred to the poor, Black, and lead-poisoned children of Baltimore as "little rodents."

Silbergeld considers such comments racist and immoral. "These defenses avoid the obvious," she writes. "Only lead—not inadequate mothers, stupid children, or blighted cities—can cause lead poisoning."[6] Indeed, lead continues to contaminate water infrastructure and older housing stock, especially in poor neighborhoods. Across the United States some 9 million lead pipes still carry drinking water to people's homes; in Baltimore City and Baltimore County, water officials lack reliable records about what some 78% of older water pipes are made of and are asking residents for help identifying their pipes using magnets, pennies, and keys, the *Baltimore Sun* reports.[7]

Or take food. In Baltimore, close to a *quarter* of the city's 650,000 residents live in neighborhoods that are "food insecure," meaning they have no access to nutritionally dense food. There may be plenty of *calories* for sale, in the form of fast food or cheap snacks and sodas sold out of liquor stores, but there are no supermarkets, or produce stands, or farmers selling their produce. On many city blocks, diabetes clinics are as common as burger joints.[8]

Or take trees. Baltimore is also near the top of the list of American cities with the greatest summer temperature disparities between wealthy neighborhoods (with dense tree cover) and poor communities (with no tree cover). As in other cities, the coolest neighborhood in Baltimore has *10 times* more tree canopy than the hottest neighborhood; on a hot day in August 2018, the Howard Center for Investigative Journalism reports, there was an *8°* Fahrenheit difference between the coolest and hottest neighborhoods in the city.[9]

So, a basic question: why do we have communities with lots of good food and lots of mature trees, and other communities—just blocks away—with no good food and no trees? This is a question for Master Naturalists seeking to understand environmental justice, and it requires a measure of investigation that can bear painful but important lessons. Indeed, one of the challenges of all environmental work can be the effort needed to understand the *context* of the problems we are working to repair.

Ecologists are trained this way: when they see a stand of dead ash trees, they understand that the degradation of forests doesn't "just happen." Trees don't just die, especially in numbers. Faced with an ecological collapse, an ecologist—before proposing a fix—will search for, and generally find, a root cause. In the case of Baltimore's Stillmeadow forest, for example, the primary suspect for the death of scores of ash trees was the invasive emerald ash borer. An ecologist would know that the ash borer arrived in North America through very specific—and traceable—vectors. As with countless other invasive pests enjoying the efficiency of a globalized shipping economy, the insect likely came to North America on a cargo ship. The same globalized economy that gives Americans access to inexpensive consumer products also exposes us to serious ecological threats. We don't get one without the other.

Environmental justice work requires a similar awareness, and a similar reckoning. In understanding not just surface appearances (a single degraded tree) but deeper, structural questions (a degraded forest system), we can work toward a more robust and lasting regeneration, of both our ecological systems and our human communities.

So, again: why do certain neighborhoods have robust tree cover, ample parklands, plenty of nutritious food, and little exposure to industrial pollution? And why do other neighborhoods, perhaps just miles (or blocks) away, have no tree cover, no parkland, no access to nutritious food, and immediate exposure to toxic pollution? Similarly, why are some rural communities home to robust forests, clean rivers, and state or county parks, and others exposed to industrial water and air pollution caused by giant chicken farms? And how have we, as Maryland residents living (for the most part) in a single watershed, become comfortable allowing such dramatic disparities between places that are "safe" and "beautiful," and others that are "unsafe" and degraded?

As with forests degraded by invasive pests, these "social" circumstances did not "just happen"—they are the result of many years, and many decisions, made by specific people and specific institutions with specific values and goals in mind. As in ecological systems, in other words, human communities have thrived, or suffered, because of specific (and traceable) pressures.

So, let's take a look at how all this has come to be and how we might work to continue to repair some of the damage and support healthier systems.

A good place to start is to go all the way back to the moment when European values and European land use practices arrived on North American shores. From the moment of first contact, Indigenous communities on Maryland's Eastern Shore (see figure 4.1) were "flooded out by hordes of English squatters," who forced them off their land to make room for European agricultural practices, the historians Helen Rountree and Thomas E. Davidson write. From the mid-1600s, settler militias attacked and destroyed Indigenous villages, forcing native people off their land and pushing many people west, where they were often absorbed into other Native communities. Communities that had thrived for centuries along Maryland's rivers, forests, and bays were overwhelmed not just by force of arms and European diseases (often used as a kind of biological warfare) but also by an economic system built not on *regeneration* but on *extraction*— pulling crops from the land, shellfish from the Bay, and trees from the forests—often for export back to Europe, and in the service of channeling wealth to a small set of very prosperous landowners.

"The English who migrated to the Chesapeake region saw the land first as being owned by their monarch and second as being a place for themselves

FIG. 4.1. An engraving (circa 1671, artist unknown) of a Susquehannock village featuring a ring of wigwams within a fenced wall. Indigenous peoples are depicted at the center of the village and also working outside of the fence. In Maryland, the Susquehannock formed villages in what would later become Allegany, Cecil, and Harford Counties. They would, by the 1600s, center primarily on lands along the Susquehanna River. Illustrator unknown. Courtesy of the Maryland Center for History and Culture [SVF], https://www.mdhistory.org /resources/sasquesahanok/.

to practice intensive farming in the attempt to become wealthy," Rountree and Davidson write. "From 1705 in Virginia and 1717 in Maryland, Indians were barred along with other nonwhites from testifying in court against whites, a prohibition that must have had serious repercussions for the tribes whose land was being encroached upon by squatters and who now could only go to the governor in each colony's capital for redress."[10]

It is worth remembering that the colonization of North America was carried out—at least in part—because the European landscape itself had become exhausted by these very extractive practices. Just for starters, the English had so depleted their native forests that—in order to maintain their imperial navy—they needed a new source of timber. The history of the early American colonies is, in a way, a history of trees.

Within just a few centuries, a famously fertile and complex mid-Atlantic ecosystem had begun to teeter: estuaries so heavily mined for fish and shellfish that populations collapsed to the brink of extinction; forests clearcut for lumber, to create tobacco plantations, or (later) for suburban housing development. By the middle of the 20th century, ecologically diverse Indigenous cropland had long since been transformed through the explosive expansion of industrial chicken farms, mono-cropped corn and soybeans, and the intensive use of synthetic fertilizers. The result? An agricultural landscape now controlled by a tiny number of companies, and creating food products that are loaded with cheap calories but very little nutrition. Considered at the time scale of human civilization, this happened very quickly.

So, to begin: Indigenous land theft, and extractive agriculture.

The region's plantation-based (and later industrial) farming methods required a great deal of human labor, and from the earliest days wealthy landowners were always in search of the cheapest labor possible. At first, this came in the form of indentured (that is to say, poor) whites from Europe, who booked passage across the ocean in exchange for (typically) seven years of hard labor. But soon this quest for cheap labor turned to enslaved Africans. Although Maryland never joined the Confederacy, its antebellum economy was fundamentally dependent on—indeed built upon—enslaved labor. In the 1660s, less than a quarter of Maryland's bound laborers were enslaved Africans, according to the National Park Service's Ethnography Program. But within just 20 years, the number was up to one-third, and by the early 1700s, fully 75%. Most worked on plantations in Calvert, Charles, Prince George's, and St. Mary's Counties, but even in Baltimore County, just a few miles from downtown, the 25,000-acre Ridgely Plantation enslaved almost 400 people, making it one of the largest industrial slave operations in the country. Today, as we shall see below, most of this work is done by (frequently) undocumented immigrants, themselves subjected to low pay, injurious working conditions, and little political recourse.[11]

So, second: industrial farming, slavery, and ongoing labor exploitation.

In other words, the remaking of the mid-Atlantic landscape from a functioning ecological system that supported a wide variety of Indigenous communities into a profit-producing collection of "natural resources" extracted and controlled by a small number of wealthy landowners demanded two profound acts of force: the taking of the land itself, and the taking of people to work the land. Both required violence (or the threat of violence), and both left scars on our land and our people that remain deeply embedded in our collective history and our contemporary culture and politics. So disquieting are these stories in some circles that in recent years a number of states—notably Florida, Texas, and Alabama—have begun restricting the teaching of this very history. To say the very least, the repression of what is fundamental to our country's history is unlikely to help us heal.

Maryland, thankfully, has thus far resisted this overt effort at historical erasure. But as elsewhere, the double-edged original sins of Indigenous land theft and chattel slavery has nonetheless become overwritten by a powerful local (and national) mythology that resists acknowledging our painful past while simultaneously continuing to celebrate private land ownership and displays of material wealth.

During the 18th and 19th centuries, wealthy white landowners in Maryland and across the agricultural South promoted a kind of pastoral ideal: the truly "American" way to live (as opposed to the outdated urban "European" way) was not in a cosmopolitan city but on a country estate, surrounded by well-maintained farmland, clear blue skies, and rolling, forested hillsides. Bustling, industrial, and ethnically diverse cities—New York, Philadelphia, Baltimore—were widely castigated as dirty, chaotic, and overflowing with poor (and often non-white) people, just as they were in the "old" cities of Europe that the first colonists (many of them poor and white) had fled to come here in the first place.

The most famous proponent of this early American vision was Thomas Jefferson, a Virginia resident, one-time European ambassador, and the country's third president. Although Jefferson admired the European statesmen and scholars he befriended in Paris, much of the city disgusted him. To

Jefferson, industrial cities "belched smoke into the air and chewed up the bodies of the many to make profits for the few," Carl Zimring writes in *Clean and White: A History of Environmental Racism.* "The poverty and filth of large cities disgusted him when he traveled in Europe, and that disdain stayed with him when he returned to America."

Jefferson's vision was for a simpler, more bucolic country, "where all farmers owned land and shared comparable shares of wealth, political power, and stewardship over thriving natural wonders," Zimring writes. In the small agricultural communities of this ideal, Jefferson boasted, you will seldom meet a beggar, and when you do, they "are usually foreigners, who have never obtained a settlement in my parish. I never yet saw a native American begging in the streets or highways."[12]

The fact that Jefferson used the term "native American" to refer to white colonists tells you all you need to know about his unwillingness to acknowledge the colonial purge of Indigenous communities, let alone the hundreds of people he enslaved on his own plantation. Indeed, the dissonances radiating from Jefferson's vision are still among the most deeply felt in our country's national discourse and remain at the very root of environmental justice work.

In the 20th century, and especially after the economic boom following the Second World War, this pastoral mode of thinking morphed into the mythology of "the American Dream": neatly appointed suburban homes surrounded by trim lawns and—critically—set apart from all the social and environmental "problems" of the city. Highways were built, subdivisions were constructed, and those who could afford to leave older cities fled to these new communities in droves and promptly set up systems of local taxation that kept their wealth (notably through their school systems) very close to home.

The Jeffersonian ideal is a narrative that has created a deep hold on the American imagination and has, over the years, only solidified: we remain a nation split between ethnically diverse cities that—in the Jeffersonian view—are "dirty," "chaotic," and (presumably) "dangerous" and suburban communities that are "clean," "orderly," and (presumably) "safe." Just to

take one recent example, during his own presidency, Donald Trump followed the established pattern—blaming victims rather than systems—
when he called Baltimore a "rat and rodent-infested mess" where "no
human being would want to live."[13]

Maryland's own suburban American Dream has led to the conversion
of tens of thousands of acres of biodiverse forests and meadows into road
networks, subdivisions, and shopping malls, creating an explosion in car-
dependent consumerism that (we now understand) has contributed to
myriad unintended environmental consequences, from deforestation to
climate change to plastic pollution in our oceans. Chemical lawn fertilizers, used obsessively by suburban homeowners, have leached into waterways, creating algae blooms downstream that destabilize aquatic life in
the Chesapeake.

But "white flight," as it has come to be known, has also had other consequences, as wealthy homeowners took their economic (and political) clout
out to the suburbs with them. With money flowing out of urban communities and into the suburbs, dense urban populations were left with broken
infrastructure, crumbling housing stock, and collapsing school systems—
and nowhere near the wherewithal to ward off the polluting industrial sites
that filled in the gaps.

It's not just Baltimore City that has lost population, though Baltimore
did lose fully 30% of its population over the last 50 years. Between 2000 and
2009, even as Maryland's population increased 7.6%, nearly a quarter of the
state's 157 municipalities *lost* population, and another 40% gained fewer
than 100 people. Cumberland, Oxford, and Snow Hill all lost population.[14]

"Where did all those people go that used to live in cities? They didn't
move to Florida; they moved out into the counties, at a tremendous cost of
infrastructure," Gerald Winegrad, a 16-year veteran of the state legislature
who went on to teach public policy at the University of Maryland, once told
me. Sprawl is exacerbated by population growth and by policy. "The myth
that Baltimore lost population because of crime, and that everyone else is
thriving, is a bunch of horse manure. Local governments make all the land

use decisions, and they haven't cared about sprawl so long as it maximizes their tax coffers."[15]

In the 20th century, through the practice of "redlining," banks and government housing policies effectively prevented Black families from receiving home equity loans. With no access to home mortgages, Black families were prevented from accumulating the multigenerational wealth that white families enjoyed through the appreciation of housing prices. To this day, home *renters* have nowhere near the same chance to accumulate wealth as home *buyers*, and the average white household has roughly 10 times the wealth of the average Black household.

The upshot? Black families, and urban Black communities, grew poorer, and suburban white families grew richer. As the decades passed, and cities lost wealth, affluent suburban families supported policies—like using property taxes to pay for their own public schools or getting tax breaks for mortgage payments—that kept their wealth recycling close to home.

Consider: Baltimore has more than 13,000 abandoned residences.[16] Its sewage system is so antiquated that the US Environmental Protection Agency (EPA) has ordered the city to expand a cleanup program for sewage backups in residents' homes. Historically, Baltimore's sewer system included dozens of outflow points "through which, when pipes were overloaded, waste flowed directly into waterways," the *Baltimore Sun*'s environmental reporter Christine Condon reports. "As the city began to close those points, the number of backups into homes increased dramatically." Heavy rains now inundate the city's decrepit pipes, causing sewage to flow backward "and spew from toilets and into basements." The city's department of public works has to send contractors out to clean the sewage from people's homes.[17]

Unsurprisingly, given the disproportionate influence of white (and typically suburban) political power, "environmentalism" for many years was something practiced and preached by affluent whites, and it focused initially on protecting wilderness areas (some might say: vacation destinations) far from the places most people actually lived. Meanwhile, city neighborhoods and the industrial sites that bordered them were considered somehow separate from worries about land protection. These sites became known as "sacrifice zones," places made ugly places by industrialization and considered doomed—but necessary to support the country's economic engines.

Writing as early as 1971, Nathan Hare, a Black sociologist and founder of the first Black studies program in the United States, noted that "white" ecology usually meant worrying that "pollution closes your beaches and prevents your youngsters from wading, swimming, boating, water-skiing, fishing, and other recreation close to home," as well as "the planting of redwood trees, saving the American eagle, and redeeming terrestrial beauty." Meanwhile "urban blacks have been increasingly imprisoned in the physical and social decay in the hearts of major central cities, an imprisonment which most emphatically seems doomed to continue. At the same time whites have fled to the suburbs and exurbs, separating more and more the black and white worlds."[18]

Many of the people left behind by all this economic outflow, those living in densely populated cities and near industrial sites, have traditionally been poor, are often non-white, and have frequently been found to have far higher exposure to toxic air and water pollutants. A widely reported study issued by the New School's Tishman Environment and Design Center showed that of the 73 waste incinerators still operating in the United States, nearly 80% are located in low-income communities or communities of color. Sometimes the only thing separating their homes from industrial pollution are chain-link fences, giving these neighborhoods the dubious nickname "fence-line communities."[19] They have also been saddled by the ominous term "sacrifice zones."

It wasn't until the 1960s and early 1970s that the convergence of the grassroots civil rights and fledgling environmental movements placed issues like toxic pollution in poor urban and rural communities on the front page of the country's newspapers. In the 1960s, Cesar Chavez—the public face of the group that would become known as the United Farm Workers movement—brought attention to California's migrant farmworkers who were vulnerable to both labor abuses and dangerous agricultural chemicals.

The toxic chemical issue further exploded into the country's consciousness with the publication of Rachel Carson's landmark *Silent Spring*, first published in the *New Yorker* magazine in 1962. Other signal moments: the shocking imagery of Cleveland's Cuyahoga River catching on fire in 1969; 20 million people rallying in the streets for the first Earth Day in 1970; the outcry over the Hooker Chemical Company's massive chemical contamination at New York's Love Canal and General Electric's 30-year contamination of the Hudson River with PCBs. By 1982, when a citizen protest erupted over the proposed dumping of 120 million pounds of PCB-contaminated soil in a landfill near a majority Black community in Warren County, North Carolina, environmental justice began developing into a bona fide movement.

In 1987, the United Church of Christ Commission for Racial Justice—using data compiled by the EPA—issued *Toxic Waste and Race in the United States*, which confirmed that race was "the most significant" variable for officials deciding where to place hazardous waste facilities—more determinant than household income, home values, or even the amount of locally generated hazardous waste. (In other words, hazardous waste was—and is—often shipped from the wealthy communities that create it to poor communities where it is dumped.) Not surprisingly, the study also showed that poor and non-white communities routinely suffered disproportionate health risks: three in five Black and Latino communities (and over half of Asian/ Pacific Islanders and Native Americans) lived in areas contaminated by these same toxic waste sites.[20]

Following this report, the United Church of Christ's director, Rev. Benjamin Chavis, coined the term *environmental racism*, which he defined as "racial discrimination in environmental policy-making and the enforcement of regulations and laws, the deliberate targeting of people of color communities for toxic waste facilities, the official sanctioning of the life-threatening presence of poisons and pollutants in our communities, and the history of excluding people of color from leadership in the environmental movement."[21]

Powerful evidence of such environmental racism began to accumulate. In 1990, the sociologist Robert Bullard published his now-classic *Dumping in Dixie: Race, Class, and Environmental Quality*, which further linked the zoning of hazardous waste facilities to historical patterns of racial segregation in the South. Two years later, the EPA published its own *Environmental Equity: Reducing Risks for All Communities*, concluding that non-white and low-income populations were disproportionately exposed to lead, air pollutants, hazardous waste facilities, and agricultural pesticides. It also led to the creation of the EPA's Office of Environmental Justice and the National Environmental Justice Advisory Council, which (at last) turned its attention to communities living near waste sites that were so dangerous they had required—a decade and a half earlier—the creation of a federal cleanup "Superfund."[22]

Yet decades later—and despite Bullard's continuing practice of what he calls "kickass sociology"—the enforcement of environmental justice standards is widely considered inadequate at best. Today, one in six Americans still lives within three miles of an active or proposed Superfund site—filled with contaminants like arsenic, lead, mercury, benzene, dioxin, and other hazardous chemicals that may increase the risk of cancer, reproductive problems, birth defects, and other serious illnesses, according to the US Public Interest Research Group. Since the Superfund program's National Priorities List was first created, less than a quarter of the more than 1,700 sites added to it have been fully cleaned up.[23]

Asthma—a disease exacerbated by air pollution—still affects 13.7% of adults in Baltimore, compared to 9% across the state and country, a recent study by the Baltimore City Health Department found. About a third of Baltimore high school students "have been told by a doctor or nurse that they have asthma, compared to about a fourth statewide," the *Baltimore Sun* reports. Baltimore City has had the highest asthma-related emergency room visits in the state. Yet in 2011, over strong public opposition, Maryland's then-governor Martin O'Malley signed a law allowing the state to subsidize industrial-scale trash burning as a "tier one" source of renewable energy—on par with wind, solar, and geothermal—even though such incinerators spew lead, mercury, carbon monoxide, particulates, and other harmful pollutants into the air.[24]

The South Baltimore incinerator has been categorized as "Baltimore's single biggest source of air pollution," the *Sun* reported.[25]

The reaction from the local environmental justice community was fierce, and for good reason. South Baltimore's Curtis Bay neighborhood has ranked in the top 10 zip codes in the country for quantity of air toxins released from 2005 to 2009, and it ranked first in the country for the quantity of air pollution from stationary sources, according to the Environmental Integrity Project; in 2012–2013, Curtis Bay–Brooklyn was one of the highest risk areas in the nation for respiratory problems.[26]

In her book *Fighting to Breathe*, Fabricant compares the ecological trauma of breathing polluted urban air to the violent death-by-choking experienced by Eric Garner at the hands of a New York City police officer. Garner's dying gasps of "I can't breathe!"—repeated 11 times and caught on videotape—became a clarion call for the Black Lives Matter movement and, Fabricant notes, a potent metaphor for city residents around the world who suffer from air pollution.

Fabricant has worked with young citizen activists like Destiny Watford, who was awarded the Goldman Prize—one of the world's most prestigious environmental honors—for her efforts to organize her Curtis Bay community, and Shashawnda Campbell, who started her own advocacy work at the age of 13 and now helps lead "toxic tours" of mountainous open-air coal pits situated right next to playgrounds. The Curtis Bay pits are part of the country's second-largest coal-export operation, taking in millions of tons of Appalachian coal and shipping it to other states (and other countries). In 2022, at a nearby coal pier, residents felt an "earth-shaking explosion, eventually traced to inadequate ventilation in a conveyor belt tunnel that led to a buildup of methane gas," the *Baltimore Banner* reported. The explosion

> shattered nearby windows and terrified the community, which had long feared such an eventuality in view of the large quantities of coal being transported in unmarked trains and the flammable fuel stored in giant tanks not far from homes, playgrounds and places of worship.

"People were scared, their kids were traumatized. You can't even imagine the panic that was happening," Shashawnda Campbell told the *Banner*. "People thought bombs went off. And when it happened, there were no officials there. None."[27]

Fabricant herself is quick to point out that wealthy communities are complicit in this toxic dynamic through their expansive cycles of consumption and waste. "Every time we throw something away, it is either buried or burned, and more often than not, there is someone living next door to wherever the burying or burning takes place," she writes.[28]

The trouble, as my University of Delaware colleague sociologist Victor Perez, notes, is that there are often great disparities between what residents of environmentally degraded communities *experience* and what they can *prove*. This is true for a variety of confounding reasons. Direct connections between industrial contaminants and specific human diseases can be exceedingly difficult to prove. Testing the cumulative impact of a wide spectrum of contaminants is almost impossible. In other words, if a person (or a community) living next to a waste incinerator, or downstream from an industrial plant, suddenly falls ill, it is very hard to isolate (and prove) that a single contaminant (or even a spectrum of contaminants) caused the illness and thus equally hard to hold a company responsible. Consider the "plausible deniability" practiced for decades by the tobacco industry: even though there was clear evidence connecting smoking and lung cancer, the cigarette industry successfully claimed its products were not at fault for *decades*.

Think how much harder it would be for a community to argue that the soot or polluted water spewing from a local factory (or incinerator) is what caused their own illness. And remember that communities living in proximity to industrial sites typically also suffer from poor housing, a lack of access to nutritious food or decent health care, a lack of tree cover, and other structural drags on their health and well-being. Research has made it clear that children growing up in urban areas—beset not just by industry but by busy roadways and limited green space and few (if any) trees—suffer disproportionately from asthma, respiratory and heart ailments, and psychological stress. "The responsibility of proving negative environmental health consequences is often laid not upon polluting industries but upon the residents of communities themselves," Perez writes.[29]

Environmental injustice is not just visited on residents of Maryland's urban areas. On Maryland's Eastern Shore, the *Bay Journal* notes, "more than 2,000 chicken houses form one of the densest congregations of their kind in the country" and—for the companies that control them—one of the most politically powerful blocs in the state. Although Maryland has enacted some of the nation's toughest water-quality regulations to protect the Bay and its tributaries, state and federal regulations for the $2.7 billion chicken industry still allow the pollution of *the air* without limits or penalties. The result: chicken complexes annually "unleash millions of pounds of ammonia into the air," the *Journal* notes. Ammonia, a major ingredient in soot, "can exacerbate a host of respiratory illnesses from asthma to COVID-19" and can also end up in our drinking water.[30]

Predictably, the people most vulnerable to this rural contamination live in low-income communities. Large-scale chicken operations are located in counties with some of the lowest wealth in the state. As two of the leading food production businesses in the state, poultry and meat processing facilities in Maryland "must acknowledge the threats they pose to environmental and public health," a recent study argued. "Low socioeconomic status communities and communities of color are disproportionately burdened by chicken and meat processing facilities across Maryland, making the state's chicken industry an environmental justice concern."[31]

This is an issue in Maryland and across the country. "For two generations now, the U.S. has effectively assimilated millions of undocumented immigrants into a caste of low-paid labor exempt from most legal protections," the journalist Hector Tobar writes. "By the U.S. government's own estimates, 41% of hired farmworkers in this country are undocumented. Nearly one in three construction workers is Latino. Despite this evident dependence on a workforce from Latin America, 'illegals' has become a noun and a freely used slur."[32]

The poultry industry is also very hard on the workers inside chicken-processing plants; among other injuries, carpal tunnel syndrome is common among workers expected to gut 50 chickens a minute. "The industry taps into marginalized and vulnerable populations," a 2016 study by Oxfam America reports. "Of roughly 250,000 poultry workers, most are people of color, immigrants, or refugees, with a significant number of women. Many poultry workers do not have a platform to speak out about the realities of life inside the poultry plants. They are disenfranchised and intimidated, and often end up injured or disabled and on the street."[33]

If finding definitive proof of negative individual health outcomes from environmental contamination can be difficult, a powerful (and effective) strategy for Master Naturalists interested in supporting environmental justice work can be found in something much less statistical: working with community leaders to collect data, but also to collect stories. A critical component of environmental justice work demands that those seeking to help in the restoration of communities or landscapes enter these new relationships *in support* of existing communities. Master Naturalists can offer support and ecological expertise, but it should always be in partnership with community leaders who are responsible for final decisions for their particular projects.

Community groups are typically far more deeply familiar with the nuances of their neighborhoods than outside "experts" can ever be. As Morgan Grove and Steward Pickett report in this volume's chapter on urban ecology, the Baltimore Ecosystem Study—the longest-running urban ecological research of its kind in the world—has set new standards for evaluating the health and resiliency (and not just the degradation) of urban ecosystems. In addition to studying ecology *in* the city, and the ecology *of* the city, they write, the project's scientists are committed to studying ecology *for* the city. This effort urges (and trains) scientists to work in close, respectful, and humble collaboration with urban communities that have spent decades (and even centuries) beneath the attention of traditional ecological science.

At the University of Maryland, Dr. Sacoby Wilson has established the Center for Community Engagement, Environmental Justice, and Health to address the "pressing regional, national and global environmental justice and health issues that impact low-wealth Black, Indigenous and People of Color (BIPOC) populations and communities." Through research, advocacy, and educational programs, Wilson's center seeks to "step up to help those most impacted by racism, colonialism, state-sanctioned violence, and state-sanctioned environmental oppression."[34]

For 25 years, the Johns Hopkins Center for a Livable Future has been conducting research, creating maps, and working with students, educators, and policymakers to build more resilient and equitable food systems.[35]

Meanwhile, on Maryland's Eastern Shore, Donna Abbott, chief of the Nause-Waiwash Indians, a small band of the greater Nanticoke Nation, struggles to hold on to tribal traditions even as her coastal ancestral lands are being threatened by sea level rise. For years, many in her tribe (and in tribes up and down the East Coast) have had to suppress or deny their native heritage, preferring to "hide in plain sight" rather than subject themselves to the prejudices they have put up with in the centuries since their ancestral lands were taken from them. "Nause-Waiwash descendants didn't organize their history around a name until the late 1980s," *USA Today* reports, and hundreds of tribal members are dispersed along the Atlantic seaboard. Now Abbott, a tribal member who "never saw herself as a leader," speaks at schools and events as its chief, invites people out trapping, and helps host an annual Eastern Shore Native American Festival each September.[36]

Protecting her native lands is only becoming more challenging, given that the Chesapeake's salt-infused water is expected to rise by over two feet by 2050. Climate change, worsening storms, and erosion have already led to the loss of a growing number of Chesapeake Bay islands. Saltwater intrusion threatens to ruin farmland and has already turned once-vibrant forests into collections of "ghost trees." Up and down the coast, dense thickets of invasive phragmites have displaced native species. More than 400,000 acres of uplands within the Chesapeake Bay region will convert to marsh by 2100, predicts Matt Kirwan, a professor at Virginia Institute of Marine Science who studies coastal landscape evolution and has Indigenous family roots along the region's coastal marshes.[37]

Another place environmental justice storytelling is happening is the Environmental Justice Journalism Initiative, a Baltimore-based nonprofit, founded by Baltimore journalist Rona Kobell and sustainability and policy veteran Donzell Brown, that helps train young people to tell their own stories about environmental inequities in their communities. A recent project: reporting on eight floating wetlands built into the Middle Branch watershed to "eat up toxins in the water by creating oxygen and filtering nitrogen and phosphorus that is in runoff."[38]

And then there is food justice, a monumental and complex problem that requires ingenuity and investment from all quarters, including Master Naturalists. In Baltimore City, the Farm Alliance of Baltimore maintains a robust network of urban farms that do double duty: expanding green spaces in the city and assisting farmers in their efforts to feed people in their own communities. Among their showcase projects, BLISS Meadows and Backyard BaseCamp, founded by the visionary Master Naturalist (and pediatric nurse) Atiya Wells, use a 10-acre land reclamation project to serve as a farm and educational hub for Black people to have "a space to process trauma and exchange their stories as well as have moments together that they would be unable to elsewhere."

Farther south, the Black Butterfly Urban Farmer Academy provides a nine-month training program in sustainable farming practices for "socially disadvantaged" farmers. The farm is named in honor of Morgan State professor Lawrence Brown's seminal book about segregation in Baltimore, *The Black Butterfly: The Harmful Politics of Race and Space in America*. In Baltimore's Sandtown-Winchester neighborhood, Pastor Derrick DeWitt, the charismatic leader of First Mount Calvary Baptist Church, recently asked people in his community about their most-pressing priorities. The responses? Crime. Drugs. Unemployment. Blight.

"So I said, 'Well, what can we do to combat all of that with one swoop?'" DeWitt told the Capital News Service. His answer: build an urban farm.[39]

More than 9,000 people live in Sandtown-Winchester, home to "some of the city's poorest residents, most of whom do not own cars," the *Baltimore Sun* notes. "Yet there's not a single food store selling fresh produce or meats within a mile of where they live."[40] DeWitt has helped transform entire city blocks into a series of greenhouses and hired people returning from prison to build the Strength to Love II farm, which now provides both nutritious produce and an elevated sense of community to his neighborhood, and a source of inspiration to the entire city.[41]

Elsewhere in Baltimore, Marvin Hayes is building a composting system meant to address climate change, food insecurity, and environmental justice—all by collecting 1,500 pounds of food waste each week and converting it to rich soil amendments for area gardens. This is urgent work. Across the United States, some 99 million acres (28% of all cropland) "has been lost to soil erosion and continues to be lost at an alarming rate," according to the 2018 *Baltimore Food Waste and Recovery Strategy* study prepared by the Baltimore Office of Sustainability.

Adding compost to improve soil quality and structure is particularly helpful in cities like Baltimore, where "much of the urban soil is severely contaminated with lead and other heavy metals," the study notes. Compost-amended soil can reduce contamination of urban pollutants "by an astounding 60 to 95 percent," and protect "against the danger associated with lead in urban soils."[42] Compost is "black gold," Hayes recently told Inside Climate News. "It sequesters carbon and is used as fertilizer to grow food. We're going to spread some compost love in this city."[43]

All of these vibrant projects, and many more like them, show just how robust, creative, and urgent environmental justice communities can be. Around the region, nonprofits like Interfaith Partners for the Chesapeake (IPC) have organized faith-based communities for the planting of thousands of trees, the construction of rain gardens, and the environmental education of hundreds of congregations. Blue Water Baltimore volunteers and staff can be seen all over the city, tearing up asphalt, planting trees, and building stormwater retention gardens—often in the city's most treeless neighborhoods. Like IPC, Blue Water Baltimore considers its primary mission to be protecting water quality, but their work—and the hundreds of volunteers they rely on each year—are holistic in their approach.

At the national level, things are also beginning to change. Environmental justice is now becoming ensconced in federal policy, with leaders appointed from historically diverse communities. In 2021, Deb Haaland, appointed by President Biden to run the Department of the Interior, became the country's first Indigenous cabinet secretary. In 2022, Ben Jealous, who spent much of his childhood in Baltimore and rose to become president of the NAACP, was named the first Black executive director of a group long considered a bastion of white environmentalism: the Sierra Club. Two years later, the country's first Black vice president, Kamala Harris, accompanied by Michael Regan, the EPA's first Black administrator, visited Baltimore's historically Black Coppin State University to announce an unprecedented $20 billion in federal grants for climate and environmental justice projects.

"The climate crisis impacts everybody, but it does not impact all communities equally," the vice president said. "Here in Baltimore, you have seen your skies darkened by wildfire smoke and you have seen the waters of the Chesapeake Bay rise, threatening homes and businesses that have stood for generations. It is clear that the clock is not only ticking—it is banging."[44]

Notes

1 Robin Wall Kimmerer, *Braiding Sweetgrass: Indigenous Wisdom, Scientific Knowledge, and the Teachings of Plants* (Minneapolis, MN: Milkweed Editions, 2013), 9.

2 J. Morgan Grove, Steward T. A. Pickett, Christopher G. Boone, Geoffrey L. Buckley, Pippin Anderson, Fushcia-Ann Hoover, Ariel E. Lugo, et al., "Forging Just Ecologies: Twenty-five Years of Urban Long-Term Ecological Research Collaboration," in "Shifts in Urban Ecology: From Science to Social Project," special issue, *Ambio: A Journal of Environment and Society* 53 (2024): 826–44.

3 Grove et al., "Forging Just Ecologies."

4 Christine Condon, "South Baltimore Medical Waste Incinerator to Pay One of Largest Environmental Fines in Maryland History," *Baltimore Sun*, October 17, 2023, https://www.baltimoresun.com/2023/10/17/south-baltimore-medical-waste -incinerator-to-pay-one-of-largest-environmental-fines-in-maryland-history/. See also Adam Willis, "Maryland Finally Agrees with Curtis Bay Residents: That's Coal Dust on Their Homes," *Baltimore Banner*, December 14, 2023.

5 Nicole Fabricant, *Fighting to Breathe: Race, Toxicity, and the Rise of Youth Activism in Baltimore* (Berkeley: University of California Press, 2022), 2–3.

6 Ellen Silbergeld, "The Unbearable Heaviness of Lead," *Bulletin of the History of Medicine* 77, no. 1 (2003): 164–71.

7 Christine Condon, "Get the Lead Out: Baltimore Urges City Residents to Test Material of Drinking Water Lines," *Baltimore Sun*, December 13, 2023.

8 Caitlin Misiaszek, Sarah Buzogany, and Holly Freishtat, *Baltimore City's Food Environment: 2018 Report* (Baltimore, MD: Johns Hopkins Center for a Livable Future, 2018), https://clf.jhsph.edu/sites/default/files/2019-01/baltimore-city -food-environment-2018-report.pdf.

9 Xander Ready, Theresa Diffendal, Bryan Gallion, and Sean Mussenden, "The Role of Trees: Poor Neighborhoods in Baltimore Have Far Less Tree Canopy Than Wealthier Neighborhoods," Howard Center for Investigative Journalism, September 3, 2019, https://cnsmaryland.org/interactives/summer-2019/code-red /role-of-trees.html.

10 Helen C. Rountree and Thomas E. Davidson, *Eastern Shore Indians of Virginia and Maryland* (Charlottesville: University Press of Virginia, 1997), 210–12.

11 National Park Service, "African American Heritage and Ethnography: The Peopling of the Maryland Colony," African American Heritage and Ethnography Training Resource, 2006, https://www.nps.gov/ethnography/aah/aaheritage /chesapeakeb.htm. See also Steedman Jenkins, "Unlearning Terra Nullis: History and Memory in a Baltimore Suburb," unpublished senior thesis, Amherst College, 2023.

12 Carl Zimring, *Clean and White: A History of Environmental Racism* (New York: New York University Press, 2015), 10–11.

13 Meredith McGraw, "President Trump Heads to Baltimore, a City He Called a 'Rodent Infested Mess,'" ABC News, September 12, 2019, https://abcnews.go .com/Politics/president-trump-heads-baltimore-city-called-rodent-infested/story ?id=65570278.

14 Maryland State Department of Planning, "Plan Maryland: A Sustainable Growth Plan for the 21st Century," December 2011.

15 McKay Jenkins, "The Era of Suburban Sprawl Has To End. Now What?" *Urbanite*, March 30, 2012.

16 Adam Thompson, Cristina Mendez, and Paul Gessler, "Multi-Billion Dollar Plan to Solve Baltimore's Vacant Home Crisis Disclosed," CBS News, December 11, 2023, https://www.cbsnews.com/baltimore/news/baltimore-mayor-to -announcement-partnership-to-combat-vacant-homes-crisis.

17 Christine Condon, "EPA Orders Baltimore to Expand City Cleanup Program for Sewage Backups," *Baltimore Sun*, July 14, 2023.

18 Nathan Hare, "Black Ecology," *Trends* 8, no. 3 (July 1971).

19 Oliver Millman, "Revealed: 1.6m Americans Live Near the Most Polluting Incinerators in the US," *Guardian*, May 21, 2019. See also Rina Li, "Nearly 80% of US Incinerators Located in Marginalized Communities, Report Reveals," *Wastedive*, May 23, 2019.

20 United Church of Christ Commission for Racial Justice, *Toxic Waste and Race in the United States: A National Report on the Racial and Socio-Economic Characteristics of Communities with Hazardous Waste Sites* (New York: United Church of Christ Commission for Racial Justice, 1987).

21 Joni Adamson, Mei Mei Evans, and Rachel Stein, *The Environmental Justice Reader: Politics, Poetics, and Pedagogy* (Tucson: University of Arizona Press, 2002), 4.

22 Bill Clinton, "Executive Order on Federal Actions to Address Environmental Justice in Minority Populations and Low-Income Populations," February 11, 1994, https://www.epa.gov/sites/default/files/2015-02/documents/clinton _memo_12898.pdf.

23 US Public Interest Research Group, "Funding the Future of Superfund," December 8, 2021, https://pirg.org/edfund/resources/funding-the-future-of -superfund/. See also Yessenia Funes, "The Father of Environmental Justice Exposes the Geography of Inequity," *Scientific American*, September 19, 2023.

24 Angela Roberts, "Study: Baltimore Children Moved from High-Poverty to Low-Poverty Areas Saw Their Asthma Improve," *Baltimore Sun*, May 16, 2023. See also Baltimore City Health Department, "Asthma," https://health .baltimorecity.gov/node/454, and Aman Azhar, "Advocates Welcome EPA's Proposed Pollution Restrictions on Trash Incineration. But Environmental Justice Concerns Remain," *Inside Climate News*, January 13, 2024, https:// insideclimatenews.org/news/13012024/epa-pollution-restrictions-on-trash -incineration/.

25 Christine Condon, "A Burning Problem," *Baltimore Sun*, March 17, 2024.

26 Environmental Integrity Project, *Air Quality Profile of Curtis Bay, Brooklyn and Hawkins Point, Maryland* (Washington, DC: Environmental Integrity Project, 2012), https://www.environmentalintegrity.org/wp-content/uploads/2016/11/2012-06_Final_Curtis_Bay.pdf.

27 Aman Azhar, "Harm City: On a 'Toxic Tour' of Curtis Bay, Academics and Activists See a Hidden Part of Baltimore," *Baltimore Banner*, August 6, 2023.

28 Fabricant, *Fighting to Breathe*, 6.

29 Victor Perez, "Environmental Justice," in *The Delaware Naturalist Handbook*, ed. McKay Jenkins and Susan Barton (Newark: University of Delaware Press, 2020).

30 Jeremy Cox, "Lawsuit Targets Air Pollution from Maryland Chicken Industry," *Bay Journal*, March 1, 2021.

31 Jonathan Hall, Joseph Galarraga, Isabelle Berman, Camryn Edwards, Niya Khanjar, Lucy Kavi, Rianna Murray, Kristen Burwell-Naney, Chengsheng Jiang, and Sacoby Wilson, "Environmental Injustice and Industrial Chicken Farming in Maryland," *International Journal of Environmental Research and Pubic Health* 18, no. 21 (November 2021).

32 Hector Tobar, "The Truths of Our American Empire," *New York Review of Books*, April 18, 2024.

33 Oxfam America, *Lives on the Line: The Human Cost of Cheap Chicken* (Boston, MA: Oxfam America, 2016), https://s3.amazonaws.com/oxfam-us/www/static/media/files/Lives_on_the_Line_Full_Report_Final.pdf.

34 Community Engagement, Environmental Justice, and Health, "Our Mission," https://www.ceejh.center/our-mission.

35 Johns Hopkins Center for a Livable Future, "Center for a Livable Future," https://clf.jhsph.edu.

36 Kelly Powers and Dinah Voyles Pulver, "Muskrats, Eagles, Marsh: This Tribal Chief Sees Climate Crisis Threaten East Coast Culture," *USA Today*, September 20, 2022, updated February 5, 2023, https://www.usatoday.com/in-depth/news/2022/09/20/sea-level-rise-not-just-beachfront-problem/10222539002/.

37 Erin Kelly, "Chesapeake Bay Marsh Expert Joins VIMS Faculty," Virginia Institute of Marine Science News, November 11, 2013, https://www.vims.edu/research/topics/global_change/ts_archive/kirwan_nff.php.

38 Environmental Justice Journalism Initiative, "Our Mission," https://www.ejji.org/about-us.

39 Parker Leipzig, "Rev. Dr. Derrick DeWitt Sr. Wears Many Hats to Transform Sandtown-Winchester," Capital News Service, December 15, 2022, https://

cnsmaryland.org/2022/12/15/rev-dr-derrick-dewitt-sr-uses-his-positions-of
-power-to-transform-sandtown-winchester/.

40 "Baltimore's Food Deserts," editorial, *Baltimore Sun*, June 15, 2015, https://www
.baltimoresun.com/2015/06/15/baltimores-food-deserts/.

41 "Baltimore's Food Deserts."

42 Baltimore Office of Sustainability, *Baltimore Food Waste and Recovery Strategy*
(Baltimore, MD: Baltimore Office of Sustainability, 2018), https://mayor
.baltimorecity.gov/sites/default/files/BaltimoreFoodWaste&RecoveryStrategy
_Sept2018.pdf.

43 Aman Azhar, "Marvin Hayes Is Spreading 'Compost Fever' in Baltimore's
Neighborhoods. He Thinks It Might Save the City," *Inside Climate News*,
August 20, 2023, https://insideclimatenews.org/news/20082023/baltimore
-composting-environmental-justice.

44 Dan Belson, Hannah Gaskill, and Christine Condon, "Vice President Kamala
Harris, EPA Administrator Announce Green Tech Grants at Coppin State in
Baltimore," *Baltimore Sun*, July 16, 2023.

CHAPTER 5

Geology

MARTIN F. SCHMIDT JR.

Maryland is great place to see and learn about geology. With over a billion years of geologic history, we have a great variety of geologic processes and sequences affecting the landforms and rocks we find today.[1]

To examine this variety, we will first take two trips across Maryland: one to look at the geography (landforms), and a slower one to investigate why the landforms are shaped as we find them. For these trips, the reader is encouraged to follow along using the online maps listed in the notes at the end of this chapter, which will provide more detail than we can include here. Then we will take a third trip—this one through time—to examine Maryland's geologic history, to see how the underlying geology came to be. To help keep track of all of this, you can refer to table 5.1 for a summary.

Trip 1: What geography can we find as we travel across Maryland?

When referring to Maryland's geography, it's convenient to group areas with similar landforms. Geographers call these "physiographic provinces," and

TABLE 5.1. Summary of the Geology of Maryland

PHYSIOGRAPHIC PROVINCE	TOPOGRAPHY	PREDOMINANT ROCK CLASS(ES)	PREDOMINANT GEOLOGIC STRUCTURES	PREDOMINANT GEOLOGIC ERA OF ROCKS
Coastal Plain	Flat, low relief; more hills in Western Shore of Bay	Sediments not consolidated into rocks	Nearly horizontal layers, dipping gently toward the Atlantic Ocean	Cenozoic and late Mesozoic
Piedmont	Rolling hills and stream valleys	Metamorphic and igneous; sedimentary in Frederick Valley area	Complex folds and faults, and igneous intrusions	Precambrian and Paleozoic; Mesozoic in Frederick Valley area
Blue Ridge	Mountain ridges, varying shapes	Metamorphic	Broad upfold (anticlinorium) with faults and complex subfolds	Precambrian and early Paleozoic
Valley and Ridge	Long mountain ridges with alternating linear valleys	Sedimentary	Tighter folds, with more plunging folds than the Plateau	Paleozoic
Allegheny Plateau	High land with deep stream valleys and higher ridges	Sedimentary	Broad folds, mainly horizontal	Paleozoic

the six provinces are shown in figure 5.1. We will talk about them traveling from east to west, which—since we are in Maryland—will often mean we are climbing uphill.

Even though it is underwater, we include the Atlantic continental shelf because it is (indeed) part of the continent and geologically continuous with the Coastal Plain. It extends 70 to 80 miles offshore, and the water depth gradually increases to 400 feet deep, beyond which it drops off quickly to over 5,000 feet deep.[2]

As folks see when driving to or from Ocean City, the eastern shore of the Coastal Plain is quite flat; geographers say it is "low relief" because of the small difference between the highest and lowest places. The western

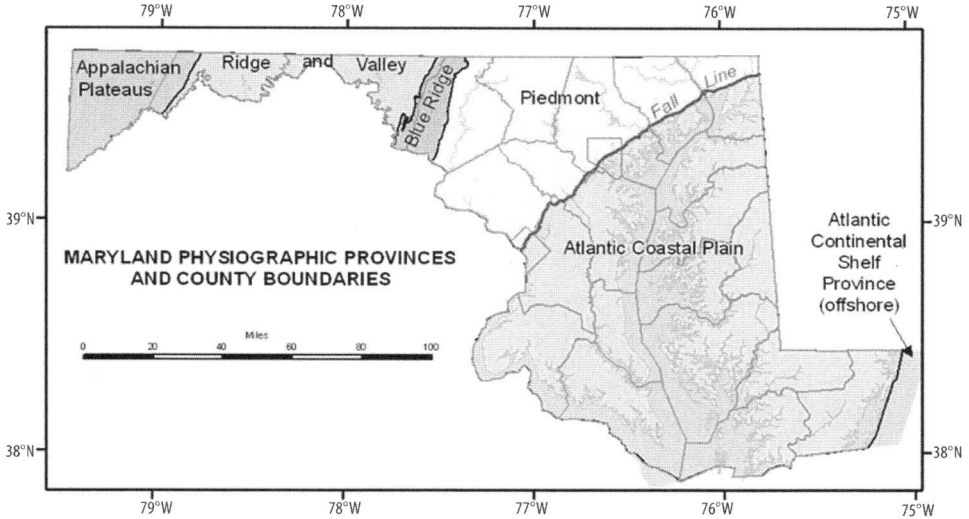

FIG. 5.1. Maryland physiographic provinces. Courtesy of Maryland Geological Survey, http://www. mgs.md.gov/geology/index.html

shore of the Coastal Plain has more relief in its ups and downs, but still stays under about 300 feet elevation.

It's a definite step up to the province of the Piedmont (literally, the "foot of the mountains"), reaching an elevation high of about 1,100 feet in northern Carroll County and Sugarloaf Mountain (at 1,260 feet) in southeastern Frederick County. The Piedmont is generally composed of rolling hills, with a descent on the western side of the province to the Frederick Valley.[3]

The Blue Ridge province has two main ridges that cross Maryland from north to south: Catoctin Mountain on the east and South Mountain on the west. The highest point in the province is 2,140 feet at Quirauk Mountain, on the South Mountain ridge. In the north, mountains fill in between these two ridges, while in the south it opens to a wide valley. One shorter ridge—seven-mile-long Elk Ridge—rises west of and parallel to South Mountain in the southern part of the province. Elk Ridge contains Maryland Heights near the Potomac River and continues as the main Blue Ridge into Virginia and West Virginia.

Continuing west, we descend to the broad "Great Valley" that is the eastern part of the Ridge and Valley province; in Maryland, this is the

Hagerstown Valley. In western Washington County and Allegany County, the province becomes a series of alternating linear mountains and valleys, trending southwest to northeast across the state. The main highways avoid the highest places, but the big (and geologically famous) roadcut at Sideling Hill shows just how substantial the mountains are.

Past Cumberland and LaVale, we start a long climb onto the Appalachian Plateaus province, a broad high area at the western end of the state. Higher mountain ridges stretch above this plateau, again trending southwest to northeast, though more widely spaced than in the Ridge and Valley, and river valleys cut into the plateau. This province includes Backbone Mountain, the highest point in Maryland at 3,360 feet.

Maryland's physiographic provinces are merely a cross section of geologic provinces that run from Pennsylvania to Georgia, so understanding them for Maryland helps us understand the geology of the entire mid-Atlantic states. (This is shown in the map that makes up figure 5.2.) Examining the *causes* of these landforms will be the purpose of our next trip.

Trip 2: Why are Maryland's landforms what they are?

For our second trip across the state, we will start in the west and move to the east. This will be a slower journey, taking the time to understand what processes caused the landforms.

First, it will help to discuss the two major factors that create landforms: *rock character* and *weathering*. Rock character includes both the *type* of rocks that exist in a place and the geologic *structure* of those rock units. Are they horizontal? Tilted? Folded? Whenever discussing rocks, it helps to clarify that *elements*—the basic units of matter—combine to form *minerals*, which are naturally occurring *chemical compounds*. Rocks are an aggregate of minerals, so a specific rock type defines (and is defined by) the physical and chemical properties of its constituent materials. The geologic structures determine where the rocks are on Earth's surface and how they will react to the other primary factor, the weather.

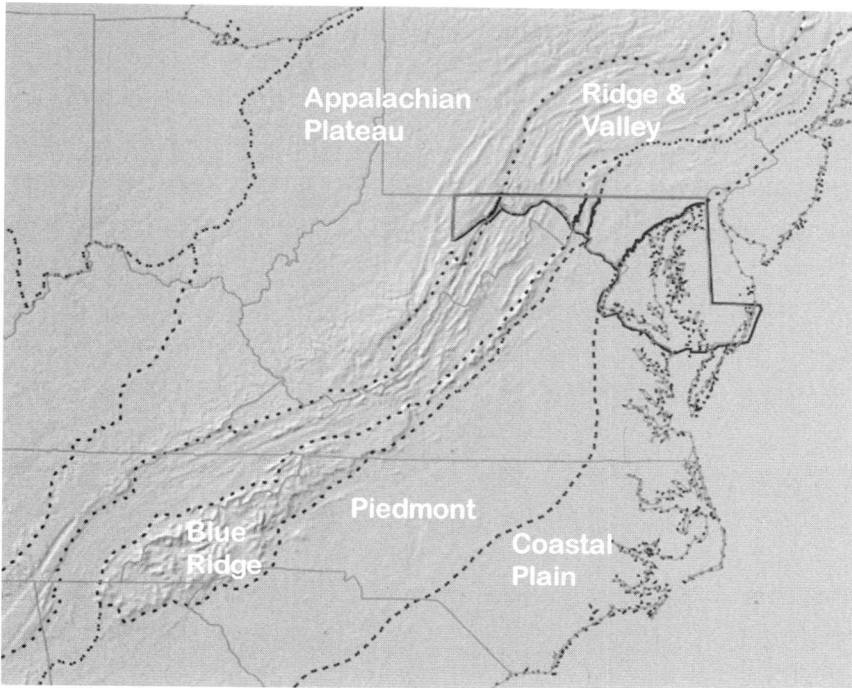

FIG. 5.2. Physiographic provinces of the mid-Atlantic states. Maryland includes all 5 major provinces. Map prepared by Martin F. Schmidt Jr., using layers available in ArcGIS Online by Esri.

Weathering is the term geologists use for the physical and chemical processes simultaneously contributing to the breakdown of rocks due to the action of the weather. *Physical weathering* is the process that turns rocks into smaller pieces of the original material. Examples include *freezing and thawing* (the same forces that create potholes in roads); the *abrasion* caused by rocks and sediments rolling down rivers and becoming smaller pieces; and the impact of tree roots, which can split rocks merely by growing and expanding inside cracks. *Chemical weathering* describes a rock's reactions with air and water, often creating entirely new minerals, which are then, in turn, affected in new and different ways.

The dual processes of weathering are more than simply an eternal cycle of turning big rocks into small rocks. By breaking rocks down and releasing their constituent elements, weathering releases nutritional materials for plants, animals, and people. In other words, weathering is fundamental to all life on Earth.

Speaking of weather: in Maryland, mountain provinces will generally be cooler than the Coastal Plain, but otherwise the entire state has a similar *temperate climate*, with revolving cold and warm seasons and a moderate amount of precipitation. Weathering processes are thus fairly consistent statewide, so it is rock *character* that will have the largest differentiating effect on landforms from one province to another.

Causes of Landforms in the Plateau and Ridge and Valley

In the big picture, Earth has been recycling all the planet's material, especially the material near the surface, since the planet formed nearly 5 billion years ago. This so-called rock cycle moves material through three major classes of rocks based on their origin: *igneous, sedimentary*, and *metamorphic*. If we pick up a rock on our travels across the western Plateau, we would identify it as one of the sedimentary rocks, so that class is the first one we need to understand.

Sedimentary rocks are made of material that has weathered out of other rocks, been deposited as sediments in one way or another, and then turned back into a new solid sedimentary rock. They are divided into two groups: *clastic*, if made of *cemented* pieces (which may be visible or microscopic), or *crystalline*, if made of crystalline masses by chemical (including biological) precipitation. Names and brief descriptions of common sedimentary rock types for these two groups are listed in table 5.2, which also shows a rock's relative *weathering resistance*, which is important in applying a basic principle about making landforms. Over time, in an area made up of two different rock types that start at the same elevation, nonresistant rock will wear away faster than resistant rock and will thus underlie a landform at a lower elevation than a resistant rock unit nearby. This is why nonresistant rocks will be found in our valleys while resistant rocks are responsible for making up our hills and mountains. We can find repeated examples of this principle as we travel across Maryland.

TABLE 5.2. Common Sedimentary Rocks and Their Weathering Resistance

SEDIMENTARY ROCKS	BRIEF DESCRIPTION	RELATIVE WEATHERING RESISTANCE
Clastics:		
conglomerate and breccia	aggregate of rounded pebbles or coarse angular fragments	usually resistant, especially if well cemented
sandstone	mainly quartz sand grains cemented together	usually resistant
siltstone and shale	very fine (microscopic) rock or mineral particles and clay; shale is in layers	nonresistant
Crystalline:		
limestone	crystalline calcite mineral ($CaCO_3$)	moderately resistant to nonresistant in a wet climate like Maryland

Knowing the types of rocks in the two western provinces, the next thing we will need to know is the *geologic structure* of the rock units. Sedimentary rocks are usually deposited in horizontal layers, but if subjected to Earth forces from *plate tectonics* (which we will discuss later), they can be *broken* by faults, or *folded*. Imagine making a cut through the land, running from northwest to southeast across these provinces, so the cut is roughly perpendicular to the lines of the ridges and valleys, and examining this cut from the side. We would find the sedimentary layers of the Plateau to be gently folded and the layers of the Ridge and Valley to be more tightly folded. For these provinces, this looks like figure 5.3.

These folds might seem to explain the mountains we find, but looking more closely we'd see that the main result of this structure is actually found in the pattern of the rocks at the surface. Figure 5.4 shows folded layers in blocks at the bottom, then flat planes at the top, which reveal the visible pattern of the rocks on the ground surface on a geologic map. Notice that these folds create alternating parallel bands of rock units at the surface; since these lay-

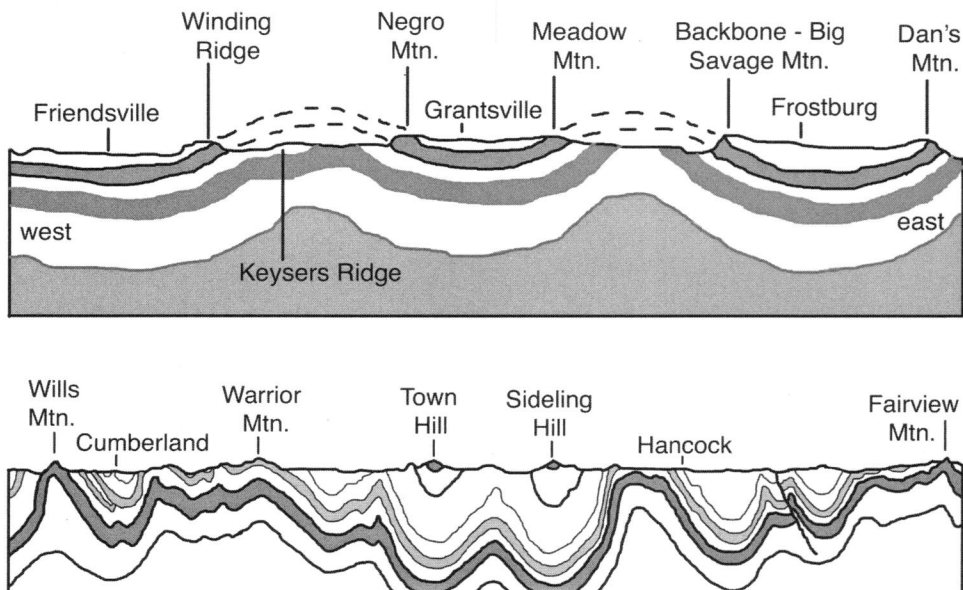

FIG. 5.3. *Top:* Diagrammatic geologic cross-section of the Allegheny Plateau Province in Maryland. This section represents a width of about 30 miles.
Bottom: Diagrammatic geologic cross-section of the Ridge and Valley Province in Maryland. This section represents a width of about 60 miles. Both diagrams are adapted from Geologic Map of Maryland, by Cloos, et.al., 1968.

ers are so exposed, this is the geologic structure on which weathering first begins to work. As suggested by the bumps on the tops of the block diagrams, the *resistant* bands of rock form *ridges*, while the less resistant rocks between them wear away to make *valleys*. In the figure, unit C is resistant in the left block, and unit D is resistant in the right block. Note that the highs and lows of the topography are not necessarily the same as the highest and lowest points of the up-and-down folds of the rock. These are the processes that create the valleys and ridges of western Maryland: the formation of the geologic structures folds rock units in certain locations, and then weathering determines the high and low landforms.

The broad folds on the Plateau result in wider spacing between the resistant rocks (often actually the same folded rock units as they come to the surface across the province), while the tighter folds in the Ridge and Valley

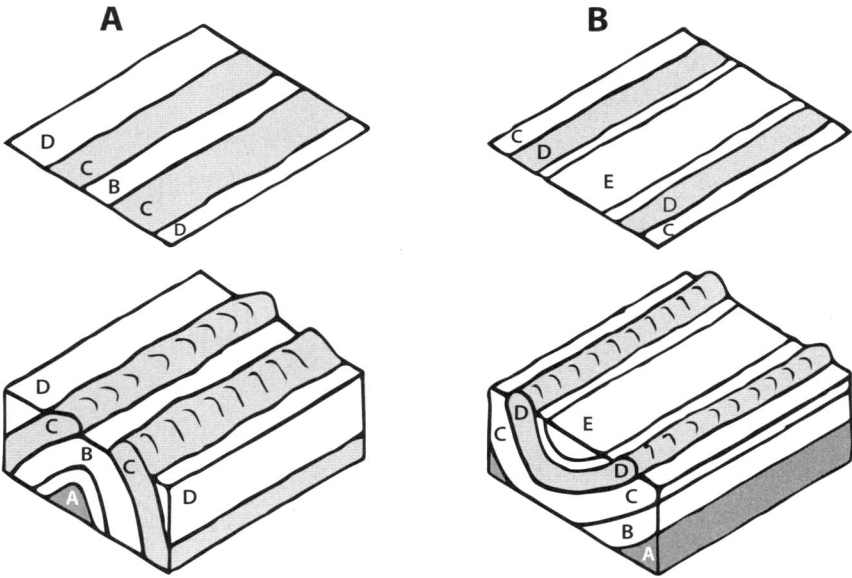

FIG. 5.4. *Top:* Pattern of rocks on geologic maps.
Bottom: Block diagrams of folded layers, with pattern of rock surfaces shown on the top.
Adapted from Martin F. Schmidt Jr.'s *Maryland's Geology* (Schiffer Publishing, 1993).

province can result in closer spacing of the landform variations. Note also in Figure 5.4 that the layers visible at the sides of the blocks are horizontal; with more intense folding these layers can be tilted, resulting in what are called *plunging folds.* These bend the rock unit patterns on the ground into S- or Z-shaped patterns, much like the zigzag pattern that would result from tilting a corrugated roof panel and sticking it in sand. Combined with the processes of weathering, the varying resistances of the rock units create winding patterns in the landforms, which we can see in places between Cumberland and Hagerstown, as well as in the Ridge and Valley province to the north and south of Maryland.

These same dynamics create the landforms of the Hagerstown Valley, but in a somewhat more uniform way. During a period of geologic history we will discuss later, many layers (mainly of limestone and shale) were deposited, then folded and faulted, and now lay beneath the Hagerstown Valley. However, since these rocks all have relatively low resistance to weathering, they have created one broad valley without much relief. The Great Valley continues north into Pennsylvania and south through Virginia; its geology is similar throughout all three states.

So, we see the landforms of the two westernmost provinces of Maryland are a result of their folded geologic structure and the ways in which those rock units have weathered. We'll continue our trip east to see how these factors play out in the next two provinces.

Causes of Landforms in the Blue Ridge and Piedmont

As we head east, reaching and climbing South Mountain of the Blue Ridge, we will see that the interaction between geologic structure and landforms continues in a similar way, but now the rocks have changed, so we need to pause a bit and see what we have.

Metamorphic rocks underlie most of the Blue Ridge and Piedmont. These rocks have been subjected to increased heat and pressure, usually the result of once being buried miles underground, and by mountain-building pressure from plate tectonics. Table 5.3 puts these into two groups: those with some type of layering are in the *foliated* group, and those that are not always layered are *non-foliated*. The foliated ones are listed in the table in order of the increasing amount of heat and pressure to which they have been subjected. For example, we say that slate is *low-grade metamorphism* while gneiss is *high-grade*. Another way of saying this is that their current state tells us a lot about their history.

The geologic structure of Maryland's Blue Ridge province is a broad up-fold spanning the entire province, with some more complexities. The region generally resembles the left-hand block diagram at the bottom of figure 5.4, with the resistant rock (labeled as C) made of the very resistant rock quartzite.

This rock makes up South and Catoctin Mountains on the west and east sides of the province, respectively. Another piece of the same quartzite, relocated by tectonic motion along a fault, also makes up Elk Ridge to the southwest. As elsewhere, once the quartzite formations were put in place by

TABLE 5.3. Common Metamorphic Rocks and Their Weathering Resistance

METAMORPHIC ROCKS	BRIEF DESCRIPTION	RELATIVE WEATHERING RESISTANCE
Foliated:	These all have layers.	
slate	very fine grained, breaks into even layers	moderately resistant
phyllite	fine grained, satin-like shine, breaks in slightly uneven layers	moderately or less resistant
schist	shiny due to visible mica grains, breaks in uneven layers	moderately resistant
gneiss	coarse grained showing layering of minerals but usually does not break in layers	moderately to highly resistant
Non-foliated:	These may be non-layered.	
quartzite	very hard, fine to coarse grained, mostly quartz	very resistant
marble	medium to coarse grained, mostly calcite mineral	less resistant than other metamorphic rocks in a wet climate like Maryland

folding and faulting, slow weathering makes them stand out well above the sedimentary rocks that have worn away to the west and east of the province.

And though most of the Piedmont is made up of metamorphic rocks, as we come down the east side of Catoctin Mountain, we find the Frederick Valley is a north-south band of sedimentary rocks, less resistant than the surrounding metamorphic rocks. This accounts for its lower elevation. As we move east onto the metamorphic rocks that will continue nearly to the Chesapeake Bay, we have to step up slightly onto what is sometimes called the Piedmont Plateau.

The geology of the Piedmont is highly complex because, over the long haul of geologic time, there have been multiple mountain ranges that have risen and eroded in this area. The rocks now on the surface were once miles underground, at the roots of those mountain ranges where they were intensely folded, faulted, and changed by metamorphism. Most of the metamorphic rocks are of similar resistances, so the multiple rock types in the Piedmont have weathered to form rolling hills and valleys without the sharper differences we saw in the western mountain provinces.

Following the weathering patterns we have already seen—geologic structures placing rock units and weathering emphasizing their differences—there are places with more *relief*. An example of a more resistant rock making a higher area is the irregular quartzite that makes Sugarloaf Mountain stand up above its surroundings. Meanwhile, areas underlain by less-resistant marble (especially the unit named Cockeysville Marble) have worn away faster than adjacent rocks to make the many valleys north of Baltimore, such as the Greenspring, Lutherville-Timonium, and Hunt Valleys. And the folding of the Piedmont rocks has created structural undulations that we can describe as *geologic domes*. These generally circular areas can show a common sequence from the center moving out: an elevated central area made of *gneiss*, then a sharp ridge of resistant *quartzite*, then a drop down to much less resistant *marble*, before rising up again to areas of *schist*. So, the hills and valleys we travel can at least sometimes be explained by going from one rock type to another over the complex Piedmont geology.

Here is where we will also need to discuss *igneous* rocks, which are those that solidify from melted rock—*magma* (if formed below the surface) or *lava* (if formed above). Though we have few of either in Maryland, they can be found in the Piedmont. Table 5.4 gives the characteristics of igneous rocks.

As we shall see, there have indeed been volcanoes in Maryland in the past, but plate tectonics over time has turned those igneous rocks into some of the metamorphic rocks we find today. Our only remaining igneous rocks are ones that have cooled underground to make granite and similar rocks. Locally rare as they are, these formations can be important for human use, and they reveal our geologic and cultural history (consider the buildings made of Ellicott City granite). Geologically speaking, most of these rocks weather in a similar way as the surrounding Piedmont rocks, and so don't remain as distinctive landforms.

TABLE 5.4. Common Igneous Rocks and Their Weathering Resistance

IGNEOUS ROCKS	BRIEF DESCRIPTION	RELATIVE WEATHERING RESISTANCE
Coarse, visible grains:	These all solidified underground.	
pegmatite	very large grains, usually quartz and feldspars	resistant
granite	mostly light-colored minerals: quartz, feldspars	rather resistant
diorite	salt-and-pepper mixture of light and dark minerals	moderately resistant
gabbro	dark minerals: pyroxenes, amphiboles	moderately resistant
Fine grained, generally uniform texture:	These all solidified at surface volcanoes; only metamorphosed forms of these are found in Maryland.	
rhyolite	mostly light-colored minerals: quartz, feldspars	generally nonresistant
andesite	mainly gray due to mixture of light and dark minerals	moderately resistant
basalt	black, made of dark minerals: pyroxenes, amphiboles	moderately resistant

There's always an exception, however, and that is the igneous rock *pegmatite*. This rock, usually formed as melted material, oozes from below into cracks in underground rocks and cools to make flat, often near-vertical, sheets of solid rock called *dikes* (they get the name because their weathering pattern is similar to the walls the Dutch use to hold back water). As it solidifies, the chemistry of this magma forms very large mineral grains; when exposed at the surface, it weathers more slowly than most other rocks. Because the dikes themselves are not large, they only rarely make mountainous ridges; more often they make waterfalls, as streams wear the dikes away more slowly than the surrounding rock. We can find such streams in the Piedmont, such as along the Jones Falls and streams in Patapsco Valley State Park.

As we have moved from western Maryland to the Piedmont, we should note both the increasing *deformation* of rock layers and the increasing amount of *pressure* shown by metamorphism. We will see why all this is true when we do our history trip. Meanwhile, we will press onward by taking a step down and east to our last province.

Causes of Landforms in the Coastal Plain

Once again, we need to check out the rock types and geologic structure to understand our easternmost province, but now we have a new twist. The Coastal Plain isn't underlain by the types of rocks found farther west; instead, it sits on mixed layers of gravel, sand, and clay that have not formed (or *consolidated*) into rocks. Right away, this suggests why there is little *relief* here: the formations are made of materials that erode easily and thus can't stand up as hills very well. In addition, the geologic structure is nearly flat-lying layers, which have been uplifted enough on the Chesapeake Bay's Western Shore for rivers and streams to cut some hills and valleys—but not on the Eastern Shore, where the horizontal layers have produced almost entirely flat land. The overall structure is a large wedge: thin at the Piedmont border to the northwest, but nearly 8,000 feet thick below Ocean City. We will see how this formed in Trip 3.

Now that we understand the underlying materials of both the Piedmont and the Coastal Plain, we can explain the step up we took in our first trip as we traveled west. Remember that the metamorphic rocks of the Piedmont are rather resistant to weathering and erosion while the Coastal Plain's sediments are easily eroded. As an analogy, consider a stream of water flowing across a paved driveway and running off the edge: the water won't erode the driveway but can erode the dirt at the edge, perhaps even (over time) creating a small waterfall as it crosses the edge. Similarly, the streams and rivers of the Piedmont flowing toward the Chesapeake watershed have created waterfalls—or, more commonly, rapids in the rivers—as they cross from the last bit of metamorphic rocks in their streambeds to the softer sediments of the Coastal Plain. This boundary, which technically separates the Piedmont and Coastal Plain, occurs on each of the streams in the region. It is known as the *fall line* (or *fall zone*, as the boundary can be hard to locate) and marks the dividing line shown in figure 5.1.

The geologic change at the fall zone has long had an impact on human cultural development. As the first European settlers moved up these streams in boats, water routes became unnavigable at these rapids, so they had to change to land transportation; it was often at these points that people built villages like Richmond, Washington, Baltimore, and (north of our watershed) Philadelphia. Later, we connected these cities with major roads, such as Interstate 95, that also follow the fall line. This is just one example of how geologic history—even things that happened hundreds of millions of years ago—have affected human decisions and actions today.

Trip 3: Traveling through time in Maryland geology

Having discovered a variety of geologic structures in our two trips across Maryland, we can now investigate how those formations developed in the first place. As before, there are some preliminary steps to be ready for the journey.

Let's first look at *plate tectonics*, which describes how the planet's outer layers are not fixed in place but are enormous, moving geologic slabs that include both continents and ocean floor. This process—slabs moving over and against each other—has created a variety of geologic interactions that have occurred in Maryland over the last billion years.

We will describe the timing of events using the approximate years before the present, but it is useful to know how time periods have been divided up in geology. These named time periods are shown in table 5.5, and the eras referred to in the table 5.1 summary.

This journey will take us through more than a billion years in just a few pages, so it will help to organize the trip into three stages according to what the continents were doing: moving toward one another (*converging*), or moving apart (*diverging*).

TABLE 5.5. Geologic Time Scale

EON	ERA	PERIOD			EPOCH	AGE
Phanerozoic	Cenozoic	Quaternary			Holocene	0.01
					Pleistocene	2.6
		Tertiary	Neogene		Pliocene	5.3
					Miocene	23.0
			Paleogene		Oligocene	33.9
					Eocene	55.8
					Paleocene	65.5
	Mesozoic	Cretaceous				146
		Jurassic				200
		Triassic				251
	Paleozoic	Permian				299
		Pennsylvanian				318
		Mississippian				359
		Devonian				416
		Silurian				444
		Ordovician				488
		Cambrian				542
Precambrian	Proterozoic					2500
	Archean					4000
	Hadean					4600

This geologic time scale shows named time divisions. The formation of Earth is at the bottom, and the present time is at the top. The age column shows millions of years before the present for the beginning of the time period on that line. Note the age steps are not uniform, so most of the time (about 88%) is in the Precambrian, while the remaining 12% can be better subdivided using what we can learn from fossils. Adapted from time scale by US Geologic Survey, https://pubs.usgs.gov/fs/2010/3059/pdf/FS10-3059.pdf.

Maryland Geologic History Part 1:
Continents Diverging

We start our time travel a bit over a billion years ago. A supercontinent called *Rodinia* formed when land (that would later become North America) collided with land that would become South America. Where these continents collided, mountains were pushed up, and our oldest Maryland rocks formed as metamorphic gneiss at the deep underground roots of this range known as the Grenville Mountains. These mountains eroded for a long time. By about 750 million years ago (hereafter abbreviated *mya*) these gneisses were exposed at the surface. Around this time, Rodinia began to *rift apart*, splitting the continent into several blocks in the area that will be central Maryland.

This splitting created volcanoes that produced basalt lava (see table 5.5), which we find today in the north-central Blue Ridge, though now slightly metamorphosed into metabasalt. As the continents and rifted *continental fragments* moved apart, and the edges subsided, a shallow ocean covered Maryland. In this water, sediments eroding off North America were deposited on top of the old gneisses. These deposits made sedimentary rocks that will later be metamorphosed, making up many of the rocks we find today in the Piedmont and Blue Ridge. In addition, a large area of *carbonate sediments*—formed by millions of years of accumulating both mud and shells made by sea creatures—piled up to make thick layers of limestone like those we find today in the Frederick and Hagerstown Valleys. These sediments formed on what is called a *passive margin* in plate tectonics, because as an ocean opens, the sediment can accumulate undisturbed along its edges for a long period—in this case it was over 100 million years, a situation we will see again later toward the end of our time travel.

Maryland Geologic History Part 2:
Continents Converging

The region's restless tectonic plates had changed their direction by 500 mya, so plates began to converge offshore east of North America. In a process called *subduction*, one section of ocean crust slid under another section. The material pushed into the Earth heated up, sending up plumes of magma that formed a line of volcanic islands on the surface similar to the Aleutian Islands today. These islands were moving toward North America, and by about 420 mya they collided with the main continent to form what geologists call the Taconic Mountains, a range running northeast to southwest across the eastern side of Maryland. In addition to pasting the rocks of the volcanic islands onto the continent, this collision also folded, faulted, and metamorphosed the sediments of the Piedmont and Blue Ridge. The continent west of these mountains was bent downward and filled in with a shallow sea, and as the Taconic Mountains eroded, these sediments were deposited in this shallow water and later became some of the sedimentary rocks we find today in Maryland's two westernmost provinces. This collision also emplaced the magmas that cooled underground to make most of the igneous rocks we find today in the Piedmont.

There were more continental fragments in the ocean east of North America, and as the converging plate motion continued, these fragments made the next collision with the continent by about 375 mya. This time it formed what geologists call the Acadian Mountains. Their formation furthered the metamorphism of rocks in the east and shed more sediments to the west to be sedimentary rocks at that end of Maryland.

While all this was going on, the far larger continent of Africa was also moving closer to North America. These two continents collided about 320 mya, forming the supercontinent Pangaea, and pushing up the Allegheny or Appalachian Mountains. The continents continued to push together for a long time—until about 260 mya—pushing up mountains at least to the height of today's Rockies or Alps and perhaps as high as the Himalayas. These forces completed the metamorphism of the Piedmont and Blue Ridge rocks we see today and compressed and folded all of the rock layers in Maryland, bringing them to their present positions. The range of deformation (from least to most) that we found in the geologic structures in Trip 2 above (while traveling from west to east) is a result of this collision and had a huge influence on Maryland geology and landforms we find today.

Maryland Geologic History Part 3:
Continents Diverging, Again

Plate tectonics runs in endless cycles. Now that Pangaea has formed—and having gone through an entire cycle—we see a supercontinent ready to split apart just as we were at the start of this history. Pangaea began to rift about 245 mya, again breaking and subsiding in multiple places before a single active rift grew to create the Atlantic Ocean. The Frederick Valley in Maryland is one of the rifts that became inactive, known as a *failed rift*, and we do find some basalt that intruded there again. The rift itself collected some of the sediments that make up the red sedimentary rocks of the valley.

As the Atlantic Ocean began to widen, the *subsided edge* of the North American continent once again became (and remains today) a *passive margin*, where sediments eroded from the high Appalachians accumulated in shallow water. These sediments make up the Coastal Plain of today. At first, the ocean's shoreline was near the current locations of Baltimore and Washington, DC, and we can still detect the thin edge of the sediments there. As the shore fluctuated and moved seaward, the sediments get thicker all the way to Ocean City. This thickening wedge continues out underwater as the continental shelf (remember that province?), where the sediments can be 15,000 feet thick. It's when we imagine putting all these sediments back on the land that we can understand just how high the Appalachian Mountains once were.

We're almost to the present day, but we still have to examine a few more important events that formed Maryland as it is today. About 35 mya, a significant meteorite plummeted to Earth in what is now southeastern Virginia. The meteorite was two miles or more in diameter and made a spectacular crater over 50 miles in diameter. Coastal Plain sediments filled the crater for tens of millions of years, which is why it wasn't discovered until the 1990s.[4]

But the meteorite impact had another long-lasting effect: About 20,000 years ago, Earth's climate cooled, and a lot of water from the oceans became frozen ice in glaciers on the land. The resulting loss of liquid sea-water moved the Atlantic to the very edge of the continental shelf, perhaps 70 miles out from Ocean City. As sea levels dropped, all the rivers in the area developed new channels on the exposed shelf and were drawn downhill to the depression formed by the meteor crater. The Susquehanna, Potomac, and rivers in Virginia all cut deeper channels (below what had once been sea level), came together at the crater, and flowed out to the ocean from there. As ice covering much of the East Coast north of Maryland began to melt, sea levels began to rise, flooding these channels and forming the Chesapeake Bay. So, the meteorite did not technically *create* the Bay; instead, it influenced the shapes of the river valleys that form the Bay. It's appropriate that the Bay's formation story comes at the end of our Maryland geology story because the underlying Coastal Plain had to be built long before the Bay could form in the province.

So, yes, lots has happened geologically to make Maryland the varied terrain it is today. In turn, the landforms affect the human-made and natural aspects of what we find on the land, connecting us to geological processes and events that happened long ago. This is an interesting perspective to keep in mind as we study the many features of the natural world covered elsewhere in this book.[5]

Notes

1 For a further exploration of Maryland's geologic history, see Martin F. Schmidt Jr., *Maryland's Geology* (Atglen, PA: Schiffer Publishing, 1993) and John Means, *Roadside Geology of MD, DE, and Washington DC* (Missoula, MT: Mountain Press, 2010).

2 GEBCO Bathmetry 2022 by ESRI in ArcGIS Online.

3 Elevations throughout this chapter come from the LIDAR elevation map for Maryland created by the Eastern Shore Regional GIS Cooperative (ESRGC) and available for GIS use at https://data.imap.maryland.gov/. There's also a list of highest and lowest elevations by county at the Maryland Geological Survey site at http://www.mgs.md.gov/geology/highest_and_lowest_elevations.html.

4 "The Chesapeake Bay Bolide Impact: A New View of Coastal Plain Evolution," USGS Fact Sheet 049-98, Coastal and Marine Geology Program, https://pubs.usgs.gov/fs/fs49-98/.

5 There are many changing internet sites on geology, from local to global, so search for those. Try starting with the Maryland Geological Survey website: http://www .mgs.md.gov/; interactive maps of Maryland: https://gisapps.dnr.state.md.us /MERLIN/index.html; interactive maps for the United States: https://apps .nationalmap.gov/viewer/; and National Geologic Map Database: https://ngmdb .usgs.gov/maps/MapView/. The Maryland Geology for Education web map can be accessed at https://mdgeoed.maps.arcgis.com/home/webmap/viewer.html ?webmap=d7cfa7a928f14d89ae9556612001033b.

CHAPTER 6

Soils

DAVID RUPPERT

We have all seen plants growing out of a crack in a rock. How could this be? After all, in geology speak, rock is consolidated, cemented, and rigid—clearly, not a place for a plant to grow. But look more closely at the crack, and you will see that it is actually an opening in the rock, a place where water, air, and nutrients can enter, be stored, and escape: in other words, a *pore*. Without pores, geological materials would have surfaces as smooth as kitchen countertops and interiors as unforgiving, impenetrable, and sterile as we know rocks can be.

Because rooted plants are just like us—taking in and expelling food, air, and water—they need a substrate that allows for these sorts of exchanges. A mere pore in a rock can, in some cases, be enough to function as *soil*.

Soil-making porosity can be created in many ways. Wind and water, freezing and thawing, crushing pressure from other rocks, intrusive tree roots: all these things can split or abrade rocks—quickly or over time—into a handful of (or a million) smaller pieces. The forces that change mineral particle size without chemical alteration are called *physical weathering*.

TABLE 6.1. Weathering: How Geologic Materials Are Made Smaller or Chemically Altered

	PROCESS							
	ABLATION		SPLITTING				MINERAL DISSOLUTION	FORMATION OF NEW SOIL MINERALS
Mechanism	wind erosion[†]	water erosion[†]	thermal expansion	freeze/ thaw	pressure relief	root wedging	radically different conditions in soil versus location of formation of original mineral	
Result	geologic mineralogy unchanged, but particle size is diminished; smallest size achieved is silt						dissolved ions moving downward or out of the soil	clays; unique mineralogy; smallest particle sizes
Category	physical weathering						chemical weathering	

[†] Wind and water do not cause the ablation of materials; rather, the solids entrained in them do.

    While rocks can be changed in size (physical weathering), they can also be changed in substance (*chemical weathering*). Minerals exposed to altered chemistry may be dissolved completely, or may be re-formed into new minerals (see table 6.1). Magma- and mantle-formed rocks start deep in the Earth, under tremendous pressures and temperatures, with a limited amount of water and an even more limited supply of gaseous oxygen. When these deep-in-the-Earth, high temperature– and high pressure–formed materials encounter conditions at the Earth's surface, they experience a completely different (colder, wetter, more oxygenated, reduced-pressure) environment. And because Earth's environmental conditions are so different from those under which most geologic minerals formed, many geologic minerals become less stable at the Earth's surface and dissolve slowly under millions of years of relentless rainwater.

    This mineral instability at the Earth's surface has three profound implications. The first is that oceans become saltier over time. The dissolution of crust- and soil-borne minerals and their passage out of the soil system to the world's oceans (where they have no escape) mean that the dissolved ion content of the Earth's ocean system has been steadily increasing since the oceans were first created.

The second implication is the formation of novel, unique-to-the-soil minerals called *clays*. Particular clays arise because of specific conditions in the soil at the time of their formation, including the mixture of dissolved ions out of which the clay will form, pH, the amount of oxygen, and temperature.

Clays are unlike geologic materials because they are *formed in the soil* and are uniquely small in shape and size (see tables 6.1 and 6.2). Such minerals tend to form (nanoscopic) *flat sheets* rather than the three-dimensionally symmetric *crystalline structure* of many geologically formed minerals. Because of this unusual structure, soil-formed clays enhance soil fertility in two ways. First, they preserve some amount of porosity, even when they fill a void created by the dissolution of a previously present mineral.

Second, because water in soil contains a wealth of different dissolved ions—and because soil-mineral formation can occur relatively quickly— clay crystals often contain many impurities, which can further compromise the crystal structure, bending and twisting the clay sheets and altering their charge away from neutrality. Unlike physically weathered geologic materials, the inherent chemical charge of clays allows them to capture dissolved ions from soil water, retarding the runoff of nutrients to the ocean and thus making the nutrients available for plants (and thus animals and humans) (see table 6.2).

A third profundity: many of the *sedimentary rocks* that give birth to modern soils themselves contain clay. This can only be because these rocks (shales, slates, most limestones, and many others) themselves were formed at least partially out of clay-containing sediments. Hence many of our modern soils are made out of materials that originally formed in soils that no longer exist, predating even the ancient rocks that contain them. Everywhere we go, we walk on deep time that has been consistently reorganized.

Soils are places of *novel reorganization*: in other words, because of a variety of processes (physical, chemical, biological), soils are *distinguishable from their parent, geological materials*. As soils age, the geologic materials

TABLE 6.2. Soil Particle Sizes and Resultant Characteristics

PARTICLE SIZE CLASS, SIZE (MM)		SAND (0.05–2 MM)	SILT (0.002–0.05 MM)	CLAY (<0.002 MM)	
Feel in hand		Gritty	Soft, velvet	Sticky, plastic, moldable	
Chemistry	Origin/chemical makeup	Pulverized rock; same chemistry as parent rock		Soil-formed Al, Si, and O platelets and trace elements (e.g., Fe)	
	Surface area, volume	Low		High	Depends on the specific mineralogy of the clay
	Inherent charge	Neutral		Negative[†]	
	Nutrient-holding ability and buffering against chemical change	Negligible		Non-zero but variable	
Porosity, Hydration, Aerationology, Aeration	Average pore size/ hydraulic conductivity[‡]	Large/high	Medium	Small/low	
	Ability to hold water . . . / against gravity	Very little; water freely drains	Yes; water held too strongly for gravity to drain		
	Ability to hold water . . . / against plants	Very little; plants can pull water out of pores of these sizes		Yes	Water in "micropores" between clays is held too strongly
	Ability to hold water . . . / long term (for plant uptake)	Very little; drains freely after wetting	Yes	No	
	Ability to provide aeration	Yes	Only when soil is partially dry	No; usually filled with water	
Erodability	Moved by wind	Usually smaller sand sizes moved locally (dunes)	Can be moved long distance (e.g., dust bowl "loess")	Resistant to erosion; too sticky to be moved unless primarily stuck to sand or silt	
	Moved by water	Increased erodability with smaller size	Easily moved by water		
Sedimentation	Drops out of still water. . . .	within seconds	within fractions of a minute to hours	within days to years in fresh water; accelerated in salt water	
	Freshwater deposition	Rivers	Slow flow	Lakes (still water)	
		Moderate flow			
	Marine deposition	Shoreline (beaches)	Near shore (mud)	Offshore	

[†] In Maryland, most Si-containing clays (the vast majority of clays) tend to be negatively charged; Al- and Fe-dominated clays (the minority) tend to be positively charged.
[‡] Hydraulic conductivity is the speed of water movement through the soil layer when saturated with water; usually measured in in/hr. Healthy soil structure increases hydraulic conductivity beyond that provided by particle size.

out of which they form are slowly transformed in structure, mineralogy, and organic content. These transformations occur at different depths and at different speeds according to the amount and nature of:

1. Organic matter and sediment additions

2. Rainfall infiltration and descent and the seasonal height of the water table

3. Soil temperature

4. The availability of oxygen and the mineralogy of the soil material that is present

5. The action of soil biota (of all sizes, e.g., bacteria through badgers)

6. The landscape setting and the *disturbance regime* (how often or how infrequently soil development is interrupted)

Depending on the uniformity of these factors—and the longer these factors are brought to bear in a uniform way—the soil will form layers (*horizons*) that are distinct from what it inherited from its original geology. Basic field observations about the presence (or lack) of certain features, such as degrees of cementation/consolidation, soil structure, clay movement, or organic matter, are enough to distinguish basic *master horizon* types from one another (see table 6.3).

Field Techniques Used to Describe Soil Profiles and Horizons

STRUCTURE AND CEMENTEDNESS

In most cases, soils develop out of rock or sediments originally provided by geology. When material still has the same structure as inherited from geology, the soil horizon is called a "C" or "R" (table 6.3). If the horizon is sufficiently free of cementation, such that it is diggable with a hand shovel, the

TABLE 6.3. Tell-Tale Soil Features, Soil-Forming Processes, and Resulting Soil "Master Horizon" Layers

DIGGABLE WITH A HAND SHOVEL?	STRUCTURE OF THE MATERIAL	PRESENCE OR LACK OF OXYGEN DUE TO POSITION OF WATER TABLE	CUMULATIVE EFFECT OF THOUSANDS OF YEARS OF RAINFALL	DECAYING PLANT MATTER		SOIL "MASTER HORIZONS"
				MIXING BY SOIL FAUNA INTO MINERAL MATERIAL	NATURE OF THE MATERIAL	
Yes; uncemented, unconsolidated[†] "regolith"; supports the rooting of plants	Created in place during soil formation; dissimilar to the structure of the originally deposited geological material; "soil horizons"	Graying because of lack of oxygen (below water table) or reddening because of exposure to oxygen (above water table)	Clay particles displaced below[§]	Addition of organic matter but little mixing	Primarily organic	O
				Addition of organic matter and mixing with mineral material	Primarily mineral	A
				Usually no addition of organic matter; Bhs horizons are an exception		E[#]
			Clay particles accumulated from above[§]			B
			Often too deep for clay deposition or chemical alteration-induced color changes, but exceptions to this rule occur			C
No; cemented, "consolidated rock"[‡]; generally does not support the rooting of plants	Substantially unaltered from the original geologically deposited material; "parent material"					R[¶]

[†] In the Coastal Plain, unconsolidated sediments stretch meters to kilometers deep. From a soil perspective, bedrock is simply nonexistent in the Coastal Plain. Hence, R horizons are not observed in the Coastal Plain.

[‡] Cemented, consolidated rock is only sufficiently close to the surface to be observed in soil profiles in the Piedmont and other geologic provinces to the west. In the Piedmont, consolidated rock may be sufficiently deep to not be commonly observed.

[§] Clay is given special mention here because of the ubiquity of its downward movement in sufficiently undisturbed soils in Maryland. There are many ways to get a B horizon, but in Maryland the accumulation of clays is a major one.

[#] E horizons are zones of loss and leaching; neither organic matter nor clays collect. They are typically lighter in color than surrounding horizons and are typically sandier and less clayey than underlying B horizons.

[¶] Not true soil horizons unless they are penetrated with some frequency by roots; they are mentioned here because they are interesting features of some soil profiles in the Piedmont and west when you are lucky enough to observe them.

material is *unconsolidated* and considered a *C horizon*. Layers that are *cemented*, with *consolidated* (undiggable) geologic materials (e.g., rocks) are considered *R horizons* (see table 6.3).

As a soil develops in a particular place, it *self-organizes*, and the structure initially provided by geology or sedimentation is altered by the shrink/swell action that accompanies drying/wetting and freeze/thaw cycles, the downward movement of clays between the structural units as they develop, and the burrowing and defecation of soil animals. In the end, the structure provided by geology is replaced with soil-developed structural units (see table 6.3).

Generally, soil structural units (*peds*) are larger when they are deeper in the ground, mostly because of reduced biological activity and more uniform temperature and moisture conditions. *Blocky* structure reflects peds that are roughly *equidimensional* in shape. Most blocky peds in Maryland are *subangular*, meaning that their edges are *somewhat rounded*; *angular* blocks, with knife-edge definition, typically only occur in clayey soils and even then are rare in Maryland.

While subangular blocks can occur in any soil horizon in Maryland, vertically oriented *prismatic* structures develop, if at all, only in the subsoil. *Granular* structure characterizes untrammeled *A horizons*, where fecal material from countless large and microscopic creatures has glued soil particles together. If left unprotected, these roughly spherical and small structures may erode downhill, leaving thin A horizons behind. Vehicular or foot trafficking can smash peds or squeeze them across one another, creating "platy" structure that might look geological but is not.

TEXTURE

Soil texture has profound implications for soil functioning and helps in the identification of certain horizon types relative to one another. The United States Department of Agriculture (USDA) defines *soil particles* as being two millimeters or less in size, and it is true that the size of soil particles determines much about how their surrounding soil layers function. But not all soil particles are the same. Most soil horizons are a mixture of multiple particle sizes, and the functioning of the whole horizon emerges from the many parts (see table 6.2).

Of course, no one measures the size of *all* the soil particles, but just examining by feel you can get a good idea of the relative distribution of sand, silt, and clay. The relative percentage share of sand, silt, and clay particles (called soil *separates*) makes up what is called *soil texture*. A *loam* is simply a soil texture (as felt by the hand) where the grittiness of sand, the moldability of clay, and the softness of silt are in relative balance. In gardening magazines, "loam" is often used to indicate a higher percentage presence of *organic matter*, which tends to increase soil *fertility*, but in the soil science world, a "loam" is merely a balanced-feeling soil reflecting only the "mineral" matter. Note that a "minerally balanced" soil does not necessarily "feel" balanced; it feels more "clayey" than it is because the inherent high surface area of clays dominates the mechanical (and chemical) interaction of soil particles.[1]

ORGANIC MATTER

The presence of *organic matter* is essential to the identification of O and A horizons and Bh horizons. Because clay-heavy soils contain much more surface area for organic molecules to bond to, a soil's organic matter content generally increases with greater clay concentrations. Sandier soils, with larger grains and thus less surface area, leave organic matter more exposed to oxidation by soil biota; 3% organic matter in sandy soils would be quite high (clayey soils can have up to 8%). Tillage and similar disturbance tend to lower concentrations of organic matter by exposing it to greater aeration and therefore greater oxidation (and breakdown) by soil biota (see tables 6.3 and 6.4).

COLOR

Generally speaking, the darker the soil, the greater the amount of organic matter. But remember that because sandy soils contain less surface area, a given amount of organic matter may make them *appear* darker; dark colors may also derive from *manganese oxides* or *iron sulfides*, and their coloration can be difficult to distinguish from that of organic matter. Orange/brown/

TABLE 6.4. Hints and Tricks to Identifying B Horizon Types in Maryland

FIELD INDICATORS/ FORMATION PROCESS	FREQUENCY	TYPICALLY FOUND		NAME	ABBREVIATION[†]
Accumulation of clay displaced from overlying horizons, usually visible as waxy coatings on the outside of peds; often accompanied by yellowing, reddening, or browning due to oxidation	Most common	throughout Maryland in subsoils undisturbed for thousands of years	Piedmont and terraces of the Coastal Plain	Argillic horizon	Bt[‡]
Slight clay increase without the formation of waxy coatings	Common	throughout Maryland in subsoils undisturbed for hundreds of years; relatively stable floodplain or mountainside landscapes	floodplains in any geologic province; mountainsides	Cambic horizon	Bw[‡]
Reddening, yellowing, or browning due to oxidation					
Development of a unique structure unlike over- or underlying horizons					
Graying, "gleying" due to lack of oxygen	Uncommon	in wetlands in any geologic province			Bg
Densified silty material; restricted water flow and rooting; large prismatic peds; very hard to excavate		in all geologic provinces; largely missing from the Mid and Lower Shore of Maryland, though present in Delaware		Fragipan	Bx
Accumulation of organic matter and Fe and Al displaced from overlying horizons, visible as a dark subsurface band		in extremely sandy subsoils of the Coastal Plain or Allegheny Plateau; undisturbed for thousands of years mostly under conifers		Spodic horizon	Bh or Bs

[†] Subscripts, like you see here, can be mixed and matched to improve a description. A gleyed argillic horizon would be designated Btg, for example. A gleyed argillic in a fragipan would be indicated with Btgx. For reasons given below, Btw horizons are never indicated.

[‡] Soil scientists generally shy away from calling Bw and Bt horizons in the same soil profile. It is one or the other. A clayey Bw horizon is seen as "on the way" to becoming a Bt. A soil with a Bt and an overlying or underlying horizon that is somewhat developed would likely have the partially developed horizon designated as a "transitional" AB, EB, or BC, etc. horizon as circumstances indicate.

yellow/red colors generally indicate the presence of iron oxides, which differ in hue depending on the mineralogy and crystal size. Green soil colors often indicate a specific iron mineral called glauconite, which is relatively rare.

To quantify soil color, scientists use the *Munsell system*, which looks at *hue* (basic color), *chroma* (hue intensity), and *value* (lightness/darkness). Interestingly, the majority of a soil's mineral material is made up of *quartz* and *aluminosilicate clays*, which are colorless (and generally white or gray). Hence soil color is usually indicative not of the bulk soil, but rather of surficial *trace coatings*. While small in mass, usually less than 5% of the soil's weight, the presence or absence of these trace coatings can be extremely instructive about conditions within the soil.

For example, where the water table is seasonably high, iron oxide typically does not persist, and soils are bleached (Bg horizons, table 6.4; albic horizons, table 6.5). This occurs because the lack of aeration under saturated conditions forces bacteria to "breathe" elements other than oxygen (see table 6.6). The relative abundance of iron in many soils makes this use of iron by bacteria very common. When bacteria support themselves via iron, which normally colors upland soils reddish (Fe^{3+}), they transform it to a colorless and soluble form (Fe^{2+}), which can flow away with groundwater. The loss of Fe results in a bleached soil displaying the unrubified native color (white-to-gray) of the silicate minerals that make up the bulk of soils. The presence of strongly "gleyed" soils often is used to indicate the presence of wetland soils. *Redox reactions* such as the above reduction of Fe^{3+} to Fe^{2+}—performed by bacteria—affect not only soil color but also soil horizon development and soil functioning (see table 6.6).

SMELL

Professionals and laypeople will often smell a dark A horizon or aerated compost for the sheer pleasure of the rich smell. This stands in contrast to a foul-smelling waterlogged compost or waterlogged mucky soil, where a lack of oxygen has forced soil bacteria to breathe elements other than oxygen and produce volatile organic molecules as waste. In waterlogged brackish wetlands, the "rotten egg" smell indicates bacterial use of sulfate SO_4 (rather than oxygen), producing (smelly) hydrogen sulfide rather than (odorless) carbon dioxide (see table 6.6).

TABLE 6.5. Hints and Tricks to Identifying Other Horizon Types in Maryland

FIELD INDICATORS/FORMATION PROCESS			FREQUENCY	TYPICALLY FOUND	NAME	ABBREVIATION
Senesced plant material that is too loose to support the roots of plants and therefore is not yet part of the soil; not a soil horizon; not to be confused with Oi			Very common	In all geologic provinces	"Duff" (forest) "Thatch" (lawns) "Residues" (farmland)	N/A
Stages of decay in O horizons	Specific identity of origin of organic material can be made		Rare	In wetlands in any geologic province; sometimes as a thin layer in forests	"Fibric" O horizon	Oi
	Specific identity of originating material cannot be made	Plant parts fragmented			"Hemic"	Oe
		Plant parts decayed into a "muck"			"Sapric"	Oa
A horizon has a sharp lower boundary	Sign of farming (mixing of soil to the level of tillage); such a sign will persist for many years after the cessation of tillage		Very common	In farmland and former farmland in any geologic province	Ap	
A horizon deep and unacidic	In Maryland, found only in oyster shell middens created by Native Americans where the alkaline shells buffer the acidifying effect of rain		Extremely rare	Bluffs above major rivers in the Coastal Plain (e.g., Potomac, Patuxent)	"Mollic epipedon"	A
Colorless or nearly colorless E horizon	Horizon not only removed of clays and organic matter but also removed of Fe oxide pigmentation; visible is the raw color of the primary minerals making up the horizon		Rare	Highly acidic leached soils likely either sandy or with a high water table	"Albic horizon"	E
C horizon made up of saprolite	Found in undisturbed residual[†] soils where moisture has dissolved some of the rock's original cementation		Likely common; rare to see due to presence at depth	Piedmont and geologic provinces to the west	"Saprolite"	Cr

[†] Residual soils are soils developed out of chemically weathered bedrock, unmoved from its original location in the landscape.

TABLE 6.6. Redox Transformations Induced by Respiring Soil Microbes[†]

	COMPOUNDS[‡]		
	OXIDIZED	REDUCED	CONTEXT
High → Oxygen concentration in the soil environment ← Low	O_2 (g)	CO_2 (g)	Aerobic respiration performed obligately by all biota save for a few bacteria/archaea; some bacteria/archaea are capable of the reactions below, thereby driving major aspects of nutrient cycling.
	NO_3^- (aq)	N_2 (g)	N loss in farm fields (negative outcome); N removal from runoff into wetlands (positive outcome).
	MnO_2 (s)	Mn^{2+} (aq)	Discrete, black manganese features appear in soil horizons within a weakly perched water table or poorly structured soil.
	FeOOH (s)	Fe^{2+} (aq)	As with Mn, but concentrations are ruddy. Consistent saturation can lead to wholesale removal. A more useful indicator because Fe concentrations typically >> Mn.
	SO_4^{2-} (aq)	H_2S (g) FeS (s)	Rotten egg–smelling H_2S or black FeS indicates brackish wetlands.
	CO_2 (g) $OM^{§}$ (s)	CH_4 (g)	Wetlands as sources of more powerful greenhouse gas CH_4. Diminished OM consumption under anoxia slows OM breakdown leading to OM accumulation in highly inundated soils.

(Right-side axis: High → Energy made available to the organism ← Low)

[†] Life forms consume organic matter and respire by placing electrons on an oxidized compound, thereby "reducing" it. The most common example is the "reduction" of O_2 to H_2O and CO_2, performed by most bacterial/archaeal species and all other types of species. But some bacteria/archaea have a trick up their sleeves: the ability to respire something other than oxygen. Above is a hierarchy of reduction reactions (oxidized compound + electrons ◊ reduced compound) that bacteria/archaea are capable of performing. Typically, specialized bacteria/archaea are required for any particular reduction process. In the event of waterlogging of the soil, after oxygen has been depleted, respiration occurs in sequence downward as oxidized compounds above become scarce. This table is but a window into the many-roomed mansion of bacterial diversity and capacity.

[‡] "aq" aqueous (soluble); "g" gaseous; "s" solid

[§] "OM" organic matter

TASTE

Yes! Some soil scientists taste their dirt! It is a way to gauge the *solubility* of a mineral. If a mineral has no taste, it does not dissolve readily. Mineral color, plus this solubility test, can go a long way in helping determine mineralogy in some cases. A "taste test" is used more commonly in drier, western landscapes, where soluble salts are more common. Soils utilized as septic system drain fields are typically not sampled in this way—at least by this author!

Soil Classification

When soil scientists classify soils, they typically do so using *B horizons*. B horizons, being subsurface, are less changeable than more surficial horizons and are therefore more diagnostic of long-term conditions and processes (see table 6.4).

In Maryland, rainfall is (on average) nearly constant from west to east and results in most upland soils being wet enough to transmit downward-moving water in most months of the year. As a result, this dominance of downward-moving water slowly draws clays out of upper horizons and deposits these clays in lower horizons—a slow process requiring thousands of years to be distinctly expressed. When well expressed, clay accumulation occurs as *waxy clay coatings* on or within peds (see Bt horizons, table 6.4). When above the seasonal high water table, Bt horizons are often ruddier than horizons above or below them. This reddish color is provided by *iron oxide clays* that accompany the *silicate clays*, which account for most of the clay accumulation. This is not a fast process; the formation of Bt horizons might take a few thousand to tens of thousands of years depending on the particulars of the site.

Bw horizons often form through the clay accumulation that create Bt horizons. But in Bw horizons, the process has been "weak"—either because not enough time has gone by (the soils are developing in a recently abandoned floodplain) or because the process of *infiltration* is weak (the soil is developing on a steep slope). Bw horizons can also be formed by oxidation (visible as soil reddening/yellowing) or the evolution of soil structure that is distinct from the structure of the original parent material. Bw horizons are usually in the process of becoming a more fully expressed B horizon of another type, but they are described as Bw when that transition process is incipient.

In pockets throughout Maryland—except for the middle and lower Eastern Shore, where they are absent—very dense, nearly undiggable, silty horizons (Bx) have developed in the subsoil. These *fragipans* restrict water flow and root penetration (see table 6.4).

Highly acidic conditions can move organic matter (and aluminum and iron atoms) from O and A horizons all the way through E horizons before these materials get stuck and precipitate in unique horizons characterized by a dark accumulation of organic matter (Bh), iron or aluminum (Bs), or both organic matter and Fe or Al (Bhs). Such soils are common in the continent's north, where granitic rocks of the Canadian Shield yield acidic sandy soils supporting conifers; in Maryland, similar horizons occur in the high, cold, sandy, coniferous-covered areas of the Allegheny Plateau. But in Maryland this horizon type is most often found in the Coastal Plain on ancient, stabilized dunes, or extremely sandy old alluvium supporting loblolly pines. As with Bt horizons, Bh, Bs, or Bhs horizons require thousands of years to develop.

Just as many different types of B horizons can be observed in Maryland, other master horizon types sometimes also have special or noteworthy expressions (see table 6.5). Because most soils have incomplete or indistinct expression of many of the characteristics provided in tables 6.3–6.5, people new to soil observation can feel at a loss when encountering their first soil profiles. This is why when going out into the world to observe soils, it would be best to do so with an experienced guide.

Soil Orders

Soil taxonomy (see table 6.7) is the system of soil classification used in the United States. As hinted above, it tends to utilize subsurface features (often B horizons, table 6.4) to make distinctions since subsurface changes require longer-term stability of conditions in order to be expressed. The given soil types are referred to as soil "orders." Note that we are discussing USDA soil taxonomy *before* we discuss recognizing the geological origins of soils. This is because USDA soil classification, to first approximation, considers the geological origin of soils as inessential. USDA soil classification is about the immediate, measurable properties associated

TABLE 6.7. Soil Orders

SOIL ORDERS	NECESSARY HORIZON
Ultisols and Alfisols: distinct downward movement of clay. Alfisols more inherently fertile than Ultisols, distinguished by a chemical test. Alfisols are usually formed from limestone (e.g., Monocacy Valley) or marble. Ultisols are found elsewhere out of acidic rocks (usually Piedmont) and highly weathered sediments (Coastal Plain). Old terraces; undisturbed (usually low slope) uplands. Thousands of years of development needed.	Bt
Inceptisols: limited weathering as observed through some alteration of the parent material. Usually hundreds of years of development needed. Newly abandoned floodplains (low terraces).	Bw or Bg
Entisol: so little weathering that the subsoil morphology cannot be distinguished from the original parent material. Floodplains, steep slopes, and newly disrupted soils. Distinguished as a soil because it can grow plant populations.	No B horizon; usually only A and C horizons
Histosols: nearly "pure" distinctive accumulations of organic materials unsubstantially mixed with mineral materials (deep O horizons). Characteristic of wetlands not dominated by erosion. Bogs and fens of the Allegheny Plateau; swamps, marshes, and other wetlands of the Coastal Plain.	Thick O horizon
Spodisols: subsurface organic matter or acidic metals descended from above over thousands of years (presence of Bh, Bs, or Bhs horizons). Formed in very sandy, acidic materials supporting conifers. Thousands of years of development needed. Isolated places in the Allegheney Plateau; more common in the Coastal Plain.	Bh, Bs, or Bhs
Mollisol: substantial accumulation of organic matter mixed with soil material with the soil itself high in nutrient concentrations. Characteristic of grassland soils of the great plains but found in Maryland in shell middens created from hundreds of years of shellfish harvesting by Native Americans. Found along the Potomac, Patuxent, and Choptank Rivers. So rare in Maryland that they are not mapped by the USDA.	Thick, nutrient-dense A horizon

with a soil and its horizons, not its geological origin. Nevertheless, the geological origin of the soil often is a powerful factor in the resulting soil type (see table 6.7). While this presentation of soil taxonomy is of the most basic nature, it begins to allow you to communicate with soil scientists in their language.

Identifying Geological "Parent" Material of Soils You Observe

So far you have learned how to distinguish soil horizons from geologic layers. With just one more field tool, you can add an extra dimension to your observational capacity: the type of geologic "parent" material out of which the soils you observe form. It turns out it's as easy as examining soil texture and the rocks present in the soils.

While a soil is distinguishable from the geological material out of which it formed, it is still a product of geological material. Just as Maryland has a great degree of geological diversity, the state's soil diversity is also impressive. And much about a soil in a particular place can be predicted by observing its geologic setting (see table 6.8):

1. Soils on mountainsides form either out of (a) rock weathering in place (*residuum*) or (b) avalanche or landslide material (*colluvium*).

2. Soils in stream valleys develop out of *alluvial* or (*riverine*) sediment. Alluvial material can be actively interacting with the stream (*floodplains*) or have been left high and dry (*terrace*) by a downward-cutting stream.

3. Soils in the Piedmont are *residual* (weathered out of the rock itself) but are much deeper than residual soils in the mountains because the Piedmont is flatter and erosion is relatively diminished.

4. Soils in the Coastal Plain develop out of alluvial or marine sediments or windblown dunal (*aeolian*) sands or silts (*loess*).

Certain (geologic) parent materials are associated with particular geologic provinces and landscape positions (see table 6.8), but the parent material of a horizon can also be identified in the field by simply observing

TABLE 6.8. Landforms, Geologic Provinces, and Soil "Parent" Materials

PARENT MATERIAL	LANDFORM TO BE EXPECTED IN	GEOLOGIC PROVINCES MOST LIKELY TO BE FOUND *IN MARYLAND*
Residuum	where soils develop directly out of bedrock, especially in areas of low relief	any except the Coastal Plain
Colluvium	on or below sloped or formerly sloped areas in any province	any; especially mountainous
Alluvium	river/stream valleys: floodplains, terraces in any province	modern river valleys in any province: younger alluvium; terraces: much older alluvium; much of Coastal Plain: old alluvium
Marine	barrier islands or former barrier island landscapes	Coastal Plain
Aeolian		
Loess	became entrained when high winds swept across a fairly barren landscape (e.g., exposed bed of Chesapeake Bay)	Coastal Plain

two things: (1) *texture* and (2) the presence, shape (angular or rounded), and *sortedness* of *rocks*. For example, do individual soil layers contain rocks of only one particular size (i.e., *well sorted*) or rocks of many sizes simultaneously (i.e., *poorly sorted*) (see table 6.9 and figure 6.1)?

Wind is not generally strong enough to move rocks, so rocks are typically missing from parent materials deposited by wind (see table 6.9). Clays also tend to be largely absent from wind-transported materials, not because they are too heavy, but because they are too "sticky" to be displaced by wind (unless they are coatings on sand or silt particles) (see table 6.2).

Rocks *transported by water* tend to be well sorted, with similar rock sizes turning up in the same layer of sediment. Water moving at high speed can pick up many particle sizes, but as the water incrementally slows down upon encountering landscapes of lower slopes, the biggest particle sizes tend to drop out of the flowing water and settle to the bottom. This simultaneous dropping of rocks of similar sizes accounts for the "sortedness" found in a

TABLE 6.9. Soil Texture, Rock Characteristics, and Means of Identification of Different Parent Materials

| PARENT MATERIAL | MOVED BY | TYPICAL SOIL TEXTURE | ROCK PRESENCE | |
			SORTEDNESS	TYPICAL SHAPE
Residuum	Material not moved	Any; depends on fabric of rocks weathering to form the soil	Usually more abundant with depth including decaying rock material; unsorted	Angular
Colluvium (landslides)	Gravity		Poorly sorted	
Alluvium (stream deposited)	Water	Any; depends on speed of water movement at moment of deposition	Well sorted	Rounded
Marine		Coarse sand when deposited near the shore; finer textures, becoming clayier, farther from shore; limier even farther out	None	
Lacustrine[†] (lake deposited)		Clays and silts		
Aeolian (dunal)	Wind	Medium sands		
Loess (long-distance wind deposited)		Fine sands and silts		
Organic	Material not moved	While some mineral material might be present, most of the material present is organic; senesced plants are the parent material for the soil		

[†] Lake-deposited materials are mentioned here for completeness, but Maryland has no mapped soils that are mentioned as lacustrine in origin. Lacustrine deposits do exist, but they are found at the bottom of tens of feet of peat materials in the Allegheny Plateau in the rare fens and bogs that are found there. This depth is sufficiently deep that they qualify as geologic and not soil materials. They may have qualified as soils tens of thousands of years ago, when they supported the vegetation that became the overlying peat.

FIG. 6.1. Roundedness and sortedness of alluvial stones, underlying granitic saprolite, Bw horizons at gullified Normanstone Run, Northwest Washington, DC, Piedmont. At this suburban location, southeast of the vice president's mansion, Normanstone Run presents not only its current self, with quartzite cobbles in the streambed, but also a previous self with similar, well-sorted quartzite cobbles in the soil behind. Whether current or past, the rounded, well-sorted cobbles make sense as alluvium, as well as the relatively large sizes of these cobbles, because at this location the slope is steep as Normanstone plunges toward Rock Creek. But (1) the uppermost elevation in the watershed lies less than a mile away and at most 100 feet higher in the landscape, hardly long enough or high enough for tumbling stones to become so rounded, and (2) the stones are of a fundamentally different type than the granitic saprolite (Cr) that visibly underlies the cobbles. Image courtesy of David Ruppert.

river's alluvial layers (see table 6.9 and figure 6.1). Alluvial rocks are also usually rounded because of the friction they encounter as they tumble against each other over their journey of (sometimes) hundreds of miles between mountains and the Coastal Plain. This is also why rocks in mountain streams, having tumbled less, will likely be more angular than rocks found in streams in the Coastal Plain.

Geologically speaking, Maryland's Coastal Plain is simply a big wedge of eroded stones and sediment that has—over millions of years—descended and accumulated from the eroding Appalachian Mountains. At times and in places, it has been subsequently picked up and redeposited as marine or windblown sediments.

Unlike the long, slow accumulation of alluvial material along the Coastal Plain, *colluvial* (landslide) materials are deposited simultaneously in massive dumps, typically along steep ridges in the mountains. This sudden dumping precludes sorting and rounding (see table 6.9).

Residual soils form in place, developing out of *unmoving resident* rock. Oftentimes the presence of rock fragments increases with depth, and some rocks—weathered of some of their cementing compounds—can be observed. Such weakly cemented rock is called *saprolite*, and sometimes the weathering has been so great the former rock may be crumbled or split by hand (see "Cr" table 6.5 and figure 6.1).

A final soil parent material to note is not geologic in origin, but biotic. When the water table is high, the exclusion of oxygen slows down the breakdown of plant materials (see table 6.6) to such a degree that organic matter can build up (see table 6.9) into deposits many feet thick, forming peaty soils or *histosols* (see table 6.7). These soils are now thought to be among the most valuable systems in the world for sequestering atmospheric carbon. When we buy peat, these soils and the wetland habitats they form are destroyed.

Starting with what geology provides, soils form over a long period of time on the terms of their landscape position, microclimate, plants and animals present, and frequency and nature of disturbance. They evolve as unique individuals, determined by, and helping to define, the particulars of landscapes.

In this chapter we focused on what to look for when observing soils in the field and related this to soil classification and geology. Whole other chapters could be written on topics such as soil biology or soils and hydrology, but materials on such topics are easier to find and priority was given here to material that is less accessible but still related to field observation. The most complete single reference on soils—worldwide—is provided by Maryland Professor Ray Weil in Brady and Weil (2017).[2]

Because soils are always forming, their expressions may occur as cryptic, especially to those lacking extensive field experience. The USDA Natural Resources Conservation Service (NRCS) and land grant universities are full of soil scientists wanting to get out into the field. Local consulting soil scientists are available through the Soil Science Society of America or the Mid-Atlantic Association of Professional Soil Scientists. As you make your first soils explorations, bring a professor, graduate student, consultant, or NRCS soil scientist along. Between that guided experience and the material in this chapter, you will be on your way to understanding why soils occur the way they do in our precious landscapes.

Notes

1 For more detailed soil resources, see https://www.nrcs.usda.gov/sites/default/files/2022-11/texture-by-feel.pdf.

2 Nyle Brady and Raymond Weil, *The Nature and Properties of Soils*, 15th ed. (London: Pearson, 2017).

CHAPTER 7

Taxonomy

GWENDA BREWER

Humans of all cultures have a basic desire to understand and organize the world around us, including naming things in our environment and recognizing patterns. These were the skills that helped our ancestors survive, as they needed to navigate different habitats, choose areas with needed resources, avoid danger, predict the weather, know what to eat and what not to eat, and communicate that information within their communities.

The modern-day representation of these ancestral tendencies is what we know as *taxonomy*, the science of *classification*, and *nomenclature*, a standardized naming system. As we study the incredible diversity in the biological world, we use the science of taxonomy to identify, describe, and categorize organisms, and we use a standardized system of naming organisms to communicate about them worldwide. Although we will focus here on classification and naming systems used by the typical Western scientific community, the science of *ethnotaxonomy* explores how categorizing and naming organisms is culturally influenced and how this can result in different taxonomic structures.[1]

The taxonomic and nomenclature systems used today got their formal start with Linnaeus (Carl von Linné) back in the early 1700s. This was a time of exploration and discovery, as ships sailed the world and scientists recognized a need to organize plants and animals into groups and name them. Like others before him, Linnaeus grouped organisms based on external or internal characteristics that they shared. The groupings were organized in a hierarchical system, with subgroups of organisms nested under layers of headings that included more and more organisms. The groupings went from Kingdom, which included the most species, to Phylum (animals) or Division (plants), Class, Order, Family, Genus, and Species.

Linnaeus took information from his investigations (and those of others) and put it all into one large document, the *Systema Naturae*, which also featured a consistent way of assigning names to species: a two-part specific epithet. The names themselves were Latinized (given a form consistent with Latin words and grammar), as Latin was the common scientific language of the time. These names were often based on distinctive features or descriptions derived from Latin and Greek words.

The structure for naming and the Latin convention of Linnaeus stuck (see figure 7.1). Species are still named using a binomial (from *bi*, "two," and *nomial*, "name") that is Latinized and unique for each species, made up of the "genus" and "specific epithet." These names are standardized and expressed by capitalizing the first letter of the genus and using lowercase letters for the specific epithet (e.g., *Turdus migratorius*). The scientific name of a species is usually underlined or written in italics, and the author's name and date of publication of the description of the species may be included in the scientific name after the genus and species (e.g., *Magnolia virginiana* Linnaeus 1753).

A modern-day requirement added to the naming of a species is that a *type specimen*, or *permanent reference specimen*, should be available (often in a museum collection). The basic organization scheme of Linnaeus, at least in terms of the names of taxonomic groupings, has also remained largely unchanged, although there have been *many* changes in the names

FIG. 7.1. Scientific names are approved by the International Commission on Zoological Nomenclature and the International Association for Plant Taxonomy so that every scientific name of an animal or plant is unique. The accepted scientific name of this Maryland dragonfly species is *Stylurus plagiatus*, which means "stylus-shaped thief," likely due to its long, slender shape and predatory habits. Photo courtesy of George M. Jett.

of species and other taxonomic levels, where organisms are included in higher-level groupings, and the relationships between the higher-level groupings themselves (see figure 7.2). For example, today we recognize six Kingdoms: Eubacteria, Archaebacteria, Protista, Fungi, Plantae, and Animalia (until relatively recently, the Eubacteria and Archaebacteria were included in one Kingdom, Monera). This change occurred due to differences in genetic and biochemical characteristics—something Linnaeus certainly could not have known!

Although we have adopted many aspects of the Linnean classification system, today we understand the world differently than in the time of

FIG. 7.2. New data can lead to taxonomic rearrangements. For example, the black vulture (chicks pictured here) had been placed into three different orders of birds until its recent placement in the new order Cathartiformes. Photo courtesy of George M. Jett.

Linnaeus. Specifically, we know how *evolution* has shaped the origin of new species. We also have access to a wider variety of characteristics to use to classify organisms. Genetic characters have largely changed the face of the science of classification and given us the focus on *systematics*: thinking about evolutionary relationships to classify organisms into groups and examine relatedness.

Evolution can be defined simply as *successive changes in the frequencies of alleles (forms of a gene) in a population*. The major mechanism of evolution, the process of *natural selection*, was described in the mid-1800s by Charles Darwin and Alfred Russel Wallace. In the evolutionary process, adaptation occurs through time as organism traits become better adjusted to environmental demands. A better "fit" between organisms and their environments results, leading to increased survival and reproduction.

The idea is that in the competition for limited resources, individuals vary and some individuals in a population are more successful and, critically, at least some of that variation in the ability to succeed is *heritable* (able to be passed down to offspring). As those individuals reproduce and pass that better competitive ability to their offspring, those individuals in turn are more successful and pass down the ability to their offspring. In time, there is a shift in the characteristics of the population toward those that make organisms more successful in that particular environment.

With our current understanding of molecular genetics, we see things a little differently than Darwin and Wallace did: we can measure changes at the genetic level that have taken place as a result of evolution by natural selection. "Survival of the fittest," a phrase used by Darwin, doesn't quite capture evolution by *natural selection*—the key is successful reproduction, not just survival.

So, with a background of evolutionary theory, how do modern-day taxonomists (systematists) work to determine species and classify organisms? Deciding what is and is not a *species* is a difficult challenge for taxonomists as there is no single, universally accepted definition of a species. The most common definition, known as the *biological species concept*, involves a look at which organisms can actually (or potentially) *interbreed successfully in nature*. This definition, however, can be difficult to apply when organisms can reproduce *asexually* or can *hybridize* if they come into contact.

Just like with classification in general, determining species has become less straightforward with our ability to use molecular characters. For example, some researchers define a species as the smallest diagnosable group of individual organisms within which there is a *parental pattern of ancestry and descent* (phylogenetic species concept). The biological species concept is appealing because it seems to rely more on what the animals are telling us in nature. It is not as firmly based in evolutionary theory, however, and it is difficult to tell if groups are truly isolated reproductively in the field. The phylogenetic species concept, although relying on patterns of genetic relatedness, can lead to many groups being identified as separate species, and the biological relevance of these groups can be questionable. These two species concepts are still the subject of scientific debate and keep the field of taxonomy very lively indeed.

The classification of species or of a higher taxonomic level into a particular taxonomic group can also be challenging but tends to follow a more standardized pathway. The first step in this process is to figure out what characteristics should be used to best reflect the evolutionary history of organisms, consistent with the modern-day systematics approach. This requires a focus on *characters* that are shared due to *common ancestry*. In other words, there is a direct genetic link and pattern of relatedness.

Traits of organisms may be similar due to ancestry or relatedness, but traits may also be similar when organisms respond to the same environmental pressures in the same way through time. For example, traits that convey a particular function can evolve independently in distantly related species, known as *convergent evolution*. An example of convergent evolution is webbed feet in beavers and ducks. These animals share the trait "webbed feet" not because they are closely related but because they have both developed or converged on a characteristic (webbed feet) through natural selection that makes them more successful in aquatic environments. The use of traits that are present because of convergent evolution, called *analogous traits*, will not lead to a classification that relies on shared ancestry and would not be accepted as valid. In other words, despite similarly webbed feet, we would not place these animals into the same taxonomic group as if they were closely related species.

The kinds of traits or characters that systematists find most useful are called *homologous traits*: those that are shared due to common ancestry but that are also "derived" or not *primitive*. In other words, they show some level of *specialization*. These may be anatomical, morphological, developmental, genetic, biochemical, or behavioral characteristics of living organisms or features of fossilized organisms. Once these traits have been identified, they are put together into branching diagrams on the basis of how many of the homologous characters the organism groups share.

The diagram is constructed so that it explains the data in the simplest way possible. These branching diagrams are called *cladograms* as they show the pattern of relatedness between groups of organisms (*clades*) based on the characteristics that they share (see figure 7.3). Clades represent a single

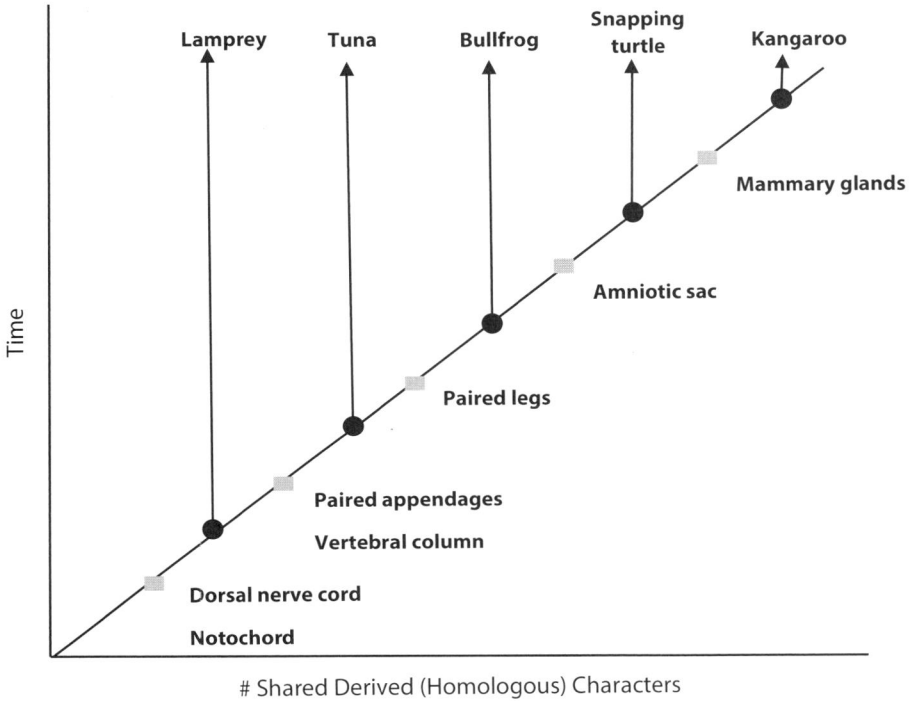

FIG. 7.3. Branching diagrams like this one are called *cladograms* as they show the pattern of relatedness between groups of organisms (*clades*) based on the characteristics that they share. Although cladograms are in some ways similar to "family trees," they do not show all of the intermediate ancestors that were part of the evolutionary history of the group. Figure courtesy of Gwenda Brewer.

common ancestor and all of the groups that have descended from that ancestor. The cladogram shows a set of hypothesized evolutionary relationships between the organisms included in the diagram. Although cladograms are in some ways similar to "family trees," they do not show all of the intermediate ancestors that were part of the evolutionary history of the group. In a cladogram, each organism group has the characteristics of the node it branches off from and also all of the nodes to its left. We can conclude from the cladogram in figure 7.3, for example, that kangaroos are more closely related to snapping turtles than bullfrogs because they both have an amniotic sac, but the bullfrog does not. Bullfrogs are more closely related to tuna than lamprey because bullfrogs and tuna share the character "paired appendages," but the lamprey does not. Cladograms can be pictured using other branching patterns, like the complicated diagram in a publication on dog breeds.[2]

Taxonomy and nomenclature form the backbone of learning about the organism groups studied in the Maryland Master Naturalist curriculum, allowing us to consistently name organisms and examine the evolutionary relationships within and between groups. Thinking about how evolutionary processes have shaped the structure, life history, and behavior of organisms and looking for patterns, similarities, and differences in the natural world, as taxonomists do, can also help us to better understand how best to conserve it.

Notes

1 Benjamin T. Wilder, Carolyn O'Meara, Laurie Monti, and Gary Paul Nabhan, "The Importance of Indigenous Knowledge in Curbing the Loss of Language and Biodiversity," *BioScience* 66 (2016): 499–509.

2 Heidi G. Parker, Dayna L. Dreger, Maud Rimbault, Brian W. Davis, Alexandra B. Mullen, Gretchen Carpintero-Ramirez, Elaine A. Ostrander, "Genomic Analyses Reveal the Influence of Geographic Origin, Migration, and Hybridization on Modern Dog Breed Development," *Cell Reports* 19, no. 4 (April 25, 2017): 697–708, https://doi.org/10.1016/j.celrep.2017.03.079.

CHAPTER 8

Botany and Plant Identification

KERRY WIXTED

Our world would cease to exist without plants. Plants produce the oxygen we breathe, sequester carbon dioxide, and provide food and habitat for just about everything on Earth—including us. Some plants are harvested for commercial products, such as timber, pulp, fiber, and medicine; others provide *ecosystem services* like regulating the water cycle, controlling and slowing down erosion, cycling nutrients, and assisting with global temperature control. Fast-growing, hardy plants—like cattails and poplars—can be used for bioremediating contaminated land.

Just three plant species—rice, maize, and wheat—provide 60% of the (human) world's food energy. The largest (by mass) and oldest living organism in the world is a tree clone called "Pando," a giant quaking aspen that weighs close to 13 million pounds, occupies almost 108 acres in Utah, and may be 80,000 years old.

More fun facts: plants can count. To ensure it doesn't spend energy on something that won't result in food, a Venus flytrap (*Dionaea muscipula*) can "count" up to five. Trigger hairs inside the "trap" need to be touched at least twice in the span of 20–30 seconds to cause the leaf to close on potential prey.

Plants can sense gravity. Higher-level plants have gravity-perceiving cells (called *statocytes*) in their roots and shoots. Inside the cells are starchy organelles (called *statoliths*) that shift with gravity and signal to plant hormones to have the plant grow in a particular direction. These cells allow shoots to grow up and roots to grow down. Understanding statocytes may help us figure out how to grow plants in outer space.

In this chapter, we will provide an overview of the amazing plant diversity in our own state, including the general evolution of plants, major plant groups, some basic plant identification characteristics, and threats to native plants.

Maryland Ecology and Plant Diversity

Because of its remarkable ecological diversity, Maryland has been called "America in Miniature." Nested between the Appalachian Mountains and the Atlantic Ocean, and divided by the Chesapeake Bay, Maryland encompasses a rich tapestry of landscapes that support a wide assortment of plant species. In Maryland, around 2,918 established taxa have been documented; 71.8% of these are native and 28.2% have been introduced.[1] Just over a quarter of native plant species tracked by the Maryland Natural Heritage Program are considered rare, threatened, or endangered.[2]

To grow and thrive, plants require the right climate (including both temperature and precipitation), nutrient availability, and associated organisms, such as beneficial *mycorrhizae*. Bedrock, as it weathers, creates the soil that plants grow within and can alter the soil's pH. Plants known as *generalists* can grow in a variety of habitats; fussier plants, known as *specialists*, have a more restricted distribution. Maryland's higher western elevations tend to support species found in more northern climates; species typically found in more southern climates live along the state's Coastal Plain.

Maryland also boasts interesting *communities of plants*, groups of interacting species that coexist within a specific geographic area or habitat such as:

Serpentine savannas found in areas with serpentine bedrock contain high levels of magnesium and low levels of calcium, making it difficult for many plants to grow in those conditions. As a result, these areas have a distinct flora, adapted to survive within these harsh conditions. Soldiers Delight Natural Environmental Area in Baltimore County is one of our best remaining examples. Much of the savanna is covered by grass species as well as serpentine specialists like serpentine aster (*Aster depauperatus*).

Shale barrens are underlaid by shale bedrock. The resulting soil is typically nutrient poor and loose and gets rather hot in full sun. Maryland shale barrens exist in the Ridge and Valley physiographic region and are found on south-facing slopes with Devonian shale. Kate's mountain clover (*Trifolium virginicum*) is a globally uncommon plant that thrives in shale barrens.

Delmarva bays, concentrated in Caroline, Talbot, and Queen Anne's Counties, are small, seasonal wetlands on the Eastern Shore. They typically are meadow-like since woody plants generally cannot endure the seasonal fluctuations in water. Sedges, rushes, and grasses dominate many of these bays along with emergent vegetation. A rare winter annual, featherfoil (*Hottonia inflata*), can be found in some of these wetlands.

Seepage fens: while some seepage fens in Maryland may be referred to as *bogs*, they technically are considered *fens* because they are fed by groundwater rather than rainwater. Both habitats are acidic, nutrient poor, and peaty. Many seepage fens occur on the Coastal Plain of Maryland, and their harsh soils support species like the carnivorous northern pitcher plant (*Sarracenia purpurea*) and small sundews (*Drosera* sp.).

Evolution of Plants

Plants are believed to have evolved from freshwater green algae, and their evolution literally changed life on the planet. The four major evolutionary periods include the movement from water to land, the development of vascular tissue, the emergence of seeds, and the emergence of flower structures.

Divisions of plants are based on their physiology and their methods of reproduction. These divisions go from the most primitive to the most advanced. Plants are *multicellular* organisms that rely on *photosynthesis*, the process by which green plants, algae, and some bacteria convert *light energy*, usually from the sun, into *chemical energy* (glucose) and oxygen. As light is absorbed into green pigments (known as *chlorophyll*) in the plants, a series of chemical reactions occur to form oxygen and glucose. While carbon dioxide, water, and light are the main ingredients needed for photosynthesis to occur, mineral nutrients such as nitrogen, phosphorus, potassium, magnesium, calcium, iron, and sulfur are also needed in smaller amounts to complete the series of complex chemical reactions.

BRYOPHYTES

The first land plants were likely seedless, nonvascular *bryophytes*, which still exist today. This group includes *mosses* (Bryophyta), *liverworts* (Marchantiophyta), and *hornworts* (Anthocerotophyta). These multicellular organisms reproduce via spores and do not have true vascular tissue, which prevents them from having *roots* (to grow down) or *stems* (to grow up). Their small, sprawling *growth habit* allows them to persist in areas with little to no soil, such as rocks and bark.

Bryophytes have two main stages in their lifecycle: the *gametophytes* and the *sporophytes*. Gametophyte stage plants are leaflike, photosynthesize, and can undergo sexual reproduction. If this stage is successful, a sporophyte will form at the top of the gametophyte and will produce spores for *asexual reproduction*. Asexual reproduction essentially creates clones of the parent, but sexual reproduction requires at least two parents to provide genetic information to create a unique offspring.

Most bryophytes require the use of a microscope to identify them down to species. However, you can easily separate the three main groups out by physical characteristics. *Mosses* have stemlike and leaflike structures; their sporophyte has a stalk and capsule. *Liverworts* have *foliose* (leaflike structures arranged in two rows) or *thallose* (large, flat rubbery leaflike structures). Sporophytes have a terminal capsule borne on a stalk, known as a *seta*. *Hornworts* resemble thallose liverworts; their sporophytes are horn shaped and split longitudinally.

One interesting group of bryophytes are the *sphagnum mosses*, which can absorb up to 22 times their own weight in liquid and have medicinal, antiseptic properties. Humans have been using sphagnum mosses for thousands of years to line cradle boards, create baby "diapers," and dress wounds. When cotton was scarce during World War I, "moss drives" were held to collect sacks of it. Water was extracted from the moss, and the dried material went to war hospitals. Today, around 37 species of sphagnum moss can be found in Maryland, mostly in acidic environments that the sphagnums help to acidify.[3]

PTERIDOPHYTES

Pteridophytes include *ferns* and *fern allies* (*horsetails* and *club mosses*). Their evolutionary advantage lies in their development of vascular tissue (*xylem*, *phloem*, etc.). Pteridophytes have special vertical tubes that transport water (via the phloem) and food (via the xylem) that allows them to grow much larger than bryophytes. Like bryophytes, they reproduce using *spores* instead of *seeds*; some have spore-bearing structures on the back of their leaves, others produce spores on their stalks.

Around 74 species of ferns have been documented in Maryland, and with the ongoing expansion of molecular systematic research, the number of fern species described continues to grow. The most commonly seen fern species in Maryland are the Christmas fern (*Polystichum acrostichoides*) and the sensitive fern (*Onoclea sensibilis*). Christmas fern is one of our few evergreen ferns, sporting leathery fronds with Christmas stocking–shaped *pinna*, or leaflets. Christmas fern foliage was traditionally used for creating holiday wreaths. Sensitive fern is *deciduous* and likes to grow in moist to wet environments. The pinnae are wavy; the fern reproduces via short, thin structures called *fertile fronds*.[4]

Club mosses are another large group of pteridophytes. Formerly known as *lycopodiums*, club mosses are evergreen, with needlelike or scalelike leaves. Club mosses are often seen in colonies attached horizontally by either aboveground *runners* or belowground *rhizomes*. Many have sporophyte structures known as *strobili* that appear slender and cone-like. The oily and highly flammable spores were once used as flash powders for early photography and for firework materials. Evergreen club mosses have also traditionally been collected for holiday decorations, a practice that—given their slow-growing nature—makes the practice unsustainable.

SPERMATOPHYTES (GYMNOSPERMS AND ANGIOSPERMS)

Spermatophytes, like pteridophytes, have vascular tissue but have the added adaptation of producing seeds instead of spores. Seeds are advantageous because they have both stored energy and a protective coat surrounding the plant embryo, so they survive and germinate even in unfavorable conditions. Spermatophytes can be separated into two groups: *gymnosperms* (which produce "naked" seeds) and *angiosperms* (flowering plants).

Gymnosperms

Gymnosperms have seeds that *lack an external covering*. In Maryland, gymnosperms include our conifer species (pines, spruces, cedars, etc.) and the non-native gingko (*Gingko biloba*). Conifers typically have scalelike leaves or waxy-coated needles that help decrease water loss in the plant. Not all conifers are evergreen: the eastern larch (*Larix laricina*), a deciduous conifer, produces gorgeous golden foliage in the fall before the needles drop. Larches are endangered in Maryland and are only found in a few boggy areas in our mountainous west. Table 8.1 covers common types of conifers in Maryland.

The eastern red cedar (*Juniperus virginiana*), a species of juniper, is common in Maryland. It grows fast and is considered a *pioneer species*, as it is one of the first tree species to colonize cleared areas. It is highly adaptable, can grow in poor and rich soils, and can even withstand salt spray. Cedar waxwings enjoy the fruits of eastern red cedars, as do other bird species like robins, catbirds, and blue jays. The cedar's aromatic oils are an effective insect repellent, which makes the wood popular for protecting clothes closets against moths; its rot-resistant wood makes it popular with furniture (and garden) builders.

TABLE 8.1. Common Types of Conifers in Maryland

TYPE	LEAVES	CONES	OTHER NOTES
Cedar (*Chamaecyparis, Juniperus, Taxodium,* and *Thuja*)	Scalelike (imbricate) leaves	Variable	Eastern red cedar (*Juniperus virginianus*) is the most common wild species in Maryland; most other species in this group are commonly planted.
Fir (*Abies*)	Soft, flat needles	Stand upright	Rare in Maryland unless planted.
Hemlock (*Tsuga canadensis*)	Two-ranked needles	Small and egg-shaped	Susceptible to invasive hemlock wooly adelgid.
Pine (*Pinus*)	Needles in clusters of 2–5	Variable; size and presence/absence of prickle can help with ID	Largest group of conifers in Maryland. Many species are economically important for timber production.
Spruce (*Picea*)	Square, sharp needles	Hang downward	Uncommon in Maryland except in western counties. Non-native Norway spruce (*Picea abies*) commonly planted.

Angiosperms

Angiosperms, formerly known as *Magnoliphyta*, are the largest group of plants, accounting for around 80% of the world's extant plant species. Angiosperms include all flowering plants, from grasses to oaks, and are distinguished by their vascular tissue and their production of flowers and fruit. Typically, angiosperms are separated into two divisions: *monocots* and *dicots* (see table 8.2).

About 99% of angiosperms are photosynthetic; the remaining 1–2% are *parasitic* for part or all of their life cycle. For example, *mycohetero-trophs* are plants (like many orchids) that obtain all of their energy needs by parasitizing fungi. All plants in the Monotropoideae subfamily are

TABLE 8.2. Divisions of Angiosperms

MONOCOTS	DICOTS
One cotyledon (seed leaf)	Two cotyledons
Parallel leaf veins	Reticulate leaf veins
Flower parts in threes	Flower parts in fours and fives
Secondary growth (woody tissue) absent	Secondary growth can be present
E.g., grasses, lilies, orchids	E.g., asters, shrubs, trees

mycoheterotrophic, including ghost pipes (*Monotropa uniflora*), pinesap (*Monotropa hypopitys*), and sweet pinesap (*Monotropsis odorata*). Other plants like oak mistletoe (*Phoradendron leucarpum*) parasitize host plants for their nutrients. Mistletoes are spread via bird feces on trees; as the plants develop, they use modified rootlike structures known as *haustorium* to attach to their host and steal water and nutrients.

Angiosperms can be classified by examining a plant's *morphology* (its leaves, flowers, etc.) or its *physiological* traits, such as its life cycle.

There are three main life cycles observed in plants:

1. *Annuals* are plants, such as petunias, that grow, mature, flower, produce seed, and die all in a single season.

2. *Biennials* take two years, or a part of two years, to complete their life cycle. During the first season, biennials, such as garlic mustard (*Alliaria petiolata*), grow as a *vegetative plant* that overwinter as a hardy rosette of basal leaves. During the second season the plant flowers, produces seeds, and dies.

3. *Perennials* live for more than two years. *Herbaceous perennials* have soft, non-woody stems, while *woody perennials* have woody stems. Woody perennials can either keep their leaves throughout the year or seasonally shed them. The dropping of leaves (known as *abscission*) is triggered by hormonal shifts

caused by environmental factors like day length and/or stress. Hormones trigger cells to weaken at the point where the leaf attaches to the stem (known as the *abscission zone*), causing the leaf to fall off.

Basic Plant Identification Characteristics

Scientists further group plants into Phylum, Class, Order, Family, Genus, and Species.

Families are large groups of plants that have similar flowering and fruiting characteristics. Knowing a plant's family can help with successful (and easier!) plant identification. For example, the plant family Lamiaceae, also known as the Mint family, has square stems, opposite leaves, and an aromatic scent. Within a family, plants can be further divided into *genuses* (or *genera*). For example, beebalms are in the Mint family and the *Monarda* genus. Members of a plant genus can be further subdivided into *species*, typically according to their similar morphological characteristics. A *species* is a group of organisms that can interbreed and produce viable (reproductive) offspring. Some species can be further divided into subspecies.

One of the easiest ways to learn plant identification is to learn morphological *family characteristics* to help narrow down your search. A few plant families and their main characteristics can be found in table 8.3.

If possible, try to compare your field observations with plant species you already know in your area. Taking note of the habitat, location, and other surrounding plant species can also provide valuable context for identification. In addition to learning plant families, try consulting field guides, websites, or smartphone apps, all of which provide detailed descriptions, images, and keys to help you narrow down your options. Apps like iNaturalist (free), PictureThis (paid), and FloraQuest (paid) can help, as can websites like the Maryland Biodiversity Project and the Maryland Native Plant Society. The 2021 *Vascular Plants of Maryland, USA: A Comprehensive Account of the State's Botanical Diversity* has the most up-to-date list of plant species found in Maryland.

TABLE 8.3. Some Common Plant Families and Their Characteristics

COMMON FAMILY NAME	SCIENTIFIC FAMILY NAME	LEAVES	FLOWERS	FRUITS	OTHER CHARACTERISTICS
Mustard family	Brassicaceae	Alternate, simple or compound; sometimes have basal leaves	Petals 4, not fused, forming a cross + from above	Silique (dry capsule)	Many are aromatic when crushed
Mint family	Lamiaceae	Opposite, simple	Petals 5, fused w/ 2 up and 3 down	Capsules w/ 4 nutlets	Aromatic; stems are square
Oak family	Fagaceae	Alternate, simple, often lobed	Catkins, wind pollinated	Nut (acorn in oaks), surrounded by the cupule	Trees
Legume family	Fabaceae	Alternate, compound	Petals 5, bilaterally symmetrical	Bean (pods)	Have symbiotic relationship with nitrogen-fixing bacteria

Field guides are an excellent way to use *dichotomous keys* to identify plants. Some recommended field guides include *Wildflowers in Field and Forest: A Field Guide to the Northeastern United States*; *Field Guide to Ferns and Their Related Families of Northeastern and Central North America: Peterson Field Guides*; *Newcomb's Wildflower Guide; Common Mosses of the Northeast and Appalachians*; and *A Field Guide to Eastern Trees: Eastern United States and Canada, Including the Midwest*.

LEAVES

As the main site of photosynthesis, *leaves* are vitally important to most plants. Leaves have pores called *stomata* that allow for the exchange of water and gas during both photosynthesis and *respiration*. Plants that need to conserve water (like the evergreen rhododendrons) often have leaves equipped with a thick, waxy outer layer called a *cuticle*. The broad, flat part of the leaf

is called the *blade*; the major vein that runs down the center of the leaf is
called the *midrib*, which provides support for the leaf and connects to the
leaf's vascular network. Sometimes, plants will have *hairs* or *glands* along
the midrib, which can help in identification. The stalk of the leaf is called
the *petiole*. Leaves that lack a petiole are known as *sessile*, but if the leaf lacks
a petiole and surrounds the stem, it is called *perfoliate*. Generally, there is
an *axillary bud* where the petiole attaches to the plant's stem. This bud
can later grow into a leaf or a flower shoot. Looking for a bud at the petiole attachment point will help you determine if your leaf is *simple* or *compound*. Simple leaves have a single leaf blade on a petiole, while compound leaves are made up of multiple leaflets arranged on a single petiole
(see figure 8.1).

When examining leaves, you should be aware of several key characteristics, including:

Leaf arrangement: Leaves position themselves to get the maximum
amount of sunlight, and determining *leaf arrangement*—how the leaves are
positioned along the stem or branches of a plant—is often a valuable first
step for identifying plants. Note that leaf arrangement can sometimes vary
within a single plant species due to factors such as age and environmental
conditions. Some plants have multiple leaf arrangements; mustards like

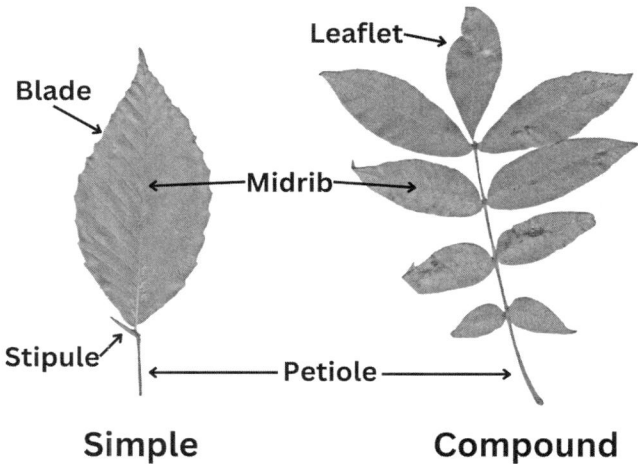

FIG. 8.1. Simple vs. compound. Image courtesy of Kerry Wixted.

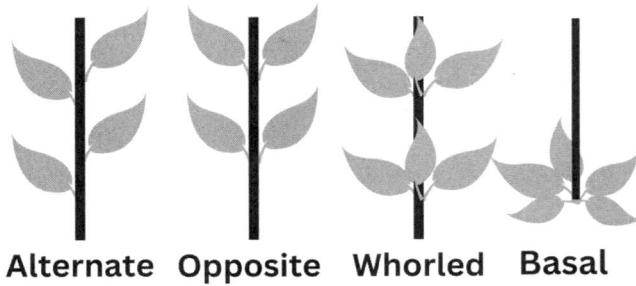

Alternate   Opposite   Whorled   Basal

FIG. 8.2. Leaf arrangements. Image courtesy of Kerry Wixted.

hairy bittercress (*Cardamine hirsuta*) have stem leaves that are arranged alternately and a basal rosette.

The most common types of leaf arrangements (see figure 8.2):

- Alternate arrangement: One leaf is attached at each node along the stem, alternating sides with each node. Because the leaves alternate, they do not directly face each other. This is the most common type of leaf arrangement in Maryland plants and can be seen in everything from oaks to asters.

- Opposite arrangement: Two leaves are attached at each node, directly opposite each other on the stem. In many cases, the leaves will appear to be paired. A mnemonic device to help remember common woody plants with oppositely arranged leaves is MADCapAHorse: Maples, Ashes, Dogwoods, Caprifoliacae (like honeysuckles), Adoxaceae (viburnums), and Horse chestnut. (Note: viburnums have been moved from the Caprifoliacae family to the Adoxaceae family, and some dogwoods like alternate-leaved dogwood do not have opposite leaves.)

- Whorled arrangement: Three or more leaves are attached at each node in a circular or spiral pattern around the stem. A local example: Culver's root (*Veronicastrum virginicum*) or the invasive hydrilla (*Hydrilla verticillata*).

- **Basal rosette arrangement:** Leaves are clustered in a circular arrangement at the base of the plant, often forming a rosette shape. Plants with this leaf arrangement include many of our mustards.

Leaf shape: Leaves come in various shapes and are adapted to different environments and growing needs. Photosynthesis, light availability, temperature, and other factors can influence leaf shape and size. Oaks commonly have *sun leaves* that grow in full sunlight and have a thicker cuticle, and *shade leaves* that are often thinner and wider.

Below are just a few examples of some common leaf shapes; there can be wide variability within each shape category. Figure 8.3 shows examples of some simple and compound leaf shapes.

- *Simple:* Leaf has a *singular blade.* Simple leaves can have different shapes, including:

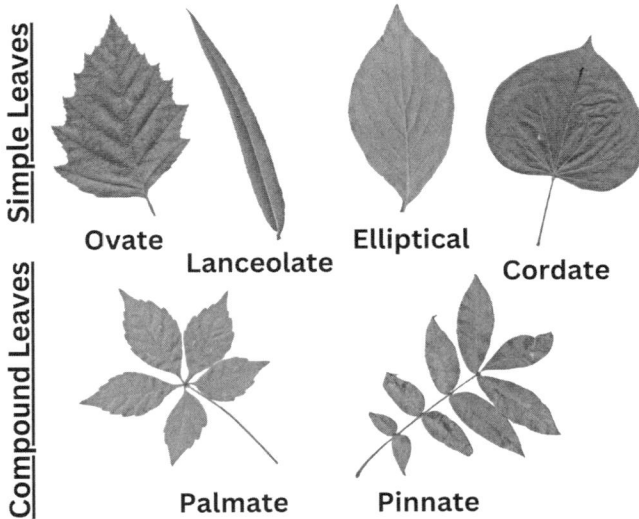

FIG. 8.3. Leaf shapes. Image courtesy of Kerry Wixted.

- **Ovate:** Egg-shaped, with a rounded or pointed tip and a wider base. Example: birch leaves.

- **Lanceolate:** Long and narrow, wider in the middle and tapering to a point at both ends. Example: willow leaves.

- **Elliptical:** Oval-shaped with equal width throughout and tapered ends. Example: beech leaves.

- **Kidney-shaped (reniform):** Leaves wider than high and shaped like a kidney. Example: garlic mustard leaves.

- **Linear:** Very long and narrow, with parallel sides and no significant tapering. Example: needlelike leaves of pine trees.

- **Heart-shaped (cordate):** Wide at the base with a notch or lobes at the top, resembling a heart. Example: eastern redbud leaves.

- **Oblong:** Rectangular in shape with slightly rounded ends. Example: some oak leaves.

- **Spatulate:** Rounded at the tip and gradually narrowing toward the base, resembling a spatula. Example: spoon-shaped leaves of succulents like jade plants.

- **Round (orbicular):** Circular or nearly circular in shape. Example: water lily leaves.

○ *Compound*: Leaf is divided into *multiple leaflets*, each resembling a small leaf, along a common petiole. Compound leaves can also be further broken down into different categories, including:

- **Palmate:** Divided into several distinct leaflets spreading out from a central point, resembling fingers of a hand. Example: buckeye leaves.

- **Pinnate:** Compound leaf with a single midrib with smaller veins branching off on both sides. Example: hickory leaves.

- **Bipinnate:** Leaf is divided into smaller leaflets, and each leaflet is further divided into even smaller leaflets. Example: acacia leaves.

- **Trifoliate:** Consists of three leaflets attached to a single petiole. Example: clover leaves.

When identifying plants, practice examining the *leaf margin, venation,* and other characteristics. Some leaves may also present *intermediate* or unique shapes that don't neatly fit into any category, so look at the overall plant to make sure you aren't just examining a weird leaf.

Leaf margin: The margin refers to the edge of the leaf. It can be *entire* (with a smooth or unbroken edge), *lobed* (with shallow or deep indentations that don't reach the midrib), *spiny,* or *toothed.* Examine the consistency of the margin along the leaf's length. Toothed margins can be further divided into categories based on the size and direction of the "teeth." Keep in mind that leaf margin characteristics can vary within a single species, especially due to factors like age, environmental conditions, and genetics. Figure 8.4 illustrates some leaf margins.

  ◦ *Toothed/serrate margin*: Edge with evenly spaced teeth. Variations include:

- **Dentate margin:** Large, pointed teeth that point outward, like the teeth of a saw blade. Example: leaves of the hazel tree.

- **Crenate margin:** Rounded teeth that are more spaced out than serrations. Example: jewelweed leaves.

- **Serrulate margin:** Tiny, fine serrations along the edge. Example: leaves of some types of roses.

Entire
Rhododendron

Serrate
Black Cherry

Lobed
Post Oak

Doubly Serrate
River Birch

FIG. 8.4. Leaf margins. Image courtesy of Kerry Wixted.

- Double-toothed/doubly serrate margin: Larger teeth with smaller teeth between them. Example: leaves of some types of birches.

Venation: Refers to the pattern of veins on the leaf. There are two primary types: *pinnate* (a single main vein with smaller veins branching off) and *palmate* (several main veins spreading out from a single point). Red maple (*Acer rubrum*) leaves are a good example of a local plant that has palmate venation. Monocots like grasses and orchids can also have *parallel* veins.

Leaf texture: Leaves can be smooth, rough, fuzzy, waxy, or prickly. Take a closer look with a 10x hand lens to see if you can discern any hairs, glands, or other structures on the surface of the leaf. It can get hard: in the botanical world, there are over 30 terms just to describe the types of *pubescence* (hairiness) in plants! Hair can protect the leaf from herbivore predation and sun exposure and can help with thermal insulation. Here are a few examples of pubescence:

- **Ciliate:** Has fringe-like hairs along the margin or midrib.

- **Hirsute:** Has coarse hairs.

- **Tomentose:** Has densely matted, soft white hairs.

Remember that plant identification can be more challenging given variations within species, hybridization, the time of year, and regional differences. Patience and attention to detail are essential when identifying leaves accurately.

FLOWERS AND INFLORESCENCES

For angiosperms, *flowers* are vital for reproduction. Most flowers have specific colors, shapes, and/or scents that have evolved to increase pollination success. *Pollination* simply refers to the movement of pollen grains from a flower's male *anther* to a female *stigma*. Successful pollination results in the formation of *fruits* and *seeds*.

The main parts of the flower (see figure 8.5) include the *petals*, which together make up the *corolla*. Inside the flower are the male *stamens* and the female *pistil*. The stamens are made up of the *anther*, which holds the pollen, and the *filament*, the stalk on which the anther sits. The pistil contains the *stigma*, which accepts the pollen and sits atop the *style*. At the base of the pistil is the *ovary*, which contains unfertilized eggs known as *ovules*. The outermost part of the flower is made up of (typically) green and leaflike *sepals*. Sepals protect the developing bud and often enclose the petals before the flower blooms. As a unit, all the sepals together form the *calyx*. Occasionally, in plants like lilies, the sepals and the petals will resemble one another and are known as *tepals*. The flower stalk is called the *peduncle*.

Flowers that contain both male and female parts are known as *perfect*; flowers that contain only male or only female parts are known as *imperfect*. If a plant has both male and female flowers on the same plant, it is referred to as *monecious* (literally "one house"). If male and female flowers are on two different plants within a species, then they are referred to as

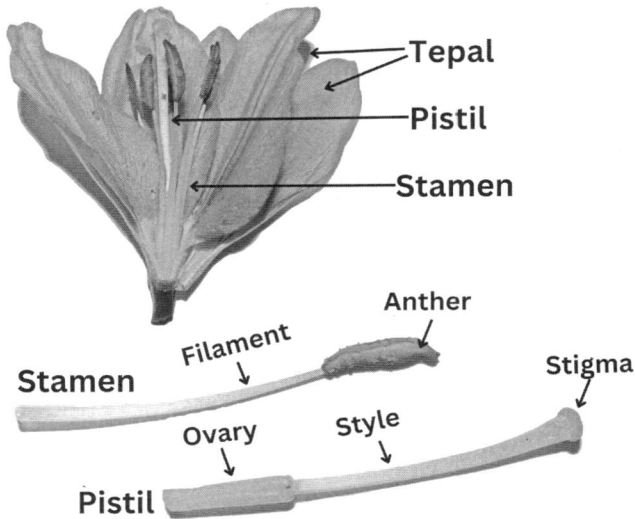

FIG. 8.5. Parts of a flower. Image courtesy of Kerry Wixted.

dioecious ("two houses"). A local example is holly. Flowers that are *radially symmetrical* (like daisies) are termed as *regular flowers*. Flowers that are not *radially symmetrical* (like violets) are considered *irregular flowers*.

Interestingly, some plants can change sex throughout their lives. For example, jack-in-the-pulpits (*Arisaema triphyllum*) begin their lives as male plants and only produce flowers with pollen. Once the plant has sequestered enough energy to produce female flowers, it changes sex. But after a poor growing season, a jack-in-the-pulpit may return the following year as a male plant, as it takes more energy to produce female flowers.

In flowering plants, the cluster of flowers is known as an *inflorescence*, and as with other plant parts, there are many forms. Inflorescence includes all the flowers on the main stalk (*peduncle*) as a single unit. If a plant has a single flower, then it is known as a *solitary inflorescence*. If all the flowers are on stalks of about the same length and arise from a common point like an inverted umbrella, then that is called an *umbel*. A plant like Queen Anne's lace (*Daucus carota*) which has multiple umbels, has a *compound umbel* inflorescence.

Sometimes, you can count the number of flower parts (petals, stamens, sepals, etc.) to help with identification. For example, the Mustard (*Brassicaceae*) family has flower parts that come in fours. Most local mustard species also have six stamens, two of which are larger than the other four.

THE ART OF POLLINATION

Most plants reproduce aboveground with *chasmogamous flowers* that typically require a *pollinator* to assist with moving grains of pollen. Some plants, like American beech (*Fagus grandifolia*), are *wind pollinated* and disperse mass quantities of pollen in the air in hopes they will find their way to a female stigma (for some people, this pollen can cause allergies). In a small subset of plants—a local example is the aquatic eelgrass (*Vallisneria*)—pollination can be carried out by water: pollen grains coated by a substance called *mucilage* float on the water's surface on their way to contacting female flowers.

Some plants can *self-fertilize* and thus do not need wind, water, or an animal to move their pollen grains. But since the seeds produced through this process are essentially *clones* of parent plants, this form of pollination—while convenient—limits genetic diversity. Some plants use self-fertilization as a backup plan if cross-pollination doesn't occur. For example, plants in the Violet genus (*Viola*) have two types of flowers, with one kept underground and completely sealed to allow for self-fertilization. These flowers are known as *cleistogamous flowers* and often appear in the summer or fall. Once the fruits form, they spill directly below the parent plant, though some lucky fruits will be carried off by ants to germinate in a new location.

Relying on pollinators—from animals ranging from insects to mammals—for *cross-pollination* is often a plant's best strategy. But because it takes so much energy to produce lures like showy flowers, attractive scents, and/or refreshing nectar, some plants have evolved tricks to dupe their pollinators while expending the least amount of energy necessary to be pollinated.

This deception can include mimicking mating partners, imitating food, and/or using scents or visual cues. For example, pink lady slipper orchids

(*Cypripedium acaule*) lure bees to flowers with enticing colors and alluring scents. Bees slipping into the orchid flower soon find themselves stuck— but without any nectar reward! To escape, bees must pass by both the plant's pollen organs and the stigma.

Some plants have *patterns* or *lines*, often visible under ultraviolet light, that act as *nectar guides*. Of course, some pollinators have gotten smart and don't fall for these tricks; they will rob flowers of nectar by cutting holes in the side of the flower, bypassing the passages containing the pollen organs.

By examining flower traits, you often can generalize which pollinators are likely to be attracted to particular flowers (see table 8.4).

Beetles are believed to be the world's earliest pollinators, so it is no surprise that magnolias, an ancient lineage of trees, are often pollinated by

TABLE 8.4. Pollinators and Flower Traits

FLOWER TRAIT	POLLINATORS						
	BEES	BEETLES	BIRDS	BUTTERFLIES	FLIES	MOTHS	WIND
Color	Bright white, blue, or yellow	Green or white	Orange or red	Red, purple	Dark purple, brown	Pale or white	Green
Nectar guides	Present	Absent	Absent	Present	Absent	Absent	Absent
Odor	Faint, fresh	Mostly absent or foul	None	Faint, fresh	Rotten	Strong, sweet	None
Nectar	Present	Some-times present	Ample, deep in flower	Ample, deep in flower	Mostly absent	Ample, deep in flower	None
Pollen	Limited	Ample	Limited	Limited	Limited	Limited	Abundant
Shape	Shallow	Bowl-like	Funnel-shaped	Tubular with lip, wide	Shallow	Tubular without lip	Small
Example	Sunflower	Magnolia	Trumpet creeper	Purple coneflower	Pawpaw	Yucca	Ragweed

Source: US Forest Service

beetles. As you can see in table 8.4, pollen is important for beetles, but nectar isn't. Interestingly, male and female flowers on some magnolia species come out at different times on the same plant to *prevent* self-fertilization. However, since pollen is important to beetles, the lack of pollen during the female flower phase can be an impediment. To get past this, some species of magnolias actually heat themselves up (using a process called *thermogenesis*) to help disperse scented oils that attract hungry beetles. While beetles get a pollen dinner during the male flower phase, they get tricked during the female flower phase, which mimics the male phase but lacks pollen.

FRUITS, SEEDS, AND DISPERSAL

Once pollinated, plants develop seeds covered by the swollen *ovary* (the *fruit*). The design of the seeds and/or fruit (their shape, the presence of a fleshy outer layer, etc.) is often linked to the plant's *dispersal mechanisms*; the farther afield a plant's seeds are dispersed, the more they avoid local competition. Small and lightweight seeds with "wings" (as in helicoptering maple seeds) can be dispersed by wind; fruits with a fleshy coating (like blueberries) are dispersed by animals (like birds) that consume the fruit and defecate the seeds elsewhere. Note that *botanical* terms for fruits can vary from *culinary* terms. For example, a banana is technically a berry, and an orange is technically a *hesperidium*, a type of berry with a leathery rind and internal segments!

As with other plant parts, there are various types of fruits, each with its own characteristics and variations. Fruit can typically be separated into *fleshy* and *dry* (see figure 8.6):

Fleshy fruits have a soft, fleshy *pericarp* (the part of the fruit surrounding the seed) and are often eaten by animals that disperse the seeds. Types of fleshy fruits can include:

- **Aggregate fruits:** These develop from a single flower with multiple ovaries, and each ovary forms a separate small fruit. Examples include strawberries and raspberries.

Fleshy Fruits

Drupe   **Berry**   **Pome**

Dry Fruits

Achene   **Samara**   **Nut**

FIG. 8.6. Fruits. Image courtesy of Kerry Wixted.

- **Accessory fruits:** Part of the fruit is derived from tissues other than the ovary. The edible part of a strawberry, for example, is derived from the receptacle (the thickened part of the stem), not the ovary. (And strawberries can thus have aggregate *and* accessory fruits.)

- **Berry:** Seeds are embedded inside the *pulp*. Examples include bananas, grapes, and tomatoes.

- **Drupe:** A single seed enclosed in a hard, woody or stony pit, called an *endocarp*. Examples include peaches, plums, and cherries.

- **Pome:** A central core containing seeds is surrounded by a fleshy layer derived from the receptacle. Examples include apples, pears, and pomegranates.

Some plant species have higher germination success after they have passed through the digestive tract of specific animals. For example, mayapple (*Podophyllum peltatum*) does better after the fruits have been consumed and excreted by box turtles. In the western United States, chokecherry (*Prunus virginiana*) does better after being digested by black bears.[5] Of course, not all digestive tracts are equal, and many fruit eaters are seed *predators* rather than *dispersers*. White-footed deer mice (*Peromyscus leucopus*) are predators of wild ginger (*Asarum canadense*), bloodroot (*Sanguinaria canadensis*), and twinleaf (*Jeffersonia diphylla*) seeds that are normally dispersed by ants. One study found that almost 47% of the twinleaf seed crop was consumed by white-footed deer mice before the seeds could fully develop and germinate.[6]

Dry fruits: These fruits have a dry *pericarp* and can be further broken down into two categories: *Dehiscent fruits* like peas, beans, and poppies split open when they are mature to release their seeds. *Indehiscent fruits* like sunflower seeds, acorns, and grains like wheat and rice do not split open when they mature. Dry fruits are typically dispersed by wind, animal attachment, and/or by *ballistic* means. (Jewelweed is a great example: when ripe, its seeds are forcefully ejected from the parent plant). A few types of dry fruits include:

- **Achene:** A small, dry, one-seeded fruit where the seed is attached to the fruit wall at a single point. Examples include sunflower seeds and dandelion seeds.

- **Samara:** A winged fruit, often designed for wind dispersal. Examples include the seeds of maple and ash.

- **Nut:** A hard-shelled, *indehiscent* fruit with a single seed. Examples include acorns and chestnuts.

- **Capsule:** A dry, *dehiscent* fruit that splits open to release seeds. Examples include poppies and iris.

One unique seed dispersal technique is *myrmecochory*, meaning the dispersal of seeds by ants. Most myrmecochorous plants produce seeds with *elaiosomes* or fatty structures on the outside of the seed. The ants, attracted to this energy-rich food source, pick up the seed and carry it back to their nest, where they consume the elaiosome and discard the rest of the seed, effectively planting it underground. Local plant species of this type include Dutchman's breeches (*Dicentra cucullaria*) and bloodroot (*Sanguinaria canadensis*). A cluster of these plants usually marks where an ant colony exists or once existed.

WOODY PLANTS

Trees, shrubs, and woody vines have specialized secondary tissue known as *woody tissue* that serves several important functions, including structural support, water transport, and nutrient storage. Woody tissue undergoes a unique process of development and maturation, resulting in its characteristic hardness and durability.

The *outer bark* serves as a protective barrier against physical damage, pathogens, and environmental conditions. Each species has its own type of bark, and although one species may look similar to another, close inspection reveals subtle differences. Bark identification is also useful during winter months, when leaves and flowers are not present or when leaves are too high up to see. Bark can be *smooth*, *scaly*, or *furrowed* in texture. Trees like American beech (*Fagus grandifolia*) have smooth, gray bark, while shagbark hickories (*Carya ovata*) have *plated* furrows that indeed look "shaggy." Some species with *ridged* and furrowed bark include oaks (*Quercus* sp.), hickories (*Carya* sp.), some elms (*Ulmus* sp.), ashes (*Fraxinus* sp.), black locust (*Robinia psuedoacacia*), sweet gum (*Liquidambar styraciflua*), black gum (*Nyssa sylvatica*), and poplars (*Populus* sp.). Other trees, like many species of birch (*Betula sp.*), can have *peely* bark. The evolutionary reasons for peely bark layers are not well known. It's possible the exposure of the tissues under the bark might increase photosynthesis and gas exchange or help cut down on parasites, fungi, mosses, and lichens.

Underneath the outer bark is the *inner bark*, which is mostly made of the *phloem*, which transports food throughout the woody plant. Beneath the phloem is the *cambium*, the growing tissue that eventually becomes new bark. (Cambium growth is directed by hormones such as *auxins*.) Beneath the cambium is the *sapwood*, which contains most of a tree's water transport system, as well as its new wood. Finally, in the center of the woody plant is the *heartwood*, which is made of dead tissue that provides a plant's structural support.

The *growing tips* of woody plants exist on *twigs*, the smaller branches on shrubs and trees (see figure 8.7). For woody plants, twigs can serve as excellent identification tools, particularly when examining characteristics like the terminal (end) bud and/or *leaf scars*. *Buds* are small, undeveloped shoots that contain embryonic leaves, flowers, or stems. Buds can be *terminal* (located at the twig's tip) or *lateral* (found along the sides of the twig). Some species, like oaks, have multiple terminal buds, while others only have a single bud. In some plants, the bud will have multiple *scales* while others will only have one. Some buds are also *hairy* or *colored*.

Lenticels can be observed on the twigs and bark of some woody species; they are the small, raised areas on a twig's surface that allow for gas

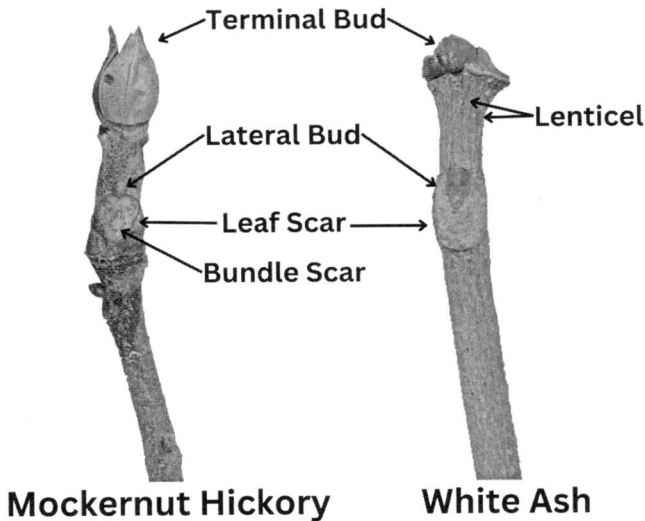

Terminal Bud

Lateral Bud

Lenticel

Leaf Scar

Bundle Scar

Mockernut Hickory   **White Ash**

FIG. 8.7. Twigs. Image courtesy of Kerry Wixted.

exchange between the internal tissues of the twig and the surrounding air. Lenticels appear as small dots, lines, or elongated structures and are more pronounced in some species than others. Sometimes, the shape of the lenticels can clue you in on plant identification. For example, many local cherry species (*Prunus* sp.) have horizontal lenticels that look like stripes or "checks" on the trunk. The raised, white lenticels are also conspicuous on spicebush (*Lindera benzoin*).

When a leaf falls off the twig, it leaves behind an identifiable *leaf scar* where it once attached to the twig. Leaf scars can be quite distinct and vary between different types of plants. Within the leaf scars are also *bundle scars*, left from where veins connected the leaf to the twig.

Another helpful identification characteristic of a twig is its central core, called the *pith*, which is composed of soft, spongy tissue and plays a role in storing and transporting nutrients. Pith can be variously *colored*, *chambered*, and sometimes *solid*.

ROOTS

A plant's *roots* absorb water and nutrients through tiny *root hairs*, help anchor the plant, and store food. Around 90% of land plants have symbiotic relationships with beneficial fungi known as *mycorrhizae*, which increase a host plant's root surface area and assist with the uptake of nutrients such as *phosphorus*. In addition, many plants in the *legume* family (Fabaceae) have beneficial *nitrogen-fixing* bacteria (*Rhizobium* spp.) that infect their roots and assist with the uptake of atmospheric nitrogen, a nutrient critical for plant growth.

Some plants have *modified stems* known as *rhizomes* that exist underground and mimic roots. Sometimes rhizomes (in plants like strawberries) form branches that send up new shoots known as *stoloniferous rhizomes*. Other plants can have large storage areas associated with their roots called *tubers* or *corms*. Similarly, *bulbs* aren't roots, either—they are short stems with roots emerging from their *basal plate*.

Notable Maryland Plants

CARNIVOROUS PLANTS

Carnivorous plants are (literally) captivating, and we have over 15 species in Maryland.[7] The carnivorous strategies in plants have evolved to meet a plant's nitrogen needs that are not met by the soil substrate it is growing within. Indeed, most of Maryland's carnivorous plants are found in fens, which are acidic and lack much nitrogen.

Carnivorous plants not only *trap* other organisms, they have a mechanism to digest and *absorb* their prey. There are two main types of traps: *active* and *passive*. Active traps, such as the iconic *Venus flytrap (Dionaea muscipula)* use leaves equipped with a *chemical reaction* or process (such as *trigger hairs*) to actively catch prey. Passive traps use sticky surfaces and strategic hairs to attract and capture prey. *Pitcher* plants, the largest carnivorous plant species in Maryland, generally occur in acidic fens and seepage wetlands. Our only native pitcher plant is the northern pitcher plant (*Sarracenia purpurea*), a threatened species; the yellow pitcher plant (*Sarracenia flava*) has been planted in several areas. Both species passively trap prey by using brightly colored tubes with the sweet scent of nectar. Hungry critters land on the pitcher, follow colorful veins in the leaf, and fall in with the aid of downward pointing hairs and slippery cells. Interestingly, the non-biting pitcher plant mosquito larvae (*Wyeomyia smithii*) has evolved to live *inside* pitcher plants; it acts as a top-level predator and feeds on parts of animals that the plant cannot digest (as well as on microscopic organisms such as *rotifers*).

Sundews (*Drosera* spp.) are tiny carnivorous plants that use sticky leaves as traps. Sundews in Maryland have spatulate-shaped or tentacle-like leaves with brightly colored glands that attract and trap prey. Insects that get trapped on the sticky mucilage secreted by the leaves' glands are gradually digested by a suite of enzymes and acids (some sundew species even bend their leaves slowly around their meal!). On average, a sundew traps and consumes about five insects per month. Maryland has five species, including the pink sundew (*Drosera capillaris*), a state endangered plant found in only three places on the Eastern Shore. The thread-leaved sundew (*Drosera filiformis*) has also been introduced to several areas in Maryland.

By far, the coolest trapping mechanism in Maryland's carnivorous plants is found in the bladderworts (*Utricularia* sp.). Bladderwort bladders are among the most complex structures in the plant world. Much of the plant exists as a fine network of *bladders* and *filamentous leaves* submerged under the water, or within a wet substrate like gravel. Because they spend much of their time below a surface, bladderworts are often overlooked until they produce showy flowers that stick out above the water or substrate surface.

Their bladders are hollow, with a lid lined with specialized trigger hairs. When closed, the bladders have a low-pressure gradient, but as water fleas or midges tap the hairs on the bladders, the lid on the bladder opens and creates a vacuum-like force to suck in the prey. Once the bladder fills with water, the lid closes, and a substance is released to seal both the door and the fate of its prey. This entire process occurs in 1/35 of a second!

Maryland has over ten different species of bladderworts, many of which can be found in ditches and open ponds. Among the most unique is the rare swollen bladderwort (*Utricularia inflata*), which resembles a floating wheel in the water when it produces its bright yellow flowers.

"POISON" PLANTS

Three notorious plant species in Maryland are poison ivy (*Toxicodendron radicans*), poison oak (*Toxicodendron pubescens*), and poison sumac (*Toxicodendron vernix*). All three contain an allergic reaction–causing, oily compound in the sap known as *urushiol*. Poison ivy thrives in urban and suburban environments and—in part because of climate change—since the 1950s has gotten *much* more potent. Increases in atmospheric carbon dioxide have increased both leaf size and urushiol content. Poison ivy isn't all "bad"—its white, waxy berries are a popular food for songbirds during fall migration and in winter when other foods are scarce. The berries are especially popular with robins, catbirds, and grosbeaks. Many birds also feed on insects hiding in the tangled vines, and small mammals and deer browse on the poison ivy foliage, twigs, and berries.

Poison oak, to date, has been documented in only about half of Maryland's counties, mostly in dry environments like *shale barrens*. It can resemble poison ivy, but its berries have hairs (hence the species epithet *pubescens*). Poison oak only grows as a small shrub and will not (like poison ivy) be found as a ground (or climbing) vine in Maryland. Poison sumac has compound leaves and is found in wetlands and peaty environments in Maryland, mostly found in the Coastal Plain in magnolia bogs of the fall zone around Washington, DC.

Threats to Native Plants

As important as they are to human well-being and biodiversity alike, Maryland's native plants face many threats, including habitat loss and degradation, invasive species, climate change, overabundant white-tailed deer, overharvest and unethical harvest, and disease.

As McKay Jenkins writes in his chapter on environmental history, the arrival of European settlers in the 1600s through the mid-1800s marked a great period of change for Maryland's forests. It is estimated that settlers had cleared almost 90% of the forests by the close of this time. While some forests were able to regenerate, other periods—especially the explosion of suburban development in the 1960s and 1970s—continued to cause a net loss of habitat. In 1991, the state legislature passed the Maryland Forest Conservation Act, a significant move for reducing the loss of forest coverage and supporting reforestation. However, many habitats remain fragmented and degraded, and some have been permanently altered. Species such as ten-lobe false foxglove (*Agalinis obtusifolia*) were lost (or *extirpated*) from Maryland's landscapes, while others like swamp pink (*Helonias bullata*) are barely hanging on.

One negative result of native habitat loss is the growing threat of *invasive species*. As Doug Tallamy notes elsewhere in this volume, invasive species pose a significant threat to Maryland's native plants. Invasive species—from bacteria to plants and animals—cause biologic, economic, or human health–related harm in their introduced regions. *Invasive insects* like hemlock wooly adelgid and emerald ash borer are responsible for significant loss of hemlocks and ashes in the state, as are *pathogens*, like the fungus (*Cryphonectria parasitica*) that leads to chestnut blight disease.

Invasive plants may alter and degrade habitats. Plants like Japanese stiltgrass (*Microstegium vineum*) are prolific breeders that create a thick thatch on forest floors and alter soil dynamics, effectively outcompeting native species. Other invasives like tree of heaven (*Ailanthus altissima*) and garlic mustard (*Alliaria petiolate*) release chemicals (a process known as *allelopathy*) that inhibit other plants' germination and/or growth. While some native plants, like black walnut (*Juglans nigra*), are also allelopathic, there are natural checks and balances that keep populations in place, and some native plants have co-evolved with black walnut over thousands of years.

Note: while *all* invasive species are "non-native," not all non-native species are "invasive." And not all pests are non-native. In areas of Maryland, high native deer populations have overgrazed sections of forests, leading to tremendous losses of some plant species and changes in forest dynamics. A deer's preference for seedlings and saplings has altered forest regeneration, particularly since they often prefer native plants (like oaks) rather than invasives. Deer can also facilitate the *spread* of invasive species: extensive deer browse can deplete native plant communities, and the resulting disturbance increases the likelihood of colonization by invasive species like Japanese stiltgrass. Deer feces can also spread the highly invasive multiflora rose.

Plant Collection

While collecting plants is a common practice, overcollection, unethical collection, and/or illegal harvesting have become problematic for some Maryland plant species. Showy, edible, and medicinal plant species are often subject to the highest exploitation pressures. Plants most susceptible to poaching in Maryland include several species of lilies, orchids, wild medicinal plants, and carnivorous plants; even species of moss have been illegally collected from Maryland forests.

As a naturalist and a consumer, it is important to be informed about local collection laws, general collection ethics, and exactly where commercial nurseries get their plants. Many public properties (such as parks) do not allow plant or seed collection, and some species, such as American ginseng (*Panax quinquefolius*), require permits for collection and sale.

In addition, even if plant collection is allowed, it is important to ask: *is it necessary?* Picking flowers and/or collecting seeds may disrupt an already threatened or failing population. When purchasing plants from nurseries, always inquire where nursery stock is sourced, and check to ensure regulations are being followed by companies selling plants.

Notes

1 Wesley M. Knapp and Robert F. C. Naczi, *Vascular Plants of Maryland, USA: A Comprehensive Account of the State's Botanical Diversity* (Washington, DC: Smithsonian Scholarly Press, 2021).

2 Maryland Natural Heritage Program, *Rare, Threatened, and Endangered Plants of Maryland* (Annapolis: Maryland Department of Natural Resources, 2021).

3 Maryland Biodiversity Project, https://www.marylandbiodiversity.com/, accessed September 1, 2023.

4 Maryland Biodiversity Project.

5 J. Auger, S. E. Meyer, and H. L. Black, "Are American Black Bears (*Ursus americanus*) Legitimate Seed Dispersers for Fleshy-Fruited Shrubs?" *American Midland Naturalist* 147, no. 2 (2002): 352–67.

6 E. R. Heithaus, "Seed Predation by Rodents on Three Ant-Dispersed Plants," *Ecology* 62, no. 1 (1981): 126–35.

7 Maryland Biodiversity Project.

CHAPTER 9

Invasive Species

DOUGLAS TALLAMY

The term *invasive species* has only recently entered our environmental lexicon and, like so many aspects of today's culture, has become curiously controversial. There is no public outcry demanding the *introduction* of new diseases or pest insects to kill our plants, new mussels to foul our freshwater ecosystems, or new species of Asian fish to wreak havoc with North American fisheries. Yet for more than a century, there has been a powerful public demand for ornamental plants that evolved on other continents, and this demand has brought us 86% of our worst invasive plants. And because invasive plants are what we argue most about, I will focus on them in this chapter.[1]

Novel Ecosystems

For as long as we *Homo sapiens* have been on the move around the globe, we have carried plants, and—to a lesser extent—animals, with us. Modern modes of transportation, international trade, and a keen desire to display unusual plants in our yards have turned what was once a trickle of introductions in past centuries into a torrent of new species entering North America from foreign lands. So great, so sudden, and so disruptive has the influx of new species been that ecologists now call the majority of today's ecosystems "novel"—literally, "new things" under the sun. They are considered "novel" because many of the species within them are just meeting each other for the first time in evolutionary history. That is, their interactions with each other are occurring without the tempering effect of long periods of co-evolution.[2]

Although some scientists are excited about the evolutionary potential of novel ecosystems, such potential will not be realized for eons, and in the meantime, many of these introductions have been devastating to native populations—and thus ecological productivity—within existing ecosystems. When a novel predator is introduced, for example, prey rarely have appropriate adaptations to defend themselves and quickly fall victim.[3] Our cute and cuddly house cats kill between 2 and 3 *billion* birds in the United States alone each year.[4] The largest extinction event in the Holocene occurred when humans colonized the Polynesian Islands some 4,000 years ago.[5] An astounding 800–2,000 species of birds were lost to human hunting, but they were also devastated by the introduction of rats to these islands. The same type of ecological ruin has occurred time and again when diseases such as chestnut blight and white pine blister rust, or non-native insects like the hemlock wooly adelgid, emerald ash borer, and gypsy moth, were introduced to North America from other lands.

But the most widespread and underappreciated consequences of creating novel ecosystems are those that occur when introduced plants replace native plant communities.

Native versus Introduced

Nearly all of us get our plants from nurseries, but the plants in most nurseries fall into two very distinct categories: they are either native to your area (that is, they share an evolutionary history with the plant and animal communities in your ecoregion or biome), or they developed the traits that make them unique species somewhere else. In Maryland, "elsewhere" is typically from East Asia, although our nurseries carry many plants from Europe as well.

In the past, we didn't care much about where a plant came from; we chose our ornamentals for the sole purpose of meeting a specific aesthetic taste. Maybe we desired a particular color or habit to complete a landscape design; or perhaps we sought a plant with a striking bloom to serve as an accent or focal point in our yard; or maybe we wanted a dense evergreen to serve as a screen; or—more often than not—we chose our plants so that our yard would look just like our neighbor's, and their neighbor's, and so on. When these were our only goals, the geographic origins of the plant mattered little. What was important was to conform.

However, if your goal is to create a landscape that enhances your local ecosystem rather than degrades it, geographic origin is the very first attribute you must consider. Plants native to your region are almost always far better at performing local ecological roles than plants introduced from somewhere else.

It is important to recognize that all invasive plants are introduced from somewhere else, but all introduced plants are not invasive. That is, they do not spread into natural areas and displace native plant communities. This is an important ecological distinction, unless, like noninvasive crepe myrtle, we plant so many of them in so many places, they might as well be invasive in terms of their impact on local biota.

Invasive Species

Invasive *species* create novel *ecosystems*, and there are more species of invasive plants, over 3,300 in the United States alone,[6] than all of the other invasive organisms combined. Invasive plants, defined as non-native species that displace native plant communities, should not be confused with fast-growing, "aggressive" native plants for one simple reason: native plants, aggressive or otherwise, have been duking it out with each other, competing for space, light, water, and nutrients, for millions of years. Over the eons, native species have evolved ways to cope with each other, and the results of their interactions define the highly diverse species composition of most native plant communities all over the world. Invasive plants, in contrast, have just arrived within a community (and "just" can be defined as within the last several hundred years, a blink of an evolutionary eye). They also have arrived without their suite of natural enemies: the insects, mammals, and diseases that keep them in check in their homeland. Invasive plants thus have an enormous competitive advantage over most native plant species, which enables many of them to run amuck across the landscape. And so, interactions between invasive plants and our native species are anything but tried and true; invasives and natives are just starting to negotiate what their future coexistence will look like, and it will take hundreds or thousands of generations for these negotiations to reach a compromise.

Japanese stiltgrass (see figure 9.1) all but eliminates plant diversity near the ground, while porcelain berry smothers other plants from the ground to the sky.

You might think monocultures of Japanese stiltgrass, autumn olive, privet, bush honeysuckle, or phragmites would easily convince anyone that

FIG. 9.1. Japanese stiltgrass. Photo courtesy of Douglas Tallamy.

fighting such invasions is a good idea. Yet questions about the wisdom of attempts to curb vegetative incursions have been raised ever since invasive ornamentals started moving from our gardens into natural areas. People concerned about the impacts of invasive plants on ecosystems have been accused of being too emotional about plant invasions, of not letting nature take its course, of ignoring the beneficial side of introduced plants, and of trying to return our ecosystems to some pristine state that has not existed for at least 14,000 years. Some criticism of efforts to control invasive plants is understandable: we spend billions each year trying to manage invaded ecosystems, and though local success is common, eradication is nigh on impossible. Wide-scale control of invasive plants is extraordinarily difficult, particularly when we continue to sell such plants in our nurseries. Fighting what many see as a losing battle might seem a fool's errand if there were not compelling reasons to do so. But there *are* good reasons to keep introduced plants off our properties.

Invasive plants can disrupt long-standing ecosystem dynamics in several ways. They can change soil hydrology, alter fire regimes, competitively exclude and hybridize with native plants, and degrade aquatic habitats. Critics of invasive control efforts argue that all ecosystems are in a constant state of flux; if an introduced plant enters an ecosystem and changes the diversity and abundance of native species, supporters of this argument say, that change is a "natural process" that should not be challenged by humans (even though humans were responsible for it). They further argue that plant invasions are not "bad" for ecosystems because there are no records of a plant invasion causing a continent-wide extinction of a native species, and that there are more plant species in North America after an invasion than before.

Are these claims valid? Let's take a close look at the evidence. First, are there really more species present after an introduced plant displaces native plant communities? The answer depends on the geographic scale being considered.[7] If we have introduced 3,300 new plant species to North America, species diversity should now be higher. And on a continental scale, it is. But ecosystems don't function on a continental scale; they function locally, and there are oodles of studies documenting the reduction (or complete elimination) of one or more plant species after the arrival of an invasive plant on a local scale.[8]

Moreover, when counting species impacted by introduced plants, we should not just count plant species; we need also to consider the animals that eat or pollinate plants and, in turn, the predators of those animals. Every time a native plant is removed from an ecosystem, or even diminished in abundance, populations of all of the animals that depend exclusively on that plant are also removed or diminished, as are the natural enemies of those species. In sum, then, at the local scale—the scale that counts ecologically—invasive plants typically shatter local species diversity, and claims to the contrary have not been supported by rigorous field studies.

What about the assertion that invasive plants have not caused any native plant extinctions on continents? The qualifier "on continents" has to be added to this statement because the threat of extinctions caused by invasive species on islands is quite high.[9] In contrast, on continents, native plant populations are larger and more dispersed, and there are more places for natives to escape invasive species; even if a native population is clobbered in one place, other populations may persist in another. However, there is one biological phenomenon associated with some plant invasions that is so pernicious, even continental scales are not protecting natives from invasive species. I speak of *introgression* or *introgressive hybridization*, where the invasive species hybridizes with a closely related native, and then—through repeated back crosses and directional gene flow—the gene pool moves closer and closer to that of the invader. This is the process by which Africanized bees have replaced the European honeybee genotypes wherever the two have come into contact, and all in just a few generations. American bittersweet and red mulberry are two examples of native plants that are rapidly disappearing from their native range in a similar manner through directional introgression with Oriental bittersweet and white mulberry, respectively.[10] Both species of native plants have been entirely replaced by their non-native congeners in Maryland. Unfortunately, the introgression is proceeding at such a rapid pace that their continental extinction seems imminent.

Even though the claim that there are no records of extinctions caused by invasive plants may not remain true for long, it is certainly true that the introduction of non-native plants has not (yet) created an extinction threat for most native species. That, however, does not mean ecosystem function has not been compromised at the site of each invasion. Besides, using global extinction as the only indication of harm is like saying the only symptom that warrants a visit to the doctor is death. When invasive plants like autumn olive, barberry, bush honeysuckle, and phragmites invade plant communities, they completely replace the local native species at that site, causing a decline in the plants on which that ecosystem has depended for eons. If introduced plants were the ecological equivalents of the native species they replaced, ecosystems would look different after an invasion, but they would be just as productive (though less stable). Introduced plants may, in fact, be equal to natives in their ability to produce ecosystem services like carbon sequestration, but they pale in comparison to natives in perhaps the most critical role plants play in nature. That is, introduced plants are poor at providing food for the animal life that runs our ecosystems.

Plants, in essence, enable animals to eat sunlight; by capturing energy from the sun and storing it through photosynthesis in the carbon bonds of simple sugars and carbohydrates, plants are the basis of every terrestrial and most aquatic food webs on the planet. But animals benefit from the energy captured by photosynthesis only if they can eat plants, or eat something that ate plants previously. And there's the rub: the group of animals that is best at transferring energy from plants to other animals is insects. Unfortunately, most insects are very fussy about which plants they eat.

The Curse of Specialization

Ah, how easy conservation would be if all plants delivered the same ecological benefits, especially to all insects. We could plant eucalyptus around the world and plant-eaters everywhere would be as happy as koalas. The nectar-filled butterfly bush so many people plant "to help the butterflies" would actually serve as a larval host for all butterflies (instead of only one species in the southwest) and deliver pollen and nectar to all 4,000 species of native bees rather than just a few generalist bees. We could rename the evening primrose moth "the every-plant" moth, and the ornamental bamboos that are consuming yards and road shoulders throughout Maryland would feed monarch butterflies as well as milkweeds do.

But alas, specialized relationships between plants and animals are the rule rather than the exception in nature, and they are far more common than generalized ones. This is particularly true for specialized relationships involving food webs: those interconnected relationships that transfer energy harnessed from the sun by plants to animals that eat plants and then to animals that eat animals. Many people refer to this transfer of energy as a food "chain," but if you were to make a diagram of a plant and then all of the species that eat that plant, as well as all of the species that eat each of those plant-eaters, the result would look far more like a very complex food "web" than a linear "chain."

The most important and abundant specialized relationships on the planet are by far the relationships between the insects that eat plants and the plants they eat. Most insect herbivores, some 90% in fact, are diet specialists, or what we call "host plant specialists," restricted to eating one or just a few plant lineages.[11] Host plant specialization has been well documented by entomologists since the early 1960s, but scientists have never been very good at talking to each other, so the importance of host plant specialization in gluing ecosystems together is still underappreciated by many ecologists, by restoration biologists, and particularly by conservation biologists. This is why proposals to reforest tropical areas of the world with eucalyptus have not been met with what should be jaw-dropping outrage. Blue gum eucalyptus from Australia and the African rubber tree are now the most abundant trees in Portugal and Puerto Rico, respectively;[12] more and more frequently, shade coffee marketed as "good for the birds" is being grown under the shade of eucalyptus, citrus, and mango, even though these plants make little to no food for birds. The specialized relationships between insects and plants are so important in determining ecosystem function and local carrying capacity that it is worth spending a little time to explain why this is so and how these relationships have come about.

Plants, of course, don't want to be eaten; they want to capture the energy from the sun and use it for their own growth and reproduction. So, in an attempt to deter plant-eaters, they manufacture nasty-tasting chemicals and store them in vulnerable tissues like leaves. These chemicals are secondary metabolic compounds that do not contribute to the primary metabolism of the plant. That is, they are not a necessary part of the everyday jobs of living and growing. Instead, their job is to make various plant parts distasteful or downright toxic to insect herbivores. Some well-known plant defenses include toxic compounds like *cyanide, nicotine, ricin, cucurbitacins*, and *pyrethrins*; heart-stoppers like *cardiac glycosides*; and digestibility inhibitors like *tannins*.[13]

But if plants are so well defended, how can insects eat them without dying? This question dominated studies of plant-insect interactions for three decades, but at this point, the answer has been thoroughly delineated. Caterpillars and other immature insects are eating machines; some species increase their mass 72,000-fold by the time they reach their full size.[14] Because caterpillars necessarily ingest chemical deterrents with every bite, there is enormous selection pressure to restrict feeding to plant species they can eat without serious ill effects. Thus, a *gravid* (read: *pregnant*) female moth attempts to lay eggs only on plants with chemical defenses their hatchling caterpillars are able to disarm.

There are many physiological mechanisms by which caterpillars can neutralize plant defenses, but they all involve some combination of sequestering, excreting, and/or detoxifying defensive phytochemicals before they interfere with the caterpillar's health. Caterpillars have typically come by these adaptations through thousands of generations of exposure to the plant lineage in question. In short, by becoming host plant specialists, insect herbivores have developed the capacity to circumvent plant defenses of a few plant species well enough to make a living while ignoring the rest of the plants in their ecosystem. For our purposes, the key point regarding host plant specialization is that it does not happen overnight; although every once in a while an insect species coincidentally possesses enzymes that are able to disarm the defenses of a plant species that it has never before encountered in its evolutionary history, it usually takes many eons for an insect to adapt to a new host plant, if it can adapt to it at all.

Does this mean insect specialists have won the evolutionary arms race with plants? Somewhat, but only in relation to the plant lineage on which they have specialized. When viewed across all lineages, plant defenses are very effective at deterring most insects. The monarch butterfly provides a great example. This species is a specialist on milkweeds that use various forms of toxic cardiac glycosides to protect their tissues. Very few insects can eat plants containing cardiac glycosides, but over the ages, monarchs have developed the enzymes that can make cardiac glycosides less toxic. They also have found a physiological mechanism for storing these distasteful compounds in their wings and blood, rendering their own bodies unpalatable to predators. And monarchs have gone one step further. Tasting bad after eating milkweed plants does not help a monarch if a bird has to eat it in order to discover its unpalatability. So, monarchs, like many other distasteful insects, advertise their bad taste with an aposematic orange and black pattern that serves as a universal warning signal to would-be predators—"Don't eat me. I taste bad."

Milkweeds are so named because, in addition to cardiac glycosides, they defend their tissues with a milky latex sap that jells on exposure to air. Insects that attempt to eat milkweed leaves soon find their mouthparts glued permanently shut by the sticky sap. Yet monarchs have found a simple but amazing way to defeat this defense; they block the flow of sap to milkweed leaves.[15] This is an example of a behavioral adaptation (as opposed to a physiological adaptation), and you can easily watch it in action right in your yard. When a monarch caterpillar first walks onto a milkweed leaf, it usually moves to the tip of the leaf and starts to eat. If any latex sap starts to ooze from the wound, the caterpillar immediately stops eating, turns around, and crawls two-thirds of the way back up the leaf. There it chews entirely through the large midrib of leaf. That simple act severs the main latex canals that move the sap throughout the leaf. With the canals blocked, all of the leaf tissues below the midrib wound become latex-free and the monarch can eat them without gumming up its mouthparts. If the monarch decides to chew through most of the leaf petiole instead of the leaf midrib, latex is blocked from the entire leaf. Incidentally, this behavior provides monarch hunters with a convenient tool for finding monarch caterpillars, for the leaf flags at the point where the monarch weakened the midrib. Any milkweed plant with a flagged leaf is or has been the home of a monarch.

The advantage of these adaptations is obvious for the monarch, but there are also disadvantages to such specialization, especially in today's world. Unfortunately for the monarch, the ability to detoxify cardiac glycosides and block latex sap in milkweeds does not confer the ability to disarm the chemical defenses found in other plant lineages. This means that of the 2,137 native plant genera in the United States, the monarch can develop (with very minor exceptions) on only one, the milkweed genus *Asclepias*. The evolutionary history of this butterfly has locked it into a dependent relationship with milkweeds, and if milkweeds should disappear from a landscape, so would the monarch.

And this is exactly what has happened across the United States in recent years. A growing culture that favors neat, turf-lined (and chemically sanitized) agricultural fields and suburban lawns, combined with a broad unwillingness to share designed domestic landscapes with milkweeds, helped reduced monarch populations 96% in just 40 years, between the 1970s and 2013.[16] Can monarchs adapt to other plant species? In theory yes, but in reality no. The monarch lineage has been genetically locked into a relationship with milkweeds for millions of years. Adaptation could conceivably modify this relationship—very slowly and over enormously long periods—but asking monarchs to suddenly (within a few decades!) switch their dependence on milkweeds to an entirely different plant lineage (say, for example, crepe myrtle) is like asking humans to learn to eat grass or oak leaves. The number of genetic changes required to make such a switch reduces the probability of it happening before monarchs disappear to near zero.

Please note that monarchs are not exceptions, either in their specialized relationship with milkweeds, or in their current plight. They are typical of 90% of the insects that eat plants; their evolutionary history has restricted their development and reproduction to the plant lineage on which they have specialized. And as we homogenized plant diversity in Maryland by replacing diverse native plant communities with a small palate of ornamental favorites from other lands, the insects that depend on local native species have declined. We have caused these declines by the way we have designed landscapes in the past. But we can and must reverse them by the way we design landscapes in the future, for such decisions will determine how well our ecosystems function.

Ecosystem Function

I talk a lot about *ecosystem function* and usually draw blank stares when I do. I use the term *function* because in some ways ecosystems are like well-oiled machines; they are built from many interacting parts that combine to perform different functions. Creating the life support systems that keep us fed, healthy, buffered from severe weather, and with plenty of clean air and water—that is, the ecosystem services on which we all depend—are some of the things functioning ecosystems do every day.

When we look more closely at how ecosystems function, though, my analogy breaks down. Machines have a specified number of parts; taking some parts away almost always impairs or destroys the ability of a machine to run, while adding more parts than the design calls for does not make the machine function better. In contrast, research over the past 60 years has shown that ecosystems are far more flexible than machines in the number of parts that run them. To me, the most fascinating result of this research has been to discover the relationship between the number of species (parts) in an ecosystem and how well that ecosystem performs its various functions.

In 1955, at the very dawn of critical ecological thinking, Robert MacArthur published a paper in which he suggested that "ecosystem productivity"—the stability and ability of ecosystems to function—was inseparably related to the number of species residing within that ecosystem. As the number of species increased, so did both ecosystem stability and productivity. MacArthur was perhaps the most brilliant theoretical ecologist of the 20th century, so for him this was an unusually simple hypothesis. Testing it, however, would prove to be extraordinarily difficult, so he didn't. He just called this relationship the "law of nature" and left it to future generations of ecologists to test.[17]

Other hypotheses followed, including the famous "rivet" hypothesis proposed by Anne and Paul Ehrlich in which each species in an ecosystem is compared to a rivet holding a plane together; and the "redundancy" hypothesis in which Brian Walker suggested that species doing the same job in an ecosystem were functionally "redundant" and thus could be lost without harm to the ecosystem.[18] These ideas too were largely theoretical exercises that went untested in the field. Slowly, however, as the resources required for long-term ecosystem studies became available and ecologists learned how to manipulate simplified ecosystems, direct measures of the relationship between species richness and ecosystem function began to appear in the literature.[19] Surprisingly, all of these studies were largely in agreement.

MacArthur had been right after all.

Like machines, ecosystems run more smoothly, longer, and more productively when they contain all of their parts. Also, like machines, some parts (species) are more vital to ecosystem function than others, and losing these parts can shut down the entire system. But unlike machines, there may be no upper limit to the number of parts that help run ecosystems; every time a new species joins the ecosystem, it runs better than it did before. Unfortunately, the opposite is also true; every time we remove a species, or diminish its numbers to the point where it can no longer perform its role effectively, the ecosystem becomes less functional and less stable. And that in a nutshell explains why replacing native plant communities with introduced plants compromises ecosystem function: not only do such plants often reduce the number of species in the ecosystem, they *always* reduce the number of interacting species. Being present but not interacting with other local species is akin to throwing a monkey wrench in a machine. The monkey wrench is a new part added to the machine, but not only does it *not* interact in a positive way with the other parts of the machine, it actually *prevents* the other parts from interacting effectively with each other.

Introduced species occupy space in an ecosystem—space once taken up by contributing native species—but they have not been present for the thousands of generations required to form the specialized relationships that run ecosystems.

Interaction Diversity

Friedrich Wilhelm Heinrich Alexander von Humboldt—an 18th-century philosopher, linguist, botanist, biogeographer, and gifted naturalist—accumulated a list of scientific contributions during his 90 years that was even longer than his name. Humboldt spent much of the time between 1799 and 1804 exploring the tropical regions of the Americas. Among other things, he was keenly interested in the way the natural world worked. Not only was he the first to propose that Africa and South America had once been attached to each other, and that human activities would change the Earth's climate if they continued unabated, but—more to our point here—he was also the first to formally recognize that it was the way species *interacted with each other* (and not the species themselves) that provides the glue that holds nature together. Humboldt reasoned that nature was not an abstract idea but a living, interconnected entity.[20] Yes, the species that constitute nature are important, but it is how these species interact and the diversity of these interactions that make nature a living force.

Sadly, like his ideas on continental drift and climate change, the importance of what we now call *interaction diversity* was forgotten soon after Humboldt's death in 1859. What replaced it is the far less subtle idea that it is the species we encounter in nature, not the myriad ways in which those species depend on each other, that is important. Recently, this species-centric view of the world has intensified as more and more species have become threatened with local or global extinction in the 21st century. And despite growing calls from ecologists to preserve entire ecosystems, what still dominates the public's perception of conservation is the plight of individual (typically "charismatic") species; our mailboxes are full of pleas to save the whales, elephants, rhinos, tigers, and jaguars. The importance of interacting suites of species has been too abstract for the public and even many scientists to appreciate.

Fortunately, this is changing, and the change is being led by tropical ecologists. While lamenting the local extinction of countless species throughout Central America, Dan Janzen, perhaps the most perceptive ecologist of the last 50 years, noted that the most insidious form of extinction was not the loss of individual species but the extinction of ecological interactions.[21] This thinking has spawned a new way of measuring nature called *network analysis*, and Lee Dyer, an ecologist at the University of Nevada, thinks such analyses will soon show that interaction diversity is an even better predictor of ecosystem function than MacArthur's species diversity.[22] After all, Dyer argues, it is interactions among species that affect all aspects of ecosystems, from primary productivity to the way populations fluctuate and the survival and reproductive success of individual species. In the few cases in which it has been studied, interaction diversity is devastated by the introduction of non-native plants.

I will close this section with some numbers, not just because they demonstrate that introduced plants reduce both species and interaction diversity, but because they hammer home *how large* these reductions are. A few years ago, my students Melissa Richard and Adam Mitchell set out to measure what happens to caterpillars when invasive plants create a "novel ecosystem." Finding habitats that were thoroughly invaded by introduced plants such as autumn olive, multiflora rose, Callery pear, porcelain berry, burning bush, and bush honeysuckle was easy. These plants now typify the "natural" areas near the University of Delaware where we did our study. The trick was finding places that were still relatively free of invasive plants. Using a combination of restored sites, and areas not easily accessed by deer (which exacerbate the spread of invasive plants), we finally found what we were looking for: four invaded sites and four primarily native sites of similar size.

Using replicated transects, we counted and weighed caterpillars at each site, once in June and again in late July. By every measure, the caterpillar community and, by extension, the community of insectivores that relied on caterpillars for food were seriously diminished when introduced plants replaced native plants. Even though there was more plant biomass along the invaded transects, there were 68% fewer caterpillar species, 91% fewer caterpillars, and 96% less caterpillar biomass than what we recorded in native hedgerows.[23]

To summarize these numbers in terms of the everyday needs of the animals that eat caterpillars, we found 96% less food available in the invaded habitats! Interactions between caterpillars and hedgerow plants were also significantly impacted by introduced plants, with invaded hedgerows supporting 84% fewer interactions and 2.3 times less interaction diversity than co-evolved hedgerows. Had we compared the dozens of species that depend on caterpillars or their adult moths in invaded and native transects—the Dipteran and Hymenopteran parasitoids, as well as the assassin bugs, damsel bugs, predatory stink bugs, spiders, toads, and birds—the impact of these plant invasions on interaction richness and diversity would have been many times larger.

Consequences

My students and I did our study in unmanaged hedgerows, what passes for "natural" areas where I live. But would we have seen the same impact on insect populations if we had conducted the study in a typical suburban neighborhood? The answer would depend entirely on the percentage of the plant life in the study landscapes that was introduced. Unfortunately, in most urban/suburban and even exurban landscapes, the majority of plants are from somewhere else.[24] My students and I have measured this in 25-year-old suburban developments in Delaware, northeast Maryland, and southeast Pennsylvania. We didn't have to work too hard because these landscapes contained very few plants at all. They were 92% lawn! But of the plants that were there, 79% (on average) were introduced not just from other parts of North America but from other *continents*. Moreover, they were largely the same species we had studied in the invaded hedgerows: Callery pear, bush honeysuckle, privet, burning bush, Oriental bittersweet, barberry, and Norway maple. Unfortunately, homeowners are still legally allowed to landscape with invasive species in most areas of the country.

This should change.

For years I have speculated about the consequences of such landscaping choices for birds with whom many of us would like to share our yards. I had to speculate because no one had directly measured what happens to bird populations in landscapes that favor introduced plants. I was pretty safe in my speculations because logic dictates that if you take away the food birds need, they won't do well. This, as the saying goes, is not rocket science. Nevertheless, I need speculate no longer. In the first study of its kind, my student Desirée Narango has measured what happens to Carolina chickadee populations and the caterpillars that support them when native plants are replaced by introduced ornamentals in suburban settings.[25]

For three years Desirée and a team of field assistants followed breeding chickadees in the suburbs of Washington, DC, during the nesting season. Using video cameras at the nests, radio isotope analyses, territory mapping, vegetation analyses, foraging observations, and citizen scientists, Desirée was able to quantify all of the variables required to model population growth of chickadees as a function of the percentage of introduced plants within the chickadees' breeding territories.

Desirée found far more than I have space to describe, but here are some of the highlights of her research. Throughout her study, parent birds foraged for food on native plants 86% of the time. Compared to primarily native landscapes in her suburban study sites, yards dominated by introduced plants produced 75% less caterpillar biomass and were 60% less likely to have breeding chickadees at all. Apparently, chickadees were able to assess the quality of the landscape before they decided whether or not to set up house in one of Desirée's chickadee boxes. If a chickadee did build a nest in a yard with many introduced plants, it contained 1.5 fewer eggs than nests in yards dominated by natives and those nests were 29% less likely to survive. Chickadees that nested where there were not enough caterpillars fed their young spiders and aphids, but these food items did not compensate nutritionally for the lack of caterpillars in their chicks' diets. Nests in yards dominated by introduced plants produced 1.2 fewer chicks and delayed chick maturation by 1.5 days compared to nests located in yards with lots of native plants.

Some of these differences may not sound very big to you, but cumulatively they are making a huge and negative impact on suburban chickadee populations. Chickadee populations achieved replacement rate (that is, produced enough chicks each year to replace the adults lost to old age and predation) only in yards with less than 30% introduced plants. Unfortunately for the chickadees in Washington, DC, suburbs, Desirée found that, on average, 56% of the plants are introduced.[26]

Although these results are not a surprise, they remove the guesswork from understanding how much our plant choices impact the life around us. Here is solid evidence that, at least for Carolina chickadees, introduced plants are not the ecological equivalents of the native plants they replace. It is hard to imagine why other insectivorous birds would not be similarly affected by introduced plants. Desirée's research helps us understand that it is the plants we have in our yards that make or break bird reproduction, and not the seeds and suet we so dutifully buy for our feathered friends, although those supplements certainly help our birds after they have successfully reproduced.[27] Her results also give us insight into what is happening beyond our yards in the natural areas that have been invaded by the ornamental plants we have brought from Asia and Europe. We can now better understand one of the factors that has caused 432 species of birds to decline at such a perilous rate.[28]

Although they depend on seeds much of the year, Carolina chickadees must find thousands of caterpillars to rear one clutch of young.

Do Birds Care If a Berry Is Native or Not?

Our studies are not the only ones that show how seriously insect populations are affected by introduced plants; dozens of other labs have found the same effect over and over again. But what about berries? Do introduced plants make up for the loss of insects by producing lots of berries that sustain birds? Mark Davis, a botanist at Macalester College, asked this interesting question in 2011 when the controversy over fighting invasive plants was at its peak. He concluded, quite logically I might add, that if birds readily eat berries from introduced plants, they must not care about the evolutionary origin of those berries.[29] As anyone who has witnessed a flock of cedar waxwings strip a European crabapple of its fruit in minutes can attest, many birds do eat berries produced by introduced plants. In fact, it is birds that make so many introduced plants highly invasive; they eat seed-laden fruit, fly some distance away, and poop out the seeds that will start another generation of that species. So, if introduced plants are providing lots of good food for birds, maybe they aren't so bad after all?

The heart of this question focuses on the nutritional value of berries produced by different species of shrubs. We can assume that all berries are equal in what they deliver to birds, but assuming things is not science. Science is all about hypothesis testing, and that's exactly what Susan Smith Pagano and her collaborators at the Rochester Institute of Technology have been up to lately. They are testing the hypothesis that berries are nutritionally equivalent regardless of whether they are produced by native shrubs or introduced shrubs.[30]

Susan focused on berries produced in the fall for two reasons: (1) nearly all of our invasive shrubs produce their berries in the fall (I can't think of any that don't), and (2) both migrating and overwintering birds depend on fall berries for the fats they need to either fuel their migration or build fat reserves for the long winter months if they don't migrate.

Susan's work has revealed a surprising and distressing pattern. Berries from introduced Eurasian plants like autumn olive, glossy buckthorn, bush and Japanese honeysuckle, and multiflora rose contain very little fat, typically less than 1%, while berries from natives like Virginia creeper, wax myrtle, arrowwood viburnum, spicebush, poison ivy, and gray dogwood are loaded with valuable fat, often nearly 50% by weight. There is variation among species, but the pattern is clear: introduced berries are high in sugar at the time of year our birds need berries high in fats.

In the fall, birds like downy woodpeckers need berries high in fat to help them make it through the winter or to give them the energy required for migration. Poison ivy berries are among the best in this regard.

Why would a bird eat a berry that does not meet its nutritional needs? I can think of two reasons. First, when an invasive shrub like autumn olive, barberry, burning bush, privet, or bush honeysuckle moves into a habitat, it typically eliminates the native shrubs that used to grow at that site. Thus, the berries produced by the invasive shrub are now the only ones available for birds. With no other option, hungry birds eat them. I can suggest a second reason birds might eat nutritionally bereft berries by asking this question: why do you eat a sugar-coated donut when you ought to be eating a garden salad? If we humans, with the most developed reasoning capacities in the animal kingdom, are unable to distinguish what is good for us and what is not, why do we expect birds to be able to bypass a sugary treat? Despite this logic, new research suggests that birds *do*, in fact, "care" whether a berry is native or not and they discriminate against introduced berries whenever they have the option. When fall migrants stop to rest and eat in a habitat loaded with invasive shrubs, they do not stay long. Instead, they seek habitat with plenty of the spicebush and arrowwood viburnum berries they need to fuel their migration.[31]

Costs versus Benefits

Every now and then a paper appears in the scientific literature that points out the ecological benefits delivered by introduced plants. The conclusion drawn by the authors is always the same: if introduced plants are doing good things in local ecosystems, perhaps we should tolerate, or maybe even encourage, their presence. This logic has been used to justify planting more eucalyptus in California (people love trees more than dry grasslands) and to discourage municipalities from spending money to fight invasive species almost everywhere. I would agree with this line of thinking on one condition: the *net* effect of the plant in question must be positive. Benefits cannot be viewed in ecological isolation; they must be compared to the costs or disadvantages associated with a particular plant as well. Demonstrating the benefits that a plant delivers is meaningless unless we also measure the ecological costs that plant brings to the system. Only in this way can we estimate whether the benefits outweigh the costs, or vice versa. If the net effect of an introduced plant improves ecosystem function, then yes indeed, let's rethink our bias against it.

Here is a typical scenario: kudzu, the Asian plant that now exclusively occupies over 7 million acres in the American Southeast,[32] serves as a host plant to our native silver-spotted skipper. The skipper, a legume specialist, has found the defensive chemicals of kudzu to be within the range of its detoxification abilities, so it can reach maturity on the nutritious leaves of this introduced species. We can put the new host association between the silver-spotted skipper and kudzu in the benefits column; kudzu is creating a food web option that did not exist before its introduction. Sticking to the food web theme, though, we now have to consider whether kudzu is having any negative impacts on local food webs. The answer, of course, is yes. When kudzu smothers young oak trees in Camden County, Georgia, for example, the oaks—host options for 454 species of caterpillars—disappear. Similarly, if black cherry is eliminated from this kudzu patch, 324 caterpillar species are lost. If willows, hickories, and maples are covered by kudzu, 247, 229, and 223 species of caterpillars are lost, respectively.[33] Such losses will hold for all of the woody plant genera lost to kudzu at this single site in Georgia.

What about herbaceous plants? Well, if kudzu smothers goldenrod, as it surely does, 94 species of caterpillars are lost. If it covers native asters, another 80 species are lost. Sunflowers lost to kudzu host 67 species of caterpillars, horse nettle 67 more species, and so on. By allowing kudzu to invade an area in Camden County, Georgia, we have gained a host plant for silver-spotted skipper but lost host opportunities for the caterpillars of over a thousand other moths and butterflies. And as with all food webs, removing food from the base of the web reverberates throughout the entire web, impacting all of the species—especially birds—that eat the caterpillars lost to kudzu. No doubt, the net effect of this kudzu invasion is not just slightly negative, it is hugely negative in terms of supporting local biodiversity.

1. The Conowingo Dam, once the largest energy-producing facility in the world, remains a source of environmental debate. It produces renewable electricity but also impedes migrating fish while holding back hundreds of millions of tons of sediment. Photo courtesy of David Harp.

2. Maryland's industrial chicken plants are a major contributor to water pollution. Photo courtesy of David Harp.

1973 developed
~ 654,500 acres

- Developed
- Agriculture
- Forest
- Water
- Wetlands
- Barren

2010 developed
~ 1,664,900 acres

- Developed
- Agriculture
- Forest
- Water
- Wetlands
- Barren

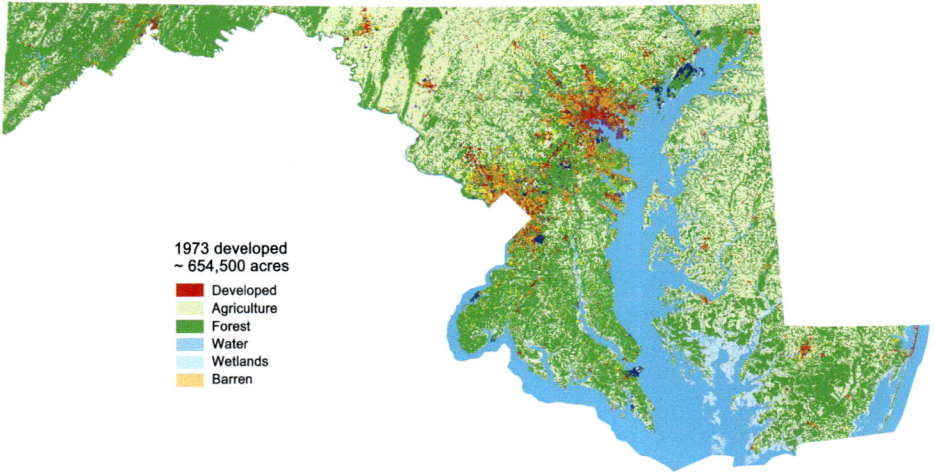

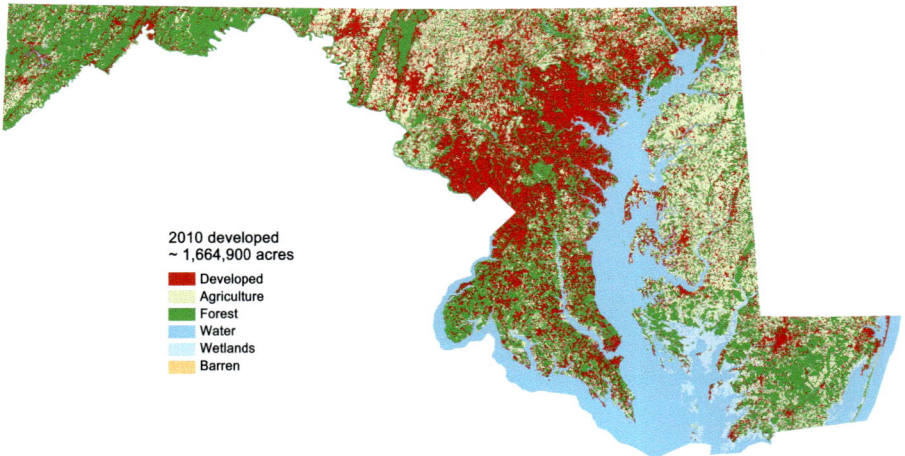

3. Maryland land use / land cover data. Developed land increased by over
1 million acres from 1973 to 2010. The number of developed acres per person has
increased by more than 80% since 1973. In 1973 (*top*), there was about one-sixth of
an acre (0.16 acres) of developed land per person (for homes, workspaces, restaurants,
retailers, schools, hospitals, houses of worship, etc.). By 2010 (*bottom*), developed land
per person had increased to about three-tenths of an acre (0.29 acres), an 84% gain.
Courtesy of Maryland Department of Planning.

4. Sunny-day flooding in Neavitt, a village near St. Michaels. Photo courtesy of Tim Wheeler.

5. Summer sunrise on Fishing Bay in Dorchester County. Photo courtesy of Tim Wheeler.

6. US Fish and Wildlife Service biologists scanning the marsh on Fishing Bay for saltmarsh sparrow, which are dwindling in the Bay as sea level rise drowns their precarious nesting habitat. Photo courtesy of Tim Wheeler.

7. "On the Deck of an Oyster Boat, Chesapeake Bay." Stereoview photograph (1912) of the deck of an oyster boat in the Chesapeake Bay. The view shows thousands of oysters that have been dredged from the ocean bottom and hauled on deck with nets by African American crew members. Photo by Keystone View Company. Courtesy of the Maryland Center for History and Culture [PP1.18.2].

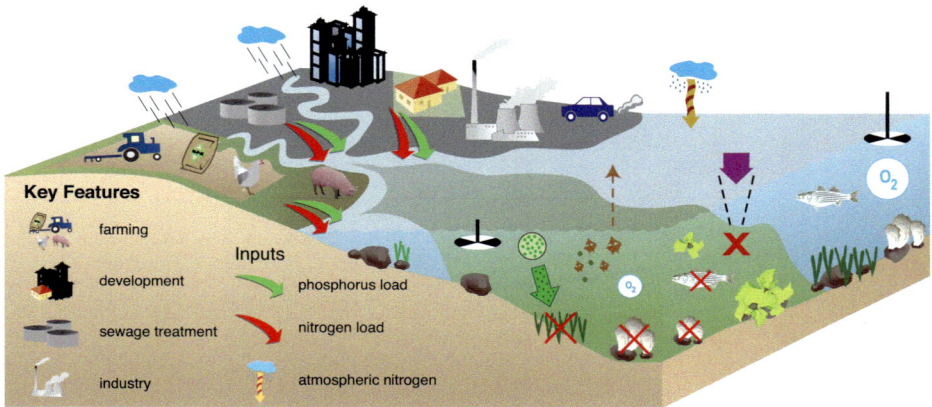

8. Sources of nutrients and eutrophication. Illustration by Jane Hawkey and Nathan Miller, Integration and Application Network, University of Maryland Center for Environmental Science.

9. Terris King and students in the Baltimore Forest School exploring the Stillmeadow Peace Park. Photo courtesy of McKay Jenkins.

10. Baltimore's Rock Rose Food Justice Project. Photo courtesy of McKay Jenkins.

11. Ultisol developed on reforested loamy sediments of the Coastal Plain (University of Maryland, College Park, Prince George's County). The thin, darkened, upper part of this soil (A horizon) has likely resulted from leaf fall and surficial rooting since the commencement of reforestation; the wavy boundary likely results from animal burrows or root channels subsequently infilled. The slightly brown lower horizon may represent an old Ap horizon dating from before reforestation, when the soil was used for agriculture, but an abrupt lower boundary created by differing organic matter content is not observed with confidence. Instead, this second horizon is designated AE owing to the slight browning due to organic matter (A) but also a lack of clay increase and lack of rubification by iron oxides (E). The simultaneous occurrence of both A-like and E-like properties indicates a transitional (e.g., AE) designation. Below the AE there are two Bts (found to be sandy clay loams), with increasing clay content, and rubification by iron oxide clays, with depth. Tree size and a lack of invasive trees in the interior of the woods indicate that reforestation occurred many decades ago and a lack of subsequent disturbance of the interior. Note that to the side of the central portion of the picture the soil has not been cleaned of sediment dripping from upper portions. Therefore, horizonation on either side of the photograph is less distinct. Note as well the (subangular) blocky peds at the foot of the profile, knocked off the soil when the face was cleaned.

Photo courtesy of David Ruppert.

12. A degrading histosol landscape, Garrett County, Allegheny Plateau. This fen, now owned by a conservation organization, was mined 40 years ago in its northern part (background) for peat, which had accumulated over the last 20,000 years to a depth of approximately 20 feet. Upon removal of the peat at one end, the resulting lake, barely visible in the background, has facilitated evaporation and a general lowering of the water table. This lowering has left the nearby unmined portion of the fen droughty and inhospitable to the survival of the unmined sphagnum. The peat in this unmined portion of the fen has begun to oxidize, leaving the present moonscape. A healthy sphagnum-based habitat persists at the far end of the fen but is under eventual threat—a legacy that persists and continues even after the cessation of mining so many decades ago. The purchase of peat-based products, ubiquitous in home and garden centers through similar processes, causes not only current but also future destruction of these irreplaceable soils and habitats. Photo courtesy of David Ruppert.

13. A downy woodpecker eating poison ivy berries. Photo courtesy of Douglas Tallamy.

14. A red-eyed vireo eating an alternate leaf dogwood berry. Photo courtesy of Douglas Tallamy.

15. The specialized skull structure of pileated woodpeckers protects their brains from injury as they hammer into trees and decaying wood to feed on carpenter ants, beetle grubs, and other insects. Photo courtesy of George M. Jett.

16. Many songbirds, like this prothonotary warbler, have a thin, pointed bill that is well suited to gleaning insects off of leaves and branches. Photo courtesy of George M. Jett.

17. Native brook trout from the Savage River. Photo courtesy of John Mullican.

18. Savage River headwaters. Photo courtesy of Ryan Cooper.

19. Female box turtle. Photo courtesy of Raymond V. Bosmans.

20. Corn snake. Photo courtesy of Raymond V. Bosmans.

21. Northern copperhead. Photo courtesy of Raymond V. Bosmans.

22. Little brown bat displaying white-nose syndrome. Although the disease has spread to bat colonies across the US, people can help by improving habitat for bats by (for example) planting native wildflower species to attract insects and by leaving dead trees standing, limiting the use of insecticides, and—for those who might visit caves where bats congregate—decontaminating gear that could be contaminated with fungal spores. Properly constructing and maintaining bat houses can be a somewhat complicated process, so be sure to conduct best practices for design and placement. Image courtesy of US Fish and Wildlife Service, https://digitalmedia.fws.gov/digital/collection/natdiglib/id/9508/rec/20.

23. A spotted salamander confidently crosses a street near Jug Bay Wetlands Sanctuary, knowing that a team of volunteers will ensure its safety. Photo courtesy of Michelle Campbell.

24. A migrating and fully grown spring peeper examines several Jug Bay volunteers and vice versa. Photo courtesy of Allison Burnett.

25. Ask a Bumble Bee! program coordinator Jenan El-Hifnawi photographs flowers and uploads them while tallying what species of bee visits them. The project explores bumblebee and carpenter bee floral preferences. Photo courtesy of Billy Clarke.

26. Tree huggers—DC tour guides try forest bathing for the first time. All ages can be captivated and find wonder in nature. Photo courtesy of Angela Yau.

Why Not Let Nature Take Its Course?

Some might actually use our kudzu example as evidence that nature is, in fact, taking its course to bring introduced plants into functional relationships with their new ecosystems. After all, the silver-spotted skipper has already adapted to kudzu (or, more likely, already possessed the necessary physiological adaptations required to eat kudzu when kudzu first arrived in Georgia). Isn't this evidence that nature is repairing itself without our help? In the long view, I suppose it may be, but unfortunately, the rate at which nature is repairing the damage we have inflicted is so incrementally slow compared to the rate at which we keep inflicting damage that true repair will not occur fast enough to prevent the loss of what we now know as nature.

The journalist Michael Pollan once asked whether there should be a statute of limitations on being "native." In other words, if a plant or animal has been in North America long enough, shouldn't we consider it ecologically equal to the organisms that evolved here? It's a good question, but it reflects how difficult it is for most people to comprehend the immense periods of time required to build evolutionary relationships. In my view, *native* is not a label a species earns after a given period of time. It is a term that describes *function*. For example, a plant should be considered a native when it acts like a native; that is, when it has achieved the same ecological productivity that it had in its evolutionary homeland, when it has accumulated the same number of specialized relationships that had been nurtured by the native plants it displaced, and when it has accumulated the same number of diseases, predators, and parasites that species that evolved in North America must endure. Time in residence is not the variable to be measured here; it is the rate at which local organisms adapt to the plant's presence.

The common reed *Phragmites australis* provides a great example. The European genotype that has displaced wetland vegetation from the Atlantic coast to the shores of Lake Michigan has been in North America hundreds of years. There is good evidence that it was used as packing material in the holds of the earliest sailing ships some 500 years ago. In Europe, phragmites supports 170 species of insects. After hundreds of years of residence in North America, only 5 insect species have started using phrag-mites as a nutritional resource.[34] Adaptation is happening, but at a glacial rate typical of evolutionary change.

And phragmites is not an exception. Very slow rates of adaptation have been recorded for a number of introduced plants. *Melaleuca quin-queneryia* has been in the Florida Everglades for over 130 years. In Australia where it evolved, *Melaleuca* supports 409 species of insects; in Florida, it only supports 8 North American insects.[35] Only 1 species has adapted to eucalyptus in California after 110 years,[36] and no species have adapted to the cactus *Opuntia ficus-indica* after 260 years.[37] In short, it takes enor-mous periods of time before introduced plants act like the natives they replace.

There are other reasons "letting nature take its course" after the intro-duction of hundreds of non-native plants is not a good idea. First, these types of introductions are anything but natural. We have perpetrated a bio-logical exchange of species across the globe so rapidly—instantaneously on an evolutionary time scale—that it has created a phenomenon that ecosys-tems have never before encountered in the history of life on Earth. There is no "natural" response to counter our meddling. Moreover, by moving in-troduced plants beyond the suite of natural enemies—the insects, mam-mals, and diseases—that keep them in check in their homelands, we have stacked the competitive deck against native plants that *do* have to contend with hundreds of herbivores and diseases. Expecting native species to suc-cessfully duke it out with introduced plants is ecologically unrealistic. And that is why natural succession from one type of plant community to another is essentially dead when invasive plants enter the scene. Most disturbances these days no longer progress from grassland to meadow to scrub to forest; instead, they become frozen in a perpetual tangle of invasive vines and shrubs, as you can see in figure 9.2. At least, that has been the case during the last 30 years. Will our native species prevail in the end? Some think they will, and I hope those optimists are right. But I wonder.

To those who claim it's a fool's errand to try to restore nature to some mythical pristine state, I say, "That's not the goal!" The natural world that

FIG. 9.2. A tangle of multiflora rose, autumn olive, and Oriental bittersweet. Photo courtesy of Douglas Tallamy.

existed before humans entered North America at least 14,000 years ago is gone, along with the Pleistocene megafauna that shaped it. The modified natural world that Europeans encountered in the 15th century is also gone, along with the Native Americans who shaped it. Since then, we have dramatically altered our watersheds, soils, forests, wetlands, and grasslands to meet first our agricultural needs and then our industrial needs. And we continue to change the land today. But none of these changes means we have to destroy (or can afford to destroy) ecosystem function. Wherever and whenever we can, we must reassemble the co-evolved relations between plants and animals and among animals themselves that enable ecosystems to produce the life support systems we all need. Introduced plants only hinder our efforts to do so.

Notes

1 Grateful acknowledgement is offered to the University of Delaware Press, in whose volume *The Delaware Naturalist Handbook* (McKay Jenkins and Susan Barton, eds., 2020) an earlier version of this essay first appeared.

2 Richard Hobbs, Eric Higgs, and Carol Hall, *Novel Ecosystems* (West Sussex, UK: Wiley-Blackwell, 2013).

3 William Stolzenburg, *Rat Island* (New York: Bloomsbury, 2011).

4 Peter Marra and Chris Santella, *Cat Wars* (Princeton, NJ: Princeton University Press, 2017).

5 Richard Duncan, Alison Boyer, and Tim Blackburn, "Magnitude and Variation of Prehistoric Bird Extinctions in the Pacific," *Proceedings of the National Academy of Sciences* 110, no.16 (April 2013): 6436–41.

6 Hong Qian and Robert Ricklefs, "The Role of Exotic Species in Homogenizing the North American Flora," *Ecology Letters* 9, no. 12 (October 2006): 1293–98.

7 Kristin Powell, Jonathan Chase, and Tiffany Knight, "Invasive Plants Have Scale-Dependent Effects on Diversity by Altering Species-Area Relationships," *Science* 339, no. 6117 (January 2013): 316–18.

8 Matthew Collier, John Vankat, and Michael Hughes, "Diminished Plant Richness and Abundance Below *Lonicera maackii*, an Invasive Shrub," *American Midland Naturalist* 147, no. 1 (January 2002): 60–72; Kathleen Knight, Jessica S. Kurylo, Anton G. Endress, J. Ryan Stewart, and Peter B. Reich, "Ecology and Ecosystem Impacts of Common Buckthorn (*Rhamnus cathartica*): A Review," *Biological Invasions* 9, no. 8 (December 2007): 925–37; Kristina Stinson, Sylvan Kaufman, Luke Durbin, and Frank Lowenstein, "Impacts of Garlic Mustard Invasion on a Forest Understory Community," *Northeastern Naturalist* 14, no. 1 (March 2007): 73–89.

9 Paul Downey and David M. Richardson, "Alien Plant Invasions and Native Plant Extinctions: A Six-Threshold Framework," *AoB Plants* 8 (August 2016): plw047, https://doi.org/10.1093/aobpla/plw047.

10 David Zaya, Stacey A. Leicht-Young, Noel B. Pavlovic, Kevin A. Feldheim, and Mary V. Ashley, "Genetic Characterization of Hybridization between Native and Invasive Bittersweet Vines (*Celastrus* spp.)," *Biological Invasions* 17, no. 10 (October 2015): 2975–88; Kevin Burgess and Brian Husband, "Habitat Differentiation and the Ecological Costs of Hybridization: The Effects of Introduced Mulberry (*Morus alba*) on a Native Congener (*M. rubra*)," *Journal of Ecology* 94, no. 6 (July 2006): 1061–69.

11 Matthew Forister, Vojtech Novotny, Anna K. Panorska, Leontine Baje, Yves Basset, Philip T. Butterill, Lukas Cizek, et al., "The Global Distribution of Diet Breadth in Insect Herbivores," *Proceedings of the National Academy of Sciences* 112, no. 2 (December 2015): 442–47.

12 Michaela McGuire, "The Eucalypt Invasion of Portugal," *The Monthly* (June 2013), https://www.themonthly.com.au/issue/2013/june/1370181600/michaela-mcguire/eucalypt-invasion-portugal.

13 Douglas Tallamy, "Do Alien Plants Reduce Insect Biomass?" *Conservation Biology* 18, no. 6 (December 2004): 1689–92.

14 O. W. Richards and R. G. Davies, eds, *Imms' General Textbook of Entomology* (Netherlands: Springer, 1977).

15 David Dussourd and Thomas Eisner, "Vein-Cutting Behavior: Insect Counterploy to the Latex Defense of Plants," *Science* 237, no. 4817 (August 1987): 898–901.

16 Lincoln Brower, et al., "Decline of Monarch Butterflies Overwintering in Mexico: Is the Migratory Phenomenon at Risk?," *Insect Conservation and Divers* 5, no. 2 (March 2012): 95–100.

17 Robert MacArthur, "Fluctuations of Animal Populations and a Measure of Community Stability," *Ecology* 36, no. 3 (July 1955): 533–36.

18 Paul Ehrlich and Anne Ehrlich, *Extinction* (New York: Random House, 1981); Brian Walker, "Biodiversity and Ecological Redundancy," *Conservation Biology* 6, no. 1 (March 1992): 18–23.

19 Oswald Schmitz, Peter Hambäck, and Andrew Beckerman, "Trophic Cascades in Terrestrial Systems: A Review of the Effects of Carnivore Removals on Plants," *American Naturalist* 155, no. 2 (February 2000): 141–53; José Rey Benayas, Adrian Newton, Anita Diaz, and James Bullock, "Enhancement of Biodiversity and Ecosystem Services by Ecological Restoration: A Meta-Analysis," *Science* 325, no. 5944 (August 2009): 1121–24; Peter Reich, David Tilman, Forest Isbell, Kevin Mueller, Sarah E. Hobbie, Dan F. B. Flynn, and Nico Eisenhauer, "Impacts of Biodiversity Loss Escalate through Time as Redundancy Fades," *Science* 336, no. 6081 (May 2012): 589–92.

20 Andrea Wulf, *The Invention of Nature* (New York: Knopf, 2015).

21 Daniel H. Janzen, "The Deflowering of Central America," *Natural History* 83, no. 4 (1974): 49–53.

22 Lee Dyer, Thomas R. Walla, Harold F. Greeney, John O. Stireman III, and Rebecca F. Hazen, "Diversity of Interactions: A Metric for Studies of Biodiversity," *Biotropica* 42, no. 3 (May 2010): 281–89.

23 Melissa Richard, Douglas Tallamy, and Adam Mitchell, "Introduced Plants Reduce Species Interactions," *Biological Invasions* 21, no. 3 (March 2019): 983–92.

24 Michael McKinney, "Urbanization, Biodiversity, and Conservation: The Impacts of Urbanization on Native Species Are Poorly Studied, but Educating a Highly

Urbanized Human Population about These Impacts Can Greatly Improve Species Conservation in All Ecosystems," *Bioscience* 52, no. 10 (October 2002): 883–90.

25 Desirée Narango, Douglas Tallamy, and Peter Marra, "Native Plants Improve Breeding and Foraging Habitat for an Insectivorous Bird," *Biological Conservation* 213 (September 2017): 42–50; Desirée Narango, Douglas Tallamy, and Peter Marra, "Nonnative Plants Reduce Population Growth of an Insectivorous Bird," *Proceedings of the National Academy of Sciences* 115, no. 45 (October 2018): 11549–54.

26 Narango, Tallamy, and Marra, "Nonnative Plants."

27 John Marzluff, *Welcome to Subirdia* (New Haven, CT: Yale University Press, 2014).

28 North American Bird Conservation Initiative, *The State of North America's Birds 2016* (Ottawa, Canada: Environment and Climate Change Canada, 2016), http://www.stateofthebirds.org/2016.

29 Mark Davis, Matthew K. Chew, Richard J. Hobbs, Ariel E. Lugo, John J. Ewel, Geerat J. Vermeij, James H. Brown, et al., "Don't Judge Species on Their Origins," *Nature* 47 (June 2011): 153–54.

30 Susan Smith, Samantha DeSando, and Todd Pagano, "The Value of Native and Invasive Fruit-Bearing Shrubs for Migrating Songbirds," *Northeastern Naturalist* 20, no. 1 (March 2013): 171–85.

31 Susan Smith, Allyson C. Miller, Charmaine R. Merchant, and Amie F. Sankoh, "Local Site Variation in Stopover Physiology of Migrating Songbirds Near the South Shore of Lake Ontario Is Linked to Fruit Availability and Quality," *Conservation Physiology* 3, no. 1 (August 2015): 10.1093/conphys/cov036; Yushi Oguchi, Robert Smith, and Jennifer Owen, "Fruits and Migrant Health: Consequences of Stopping Over in Exotic- vs. Native-Dominated Shrublands on Immune and Antioxidant Status of Swainson's Thrushes and Gray Catbirds," *Condor: Ornithological Applications* 119, no. 4 (October 2017): 800–816.

32 Irwin Forseth and Anne Innis, "Kudzu (*Pueraria montana*): History, Physiology, and Ecology Combine to Make a Major Ecosystem Threat," *Critical Reviews in Plant Sciences* 23, no. 5 (August 2004): 401–13.

33 "Native Plant Finder," National Wildlife Federation, https://www.nwf.org /NativePlantFinder/Plants.

34 Lisa Tewksbury, Richard Casagrande, Bernd Blossey, Patrick Häfliger, and Mark Schwarzländer, "Potential for Biological Control of *Phragmites australis* in North America," *Biological Control* 23, no. 2 (February 2002): 191–212.

35 Sheryl Costello, Paul D. Pratt, Min B. Rayamajhi, and Ted D. Center, "Arthropods Associated with Above-Ground Portions of the Invasive Tree, *Melaleuca quinquenervia*, in South Florida, USA," *Florida Entomologist* 86, no. 3 (September 2003): 300–323.

36 Donald Strong, John Lawton, and Sir R. Southwood, *Insects on Plants* (Cambridge, MA: Harvard University Press, 1984).

37 D. P. Annecke and V. C. Moran, "Critical Reviews of Biological Pest Control in South Africa. 2. The Prickly Pear, *Opuntia ficus-indica* (L.) Miller," *Journal of the Entomological Society of Southern Africa* 41, no. 2 (1978): 161–88.

CHAPTER 10

Birds

GWENDA BREWER

Throughout history, birds have always been among our most familiar animals. Their daytime activity means we can see them virtually anywhere we live. Even in our most urban environments, we can observe a house sparrow collecting material to build a nest or a rock pigeon cooing to its mate. Birds have also played many roles in our cultural history. They have provided food (from chicken nuggets to Thanksgiving turkeys) as well as feathers used for ceremonies, insulation (consider the down jacket!), and ornamentation (feathered hats were once so popular they almost drove some species to extinction). Birds have served as mythical symbols and omens, exotic pets, and objects of beauty and have even assisted with the ancient hunting practice known as falconry. Carrier pigeons have played a role in human communications, and the image of "the canary in the coal mine" is a powerful reminder that birds also can indicate environmental threats. Bird feathers continue to be highly valued for their important role in Native American communities in Maryland.

Biologically, birds are a diverse group of organisms in the Class *Aves*, with almost 11,000 species representing some 50 billion individuals.

Maryland is a wonderfully rich place for bird life—over 450 species of birds have been documented in the state.[1] A number of these species were "one-day wonders," but over 200 species are known to regularly breed in the state, and other species regularly pass through on migration or spend their winters here.[2] Part of the reason for this wide diversity is the variety of habitats present in the state, and—since Maryland is located in the middle of the East Coast—the presence of species at both the northern and southern edges of their ranges. Maryland also features both coastal and mountain migration corridors used by millions of birds in spring and fall each year.

But what makes a bird a bird? Characteristic features of modern-day birds include bills with no teeth, hollow bones, fused bones, bipedal locomotion, embryos that develop in a hard-shelled egg, and several other specialized anatomical and physiological features that we will discuss. But even this fundamental question has taken on new intrigue in recent years, with ongoing fossil discoveries linking modern and ancient birds to their dinosaur ancestors. Even feathers—a feature that makes birds unique among living animals—have been documented in dinosaurs.[3]

Flight

The anatomical and physiological characteristics of birds have been strongly influenced by their primary form of locomotion: flight. As obvious as this might seem, consider this: even the characteristics of modern-day *flightless* birds were influenced by flight, since they too evolved from flying ancestors.

Birds fly by *flapping flight*, in which they actively move their wings, and by *soaring* or *gliding*, in which the position of the wings and tail are adjusted but not repeatedly moved up and down. For anything that flies, including airplanes, there are four primary forces to consider: *upward lift* (and the opposing force *gravity*) and *forward thrust* (and its opposing force *drag*).

To overcome the force of gravity, birds (like airplanes) use wings with a slightly curved surface (thicker at the front and thinner at the back) that force air to flow faster over the top of their wings than the bottom. This differential—lower pressure on top of the wing compared to under the wing—is what creates lift. (Think of what it feels like when you put a slightly curved hand out of a moving car window: you feel your hand rise into the air!) Using their *secondary feathers*, which are attached to the forearm bones and extend from the body to the middle of the wing, birds also create lift by physically pushing down as they flap and by taking advantage of vertical columns of air currents rising from the surface of land or water.

Birds overcome drag by the movement of their *primary feathers*. These feathers are attached to the elongated and fused hand and finger bones in the outer half of the wing. In addition, the *alula*, a small group of feathers attached to what is left of the thumb, can be controlled to assist in takeoffs, landings, and steering. *Tail feathers* are also important in steering and landing.

During flapping flight, lift and thrust are mostly created during the downstroke of the wing (see figure 10.1). During the upstroke, birds bring their wings backward and then upward. Depending on flight speed and the shape of the wing, some lift may be produced during the upstroke

FIG. 10.1. American wigeons create lift as air flows over and under their outstretched wings and thrust as the wing is brought forward during flapping flight. Photo courtesy of George M. Jett.

(see figure 10.2). An interesting exception to this general rule is the production of lift in the upstroke by hummingbirds, whose shoulder joints allow them to move their wings in a figure eight pattern, which allows them to hover and even fly backward!

Birds create forward thrust by engaging their muscles to bring their wings forward (flapping) or simply adjusting the angle of the tips of the wings. They also take advantage of horizontal air currents ("tailwinds") to assist with forward movement (see figure 10.3). Drag, the force counteracting forward movement, is created by *friction* between the air and the bird moving through the air, and by the *disruption* in the air created as the wing travels through the air. Like bike racers in the Tour de France, some birds (such as Canada geese) reduce drag by "drafting" close behind each other, the leader breaking the trail and the followers slipping in behind; geese typically do this by flying in a "V" formation.

Wing shape and size help determine how fast birds can fly. Small birds tend to fly at about 30 mph, faster birds at 45 mph, and ducks at 50 mph. The American peregrine falcon can hit speeds of close to 200 mph during its high-velocity hunting dives. To increase speed during flapping flight, birds increase the amplitude (but not the frequency) of their wing beats. Different wing shapes also serve to provide better flapping or soaring flight and different degrees of maneuverability. For example, ruffed grouse have short, rounded wings that work well for rapid takeoff and flying through forests, whereas the American kestrel's thin, elongated wings provide speed for hunting and the ability to hover over its open field habitats. Gliding birds, such as vultures and hawks, take advantage of air currents and make subtle movements of the wingtips and tail to adjust their trajectory.

Feathers

Beautifully adapted for efficient flight, feathers allow birds to have a large, lightweight wing surface that does not require bones for support. Like our fingernails, feathers are made of a form of *keratin*. For most birds, feathers grow only from limited, linear tracts on the body. Feathers have a central shaft with barbs branching from it, and barbules branching from the barbs.

FIG. 10.2. An osprey shows how it holds its wings and tail as it lands. Some lift is still being created by placing the primary feathers sideways and by fanning the tail so that the bird does not come down too fast. On the other hand, the position of the secondary feathers has greatly reduced lift, allowing the bird to comfortably descend. Notice the flexibility in the wing at the junction between the primaries and secondaries—the bird's wrist area. Photo courtesy of George M. Jett.

FIG. 10.3. This red-tailed hawk is taking advantage of air currents to achieve forward motion. It can steer by altering the angle of its tail and by changing the positions of the separated feathers at the ends of the wings (primary feathers). Photo courtesy of George M. Jett.

These barbules have a system of hooks and notches or ridges that interlock and form a solid surface. As birds preen, they can "zip" these hooks and notches back together to maintain the solid surface. The one exception to this rule is feathers grown for insulation, like down. Down feathers grow all over the body and do not have interlocking barbs. This makes them fluffy and able to trap air.

In addition to flight, feathers provide an amazing array of functions, including insulation, communication (by accentuating mating displays), camouflage, support, water repellence, protection, and even sound production. Since feathers wear, build up parasites, or just need to be changed for breeding patterns, birds replace their feathers by *molting*—shedding the old and growing in new feathers. Most birds molt a few feathers at a time so that they can still fly, with the molt taking place before spring or before fall. An exception to this is ducks, who go through a *flightless period* when they lose all of their primary flight feathers at once before quickly growing them back. This strategy works because they can still dive to escape predators and join molting flocks, finding safety in numbers.

Birds may alternate between breeding and non-breeding plumage patterns, or (like some raptors and gulls) they may take several years to obtain the typical "adult" pattern. For example, breeding scarlet tanager males trade their brilliant red body feathers for less conspicuous yellow-green body feathers for the non-breeding period. The colors of feathers come from pigments, such as melanin (black, brown, reddish-brown) and carotenoids (red, yellow, orange), the internal structure of the feather, or a combination of the two. For example, the blue in a blue jay's feathers does not come from a blue pigment but from a combination of melanin in the feather and the internal structure of a feather, which causes us to see the reflected light as blue. If you hold a blue feather up to the light, you will see there is actually no blue pigment—just the black and dark brown pigments from melanin. Carotenoids are acquired through natural foods, and their absence or replacement by non-natives can affect plumage color.

Adaptations for Flight

Flapping flight takes a lot of sustained energy, and birds have developed several anatomical and physiological specializations that allow them to fly efficiently. These adaptations can be put into two general categories: *weight-reducing* and *power-increasing*. Anatomically, weight-reducing adaptations come from the fusion (and loss) of bones, hollow bones, a lack of teeth in the skeleton, and lighter-weight tendons (rather than bulky muscle tissue) to control the extremities. Light and strong feathers also reduce weight.

As for physiology, birds reduce weight by digesting food rapidly so that it passes out of the body quickly. They excrete uric acid rather than urea, which requires much less water to dissolve excess nitrogen. By laying eggs rather than carrying their young they reduce weight during reproduction, and their *gonads* (reproductive organs) shrink markedly in size between breeding and non-breeding seasons. Females of most bird species have one ovary and associated structures, rather than two.

To increase power for flight, birds have developed very efficient circulatory and respiratory systems. They have large, nucleated red blood cells, and a large heart relative to their body size. The heart is controlled efficiently by a series of nerve fibers, and blood flows with little resistance through stiff and smooth arteries.

A dramatic difference between birds and other vertebrates can be found in the respiratory system. Equipped with a series of air sacs in addition to their lungs, birds can direct only *oxygenated air* over the lung tissues, relieving them of the two-way air flow other animals (including us) experience during respiration, where oxygenated air passes over the lung tissues on the way in and carbon dioxide passes over on the way out. The supplemental air sacs are like thin-walled balloons inside a bird's body; they can even extend into some hollow wing bones. The flow of air follows this pathway in birds: in through the mouth and directly to the air sacs in the rear of the body, then through the lungs, then to air sacs in the front of the body, then out of the mouth. This creates a one-way, oxygenated flow of air over the lungs, which results in the extremely efficient respiratory system that allows some birds to dive to impressive depths, others to fly at high altitudes, and still others to carry out long-distance flights during migration.

Sensory Systems

The *sensory systems* (and brains) of birds also demonstrate complex specializations. For example, during flight the brain coordinates the actions of over 50 muscles. Recent studies have shown that brain size increases during the breeding season, as male singing rates increase,[4] and that species with the ability to learn and retain more sounds, like gray catbirds and blue jays, are better at problem-solving.[5]

Sight is the most well-developed sense in birds, due to their large eyes and structures that provide excellent color vision and acuity (the ability to distinguish objects at a distance). A bird's eyeballs take up a large space in their skull, and the part of the eye that we see when we look at a bird is really just a small part of the whole eyeball. Some diving species (like pelicans) have protective membranes that cover the eyes and act like goggles underwater, while other species, like pigeons, *see a different visual image with each side of their heads*. This leads to their *nystigmatic walk*, bobbing the head back and forth to get a more three-dimensional view of their surroundings.

Owls have an especially well-developed sense of *hearing*. For example, American barn owls, a Maryland species that feeds on small mammals in grasslands and marshes, can catch a moving mouse in a dark farm field. Owl ears are placed asymmetrically on either side of the head to allow better localization of sounds. They also often have well-developed *facial disks* of feathers arranged to direct sound to the ears. Added to these features are very soft feathers and comblike edges on the primaries that make them silent flyers.

The sense of *smell* in birds is not well understood, but recent studies have shown that some birds can use smell to recognize individuals, select nest materials, navigate, and avoid predators in addition to using smell to find food. Turkey vultures, for example, rely on their sense of smell to find the carrion on which they feed from up to eight miles away! Because most mammals are equipped with a very good sense of smell (and use it to identify their young), people falsely believe that birds do the same. It is not true that adult birds will smell "human" scents on their young and then reject them as some mammals will. If a baby bird is found outside of a nest, it can indeed successfully be replaced. The parents will not reject their baby if it was handled by a human.

Feeding and Digestion

Birds eat a wide variety of foods, and—as Darwin famously discovered with Galapagos finches—their bill shapes and sizes are often specialized to deal with certain kinds of foods. Local examples of these adaptations include the "dip net" of brown pelicans, the spearing bills of great blue herons, the grasping bills of belted kingfishers, the wide insect net mouth shape of tree swallows, the seed-cracking bill of northern cardinals, the straining bill of mallards, the probing bills of American woodcocks, and the wood-chisel bills of pileated woodpeckers. Examples of specialized bill shapes are pictured in figures 10.4 and 10.5.

The digestive system of birds is similar to our own, but there are some important differences related (again) to weight reduction for flight and other functions. Food first passes into the esophagus, which may have a well-developed pouch or *crop* where food is moistened or even temporarily stored. The stomach is divided into two parts: enzymatic digestion of food takes place in the *proventriculus* and mechanical digestion (remember, no teeth!) takes place in the *muscular gizzard*. The gizzard may be lined with hard, ridged plates and frequently includes grit for grinding. The small intestine functions to absorb nutrients, and the large intestine is short and resorbs water. Yellow-rumped warblers, one of Maryland's numerous winter residents, have an interesting specialization of the small intestine: they can feed on wax myrtle and digest the waxy coating on the berries thanks to microorganisms housed in their small intestine.[6]

Other Maryland birds have interesting feeding specializations that help get rid of undigestible materials like fur, insect exoskeletons, claws, and

FIG. 10.4. The American avocet, a migratory visitor to Maryland, sweeps its long, thin bill from side to side to feed on aquatic invertebrates. Photo courtesy of George M. Jett.

bones. Hawks, owls, gulls, and kingfishers, for example, cough up these materials compacted into discrete pellets. In fact, collecting and examining pellets from below roost areas is a good way to learn about what owls are eating. Tufted titmice, blue jays, and nuthatches are known to store food in caches to be eaten later. In many habitats, birds are important seed dispersers: after finding berries, they will fly to an area away from the source plant to digest their food and quickly pass undamaged seeds. This relationship is symbiotic: birds get food from berries, and plants use birds to spread their seed. Black gum, winterberry, pokeberry, and viburnums are just a few examples of local native species that benefit from seed dispersal by birds. On

FIG. 10.5. The American oystercatcher can pry open oysters and other mollusks with its chisel-shaped bill. Photo courtesy of George M. Jett.

the downside, many invasive plants like porcelain berry, English ivy, and Asian bittersweet use this same technique, which is part of the reason these species have become such a widespread problem.

Migration

As birds move through their annual cycles, they may travel long or short distances between the areas that they occupy during the breeding and non-breeding seasons. Some species spend the whole year inside the state like northern cardinals, downy woodpeckers, and tufted titmice. But of the more than 200 bird species that breed in Maryland, about one-third travel to Mexico (or farther south) for the non-breeding season, some flying as far as southern South America. Although we don't usually think of it this way, birds typically migrate to take advantage of an opportunity, like seasonal availability of food, rather than to escape bad conditions.

Long-distance migration is an amazing feat, and all of the specializations for efficient flight come fully into play to make these long journeys possible. The migration distances and flight times of species can be considerable. The red knot, a shorebird, flies 10,000 miles (twice!) each year and stops along the Maryland coast on its journey. To cross the Gulf of Mexico, the tiny ruby-throated hummingbird can fly for 26 hours straight!

Migrating birds usually stop over in one or more places along their journey for one or more days to rest and refuel, and even to wait for favorable weather conditions. One reason for Master Naturalists to encourage the planting of native plants: these are what birds have evolved to eat during their daunting migrations. And they eat intensively, especially if they are embarking on a long journey! The red knot, for example, can put on almost half of its body weight in fat in just two weeks. Guinness World Records reports that in 2022, a bar-tailed godwit flew 13,560 kilometers (8,435 miles) from Alaska to the Australian state of Tasmania without stopping for food or rest over an 11-day period, breaking the record for the longest nonstop migration by a bird.[7]

Many birds, such as our small songbirds, travel at night during migration. They may use less energy at night because it is cooler, and they can avoid daytime predators. They can also then feed during the day. On the other hand, birds—like raptors—that rely on (or take advantage of) thermals for longer movements migrate during the day.

Birds find their way to their distant destinations using a variety of cues. During the day, they can use the position of the sun as well as visual landmarks like mountain ranges, rivers, and coastlines. For night migrants, the North Star—around which the night sky rotates—is an important orienting feature. Birds also seem to have an internal map that guides them to fly in a certain direction for a certain amount of time. There is still much to be learned about bird migration, although technological advances in tracking birds have greatly increased our knowledge and will continue to do so.

Breeding and Non-breeding Periods

Looking over the full annual cycle of a bird's life, we see that for most species, more time is spent in a state of non-breeding (eight months) than breeding (about four months). For species with long migrations, the non-breeding part of the cycle may include up to several weeks of time engaged in the journey. During the non-breeding period, birds may form flocks to find safety in numbers and to learn the locations of food sources from flock members. During Maryland winters, flocks may include members of more than one species traveling together, or flocks of mostly single species, like the large concentrations of blackbirds in farm fields.

To our south, similar behavior is seen in non-breeding habitats for many of the species that migrate long distances, although some (like least and semipalmated sandpipers) may also defend territories for exclusive access to resources even in non-breeding areas. While they are breeding in Maryland, many birds defend territories to secure exclusive access to mates, shelter, nesting sites, and/or food in a particular area. Every spring, birdsong fills our fields, forests, and backyards as males sing to tell other males to keep out of their territory and to attract a mate into their territory. Unfortunately, some northern cardinals take territory defense to a new level when they perceive their reflection in a window or car mirror as a rival. In contrast, some of our breeding species, including herons and egrets, terns, gulls, and brown pelicans, nest in very close proximity to one another. We refer to these species as *colonial nesters.*

Most bird species practice a social system called *monogamy*—with one male pairing with one female. Birds use a great variety of visual and vocal displays to attract a mate and to indicate their territory boundaries. You have probably seen examples of bird displays—often emphasized by special plumage features like bright colors or long plumes—in your backyard or on nature programs. For example, the spread-wing posture of a male red-winged blackbird shows off his bright red wing patch, which may not even be visible at other times.

Birds also make a wide variety of sounds, including *songs* and *calls.* Songs are a longer series of sounds that are usually learned; calls are shorter sounds that are usually innate (not learned).

Not all species produce songs, but the study of those that do has provided interesting parallels between birdsong learning and human language learning. Birds produce sounds just as we do—by passing air along vibrating membranes—but their sound production apparatus, called the *syrinx*, is located in the chest where the trachea splits into the two bronchi. In songbirds, two vibrating membranes are typically present, and some species, like the wood thrush (a melodious bird of Maryland's forests) control these membranes separately, essentially singing with two voices at once! Although singing by female birds is less common (and is less well documented) than male singing, local examples of female singers include northern cardinal and white-eyed vireo.[8]

Bird eggs come in a variety of shapes, colors, and patterns. Egg shapes can be determined by the type of nest, with *cavity nesters* laying round eggs that use less shell, are stronger, and have a lower heat loss than *open ground* and *cliff nesting* birds that lay elongated eggs with very pointed ends so that they roll in a tight circle. The porous shell of bird eggs allows gas exchange for the developing embryo; the calorie-filled yolk provides nutrients; and the albumin (egg white) provides insulation and a water source. As the embryo develops inside the egg, liquid waste products collect in an isolated section. Female birds lay one egg roughly every 24 hours. Clutch size (the number of eggs laid during a single breeding event) varies greatly, and some species renest during a breeding season, raising more than one brood per year. In our area, for example, eastern bluebirds can have several broods in a single season, depending on when they start and available food resources.

Nests show an equally dazzling array of shapes, sizes, and placement, from more simple *open cup* nests formed with a few twigs, like those made by mourning doves, to the *woven nests* of Baltimore orioles. Nests are often lined with soft plant material or fibers; hummingbirds even camouflage their nests by attaching lichens to the outside of the nest with spiderwebs! Some species, like wood ducks, hooded mergansers, eastern bluebirds, and barn owls, have benefitted from *nest box programs* in the state. Birds considered to be *brood parasites* lay their eggs in the nests of other birds and do not build a nest or raise their own young. In Maryland, brown-headed cowbirds are the only species with this unusual breeding pattern.

Incubation patterns—who incubates and for how long—vary across bird species. In most of Maryland's species, only the female incubates and, in general, incubation begins when the clutch is complete. Eggs are turned during the incubation period so that the membranes inside of the egg do not stick to the side of the shell. In general, a short incubation period goes with a short development time and longer incubation periods go with longer development times. For example, house wrens have a 14-day incubation period and the young then develop for 12 days before leaving the nest. Bald eagles have a 40-day incubation, with a 70-day development period. Young birds (like songbirds) may be very undeveloped (*altricial*) when they hatch, or (like waterfowl) be very developed (*precocial*) when they hatch. Hawks, owls, terns, herons, and egrets fall between these two extremes.

Depending on their state of development at hatching, young birds may need their parent (or parents) to feed, brood, carry, protect, teach, and clean up after them. In some species, the young continue to be cared for by both parents even after they can fly, especially if (like terns) they have specialized feeding habits. Play, which encourages the young to practice locomotory and social skills, can also be part of a young bird's development.

Ecosystem Roles and Conservation Challenges

Our natural communities of plants and animals would not be the same without birds. As mentioned previously, birds play a role in seed dispersal and can also be important in pollination. Birds are thus an integral part of the "web of life" as both predators and prey. The nests that birds excavate, particularly tree cavity nests, provide homes for other animals that are unable to construct such nests themselves. The sensitivity of birds in general to contaminants has made them good indicators of ecosystem problems, and their sensitivity to the impacts of habitat loss, the decline of native species, and forest fragmentation can indicate a decline in ecological system function for a whole suite of species that are more difficult to study.

Given their many roles and sensitivities, we should think broadly about the main conservation challenges that we face when it comes to bird populations. Habitat loss and degradation often have the greatest impacts on birds, as well as many other species, and this is an ongoing challenge for Maryland. When land uses change—especially as human development replaces natural habitats like forests and wetlands—resources needed to feed, breed, and carry out other activities can be degraded (or entirely lost). In addition to suitable habitat for breeding, migratory birds need stopover sites and places to overwinter.

Climate change impacts on birds are a relatively new area of inquiry, but one factor that could impact birds is a shift in when and where critical insect food resources are available. For example, if the timing of insect hatches shifts earlier due to warmer temperatures, food may not be available when songbirds are feeding young unless they, too, have shifted their annual pattern.

Climate-driven habitat loss has become a serious issue, especially for beach and island nesting species and some marsh species. Introduced (and especially invasive) plants do not always provide the same value in terms of nesting habitat and food and can outcompete (and even replace) native species. Introduced predators—or native species that are out of balance—can have devastating impacts on bird populations; consider the tens of millions of birds killed by outdoor cats and the damage done to bird habitat by overabundant deer. Widespread bird mortality is also caused by collisions with cars, windows, towers, and power lines.

To assist the birds that are breeding, overwintering, or stopping over on migration in our area, it is vital that we support appropriate habitat and minimize risks to birds in our back (and front!) yards, in our local counties, and on our public lands. But it is also important—and more challenging—to continue to provide high-quality habitat for birds stopping over or spending the non-breeding season in other states and other countries. Support for conservation organizations across the full annual cycle is one way to provide support for these species.

The state of Maryland and the federal government have put in place several kinds of legal protections and programs in the state for birds and their habitats, including threatened and endangered species designations, Critical Area regulations, wetland protections, forest retention, hunting seasons, and bag limits for game species. Additional protections for native species fall under the federal Migratory Bird Treaty Act, which protects not only live birds but also limits the possession of bird parts, including feathers, and nests to those with a permit for scientific or educational purposes. A good source of information on the conservation of declining and protected bird species and their habitats, including threats and actions to address them, can be found in the current version of the Maryland State Wildlife Action Plan.[9]

Enjoying Birds in Maryland

There are many opportunities to enjoy the diversity of bird species in Maryland. Bird clubs—which can be accessed through the Maryland Ornithological Society website and through the Audubon Mid-Atlantic website—are a good place to start for those compiling bird lists, and for observation locations, field trips, and other birding information. eBird, a digital data collection system and data repository, is a great resource to discover species distributions in space and time, as well as location-specific information, including birding "hot spots." eBird is also a great place to enter your own data—through which you can help increase our knowledge of bird distributions, relative abundance, and timing of movements. The Maryland Biodiversity Project is another data repository for sightings of bird species in the state.

Identifying birds in the field can be challenging, but focusing on a few things can be helpful. Bill shape and size and feather colors and patterns—including what you see on different parts of the bird—are good external features to cue in on. Also notice how the bird moves (how it flies, walks, or swims) to see if it displays any notable behavior. Habitat and even time of year can also help to arrive at (or rule out) possibilities. Notice where birds are feeding and how they move: are the birds foraging on the ground, in shrubs, or in the trees? Do they creep along branches or tree trunks or catch insects in flight?

Sound can be extremely useful, especially during the breeding season. Since it is often difficult to see all of the birds (especially in forests), learning birdsongs can open up a whole sense of what species are present in an area. Listen for sound features when you are learning birdsongs, and try to identify what you are hearing: patterns of notes through time; pitch (how high or low the sound is and how that varies); and the quality of the sound (whistle, buzz, honk, etc.). Tricks for memorization, such as finding word phrases that capture bird sounds or syllables (like "fee-bee" for the eastern phoebe or "who cooks for you?" for the barred owl), can all help. There are many internet resources and phone applications (the Cornell Ornithology Lab's Merlin app is a particularly good one) to help with identification of birds by sight and sound, and plenty of bird quizzes to test your skills.

As you get out and enjoy the birds, whether by yourself or with others, it's worth checking on guidelines assembled by the American Birding Association to ensure that we don't interfere with—or cause harm to—birds in any way or interfere with other bird enthusiasts. These guidelines suggest that you stay back from nesting areas, roosts, and feeding sites; limit the playing of bird recordings, especially during the breeding season and under adverse weather conditions; stay on roads, trails, and paths; act in a way that does not disrupt the ability of others to observe, study, record, or photograph birds; and respect the interests, rights, and skill levels of fellow birders, as well as people participating in other outdoor activities. Following these guidelines will help us all protects birds, keep more areas open to bird-watchers, and promote a positive experience for beginners and experts alike.

Notes

1 Maryland/District of Columbia Records Committee of the Maryland Ornithological Society, *Official List of the Birds of Maryland from 1804 as of 13 October 2023*, https://mdbirds.org/records-committee/maryland-bird-records/.

2 Walter Ellison, ed., *2nd Atlas of the Breeding Birds of Maryland and the District of Columbia* (Baltimore: Johns Hopkins University Press, 2010).

3 Xing Xu, Zhonge Zhou, Robert Dudley, Susan Mackem, Cheng-Ming Chuong, Gregory M. Erickson, and David Varricchio, "An Integrative Approach to Understanding Bird Origins," *Science* 346, no. 6215 (December 2014), https://doi.org/10.1126/science.1253293.

4 Anthony D. Tramontin and Eliot A. Brenowitz, "Seasonal Plasticity in the Adult Brain," *Trends in Neurosciences* 23, no. 6 (July 2000): 251–58, https://doi.org/10.1016/S0166-2236(00)01558-7.

5 Jean-Nicolas Audet, Mélanie Couture, and Erich D. Jarvis, "Songbird Species That Display More-Complex Vocal Learning Are Better Problem-Solvers and Have Larger Brains," *Science* 381, no. 6663 (September 2023): 1170–75, https://doi.org/10.1126/science.adh3428.

6 Allen R. Place and Edmund W. Stiles, "Living Off the Wax of the Land: Bayberries and Yellow-Rumped Warblers," *The Auk* 109, no. 2 (1992): 334–45.

7 Aishwarya Khokle, "The Record-Breaking Bird That Flew from Alaska to Australia without Stopping," Guinness World Records, January 3, 2023, https://www.guinnessworldrecords.com/news/2023/1/the-record-breaking-bird-that-flew-from-alaska-to-australia-731576.

8 Karan J. Odom, Michelle L. Hall, Katharina Riebel, Kevin E. Omland, and Naomi E. Langmore, "Female Song Is Widespread and Ancestral in Songbirds," *Nature Communications* 5 (2014): 3379, https://doi.org/10.1038/ncomms4379.

9 Maryland Department of Natural Resources, *Maryland State Wildlife Action Plan 2015–2025*, publication number DNR 03-222016-798 (Annapolis: Maryland Department of Natural Resources, 2016).

CHAPTER 11

Insects

DOUGLAS TALLAMY

Many of us suffer under the misconception that human societies are sustained by human ingenuity, technology, ever growing economies, and the (occasional) Einsteins among us. Not so. We humans (and most other terrestrial species on Earth) are sustained by insects! The famed entomologist E. O. Wilson has called insects "the little things that run the world" because of the many essential ecological roles they play every day.[1] It is insects that pollinate 87.5% of all plants and 90% of all flowering plants,[2] and it is those plants that turn energy from the sun into the food that we and an unimaginable diversity of birds, mammals, reptiles, amphibians, and freshwater fish need to exist. Insects are also the primary means by which the food created by plants is delivered to these animals. Most vertebrates do not eat plants directly; far more often, they eat insects that have converted plant sugars and carbohydrates into the vital proteins and fats that fuel complex food webs.[3]

And so, it is insects that sustain Earth's ecosystems by sustaining the plants and animals that run those ecosystems. And the more plants and animals, the better: ecosystems with many interacting species are more stable, more productive, and better able to support huge human populations than depauperate ecosystems with few species. Insects also provide much of the planet's pest control in the form of millions of species of predators and parasitoids that keep food webs in balance. It is insects that rapidly decompose dead plants, releasing the nutrients they contain for use by new plant life. And by keeping the planet well-vegetated, it is insects that maintain the watersheds in which we all live, providing us with clean water and minimizing the frequency and severity of floods. As if all of that were not enough, the plants that insects pollinate sequester enormous amounts of carbon within their bodies and within the soil around their roots, carbon that would otherwise be in the atmosphere, wreaking havoc on the Earth's climate.

Indeed, humans would last only a few months if insects were to disappear from Earth. It is remarkable, then, that our cultural relationship with insects is built on disgust and animosity rather than awe and appreciation. We have created a culture in which insects and their arthropod relatives are routinely maligned and, worse, exterminated in the name of protecting crops and fighting a few disease vectors. We have declared war on all insects and we kill as many of them as possible, whenever and wherever we can. We have sponsored "National Insect Killing Week"[4] and taught our children to fear every insect they see, rather than to respect those few that might sting to defend their nest—and admire the rest.

And so, we are winning our undeclared war against insects—but at our peril. Precipitous declines in populations of the European honeybee, the 4,000 species of bees native to North America, and beautiful butterflies like the monarch and Karner blue have gotten our attention, but many other insects are disappearing utterly without notice. We have already driven three North American species of bumblebees to extinction,[5] and some 30% of the grasshoppers, crickets, and katydids of Europe are facing extinction.[6] Flying insects in Germany have declined 79% in abundance and diversity since 1989, and 46 species of butterflies and moths have disappeared from German soil altogether.[7] Similar statistics are coming to light in England and other parts of Europe and the American tropics. By killing insects, we are biting the hand that feeds us, and that has led to the most alarming statistic of all: invertebrate abundance (read: "insects") has been reduced 45% globally since 1974.[8]

It is obvious that we must form a new and more positive relationship with insects, and we must do it now. The good news is, you do not need to be an entomologist to help in this regard. As Master Naturalists, a cursory knowledge of the most diverse and numerous multicellular organisms on Earth is required but, in my view, it is more important that you understand the ecological value of insects and learn how to lead the charge in saving them.

What Is an Insect?

All insects are members of the phylum Arthropoda, organisms with a segmented body, an exoskeleton, and jointed legs. Insects can be recognized from other Arthropods like spiders, crustaceans, and centipedes by having three primary body parts: head, thorax, and abdomen and only three pairs of legs. The head bears the eyes, antennae, and mouth of the insect; the thorax is the locomotion engine and houses the wings, legs, and the muscles that work them; and the abdomen is the home of the digestive, renal, and reproductive systems. Insects are the only invertebrates that have evolved the ability to fly, and this, along with their small size, has enabled them to occupy nearly every niche on Earth and diversify into more than 4 million species—some 400 times the number of bird species. And so, a crucial question: How are we going to coexist with them in human-dominated landscapes?

Step One: Restore the Plants on Which Insects Depend

Thank goodness it is "little things" that run the world. If we needed to share our neighborhoods with big things like tigers, elephants, bison, and giraffes, we would be challenged indeed. Not only is it easy to create a world in which insects can coexist with humans, it is easy to create landscapes in which they actually flourish. All we need is more of the right plants in our landscape designs.

But which are the right plants? To answer that question, we need to decide which insects we want to help. There are a lot of insect species in the world; 3–4 million by most estimates, 164,000 of which have been described in the United States (a great many species remain undescribed!). What they all have in common is that they are directly or indirectly tied to plants, either by eating some part of a living plant, by existing solely on dead plant tissues like fallen leaves or rotting logs, by developing on the muck dead plants create when they fall into water, by eating another insect that has developed directly on plants, by eating an insect that has eaten an insect that has developed on plants, or by being a parasite on a mammal that has eaten plants, and so forth.

You get the picture: insects need plants. But the class Insecta is so large and so broad that trying to create habitat for all of them in a given space is not only impractical, it's impossible. I suggest we narrow our focus a bit by enhancing populations of the two insect groups that arguably have the greatest impact on terrestrial ecosystems: those that contribute the most energy to local ecological food webs—that is, the insects that are larger, more numerous, more edible, and more nutritious for other species than most other insects—and those responsible for most of the pollination required by plants. I speak of caterpillars, the larval stages of moths, butterflies, and skippers, as well as the group of Hymenoptera we call sawflies; and the 4,000 species of bees native to North America.

Spotlight on Caterpillars

"The early bird catches the worm," the saying goes, but that doesn't really capture it. Caterpillars are not worms, but they are in fact the mainstay of most bird diets in North America, particularly when they are rearing their young. Only thrushes like the American robin regularly feed their babies earthworms, but most thrush nestlings still feed on caterpillars.

So, birds need caterpillars to survive, and this means we ought to landscape in a way that builds the populations of such an important group of insects. For one thing, there are many types of caterpillars to work with; estimates of the number of species of lepidoptera (moths and butterflies) in North America top 14,000. This is fewer than the number (25,000) of beetle species in North America, but unlike most caterpillars, beetles are hard prey for birds to find and eat and therefore do not contribute as much to local food webs. They spend most of their lives hidden under ground, within seed pods, or tunneled deep in wood.

Beetles also sport much thicker exoskeletons, particularly as adults, and often have spiny, stiff legs that make them difficult for creatures like birds to eat and digest. Caterpillars, by contrast, are typically exposed on vegetation and their exoskeleton is thin and flexible, making most of a caterpillar digestible food instead of indigestible chitin. Think of caterpillars as little nutritious sausages and you will understand why their texture may be one of a caterpillar's most important attributes. If you have ever watched a bird feed its nestling, you know it is not always a gentle process; many adult birds forcibly stuff food down their nestling's throat, using their beak as a plunger (see figure 11.1). Insects with sharp edges can injure delicate little nestlings during such feeding bouts.

Most caterpillars are also relatively large compared to other kinds of insects. It takes 200 aphids, for example, to equal the weight of a single medium-sized caterpillar. If you are a bird looking for insects, would you choose to hunt and handle 200 aphids, or find one caterpillar? Finally, caterpillars are more nutritious than most other insects. They are high in protein, high in fats, and are the best source of carotenoids for birds,[9] particularly during the breeding season when few carotenoid-rich berries are available. It follows, then, that if birds need protein-rich prey that are high in carotenoids to raise healthy young (they do), and if caterpillars provide the best and most easily obtained source of such nutrition during the breeding season (they do), then caterpillars may not be optional components of breeding bird diets. It is also likely they are essential to successful reproduction. As with all parts of nature, there are exceptions to this generality. A few bird lineages, including finches, doves, and crossbills, can rear young on a milky substance they make from seeds, and raptors for the most part feed their young on mammals, fish, or other birds. But some 96% of North America's terrestrial bird species rear their young on insects

FIG. 11.1. It is important that food for baby birds be free of sharp edges because parent birds use their beaks to stuff insects down their offspring's throats like a plunger. Photo courtesy of Douglas Tallamy.

rather than seeds and berries (per two Peterson Field Guides),[10] and in most of those species, the majority of those insects are caterpillars or adult moths.[11] Caterpillars are so important to breeding birds that many species may not be able to breed at all in habitats that do not contain enough caterpillars.[12]

How many caterpillars are "enough"? That of course depends on which bird species we are talking about. There are "enough" caterpillars in a habitat when parent birds can find caterpillars fast enough to enable three to six nestlings to grow from egg to slightly smaller than an adult in under two weeks for most cup nesters, and a little longer for cavity nesters. This is an astonishingly fast growth rate. To achieve such rates of growth, nestlings must eat often. You and I eat 3–4 times a day (maybe 5 if we include snacks). A typical nestling, in comparison, eats a full meal 30 to 40 times a day! That means a parent (or couple) raising five chicks must bring food to the nest about 150 times a day. They are busy indeed! Most birds forage primarily within a well-defined territory surrounding the nest. For Carolina chickadees, this is about 50 meters in all directions from the nest, or an area approximately two acres in size. Thus, nesting territories must contain lots of food concentrated in a relatively small area or the nest will fail.

So how many caterpillars is that? Few people have actually sat and counted all of the prey items birds bring to their nest, but those who have had the patience to do this have recorded astounding figures. Robert Stewart, for example, made detailed records of a Wilson's warbler pair while they were feeding their young in 1973. Perhaps unsurprisingly, he found substantial differences in how hard the male and female worked at this endeavor. The male was no slouch, carrying food to the nest 241 times in a single day, but the female put him to shame; on that same day she fed the nestlings 571 times! This rate was maintained over the five days he watched the nest. Stewart did not count the actual number of caterpillars the pair brought to the nest; feeding was rapid and often a parent carried more than one caterpillar in its beak at a time. Yet even if only one caterpillar was brought to the nest each trip, the pair would have brought in 812 caterpillars per day or 4,060 caterpillars in the five days Stewart watched the nest. The chicks he observed stayed in the nest eight days before they fledged.[13]

These observations are not exceptional. Bobolinks bring food to their nests 840 times a day for 10 days in a row.[14] Sapsuckers feed their young 4,260 times, downy woodpeckers 4,095 times, and hairy woodpeckers 2,325 times.[15] All of these species regularly bring in multiple prey items per trip. Perhaps the most complete records of feeding rates were made by Richard Brewer, who counted the caterpillars that Carolina chickadees brought to their nests throughout the nesting period.[16] Brewer found feeding rates of 350 to 570 caterpillars per day, depending on the number of chicks in the nest. Over the course of a typical nesting period (16 days on average), that totals 6,000–9,000 caterpillars required to bring one nest of a tiny (3 oz) bird to fledging. A parent's job is not over, however, when the chicks leave the nest. Chickadee parents, for example, continue to feed their young for up to 21 days after fledging! No one knows how many additional caterpillars are required before young chickadees no longer depend on their parents for food.

Now let's think about an ideal neighborhood that contains not just one pair of one bird species, but thriving populations of many species. I, for one, am greedy when it comes to enjoying nature. I want Carolina chickadees in my yard, but I also want cardinals and titmice and blue jays and Carolina wrens. I want red-bellied and downy woodpeckers, white breasted nuthatches, yellow warblers, Kentucky warblers, robins, wood thrushes, ovenbirds, and indigo buntings. And I am so greedy that I also want bluebirds, catbirds, common yellowthroats, great-crested flycatchers, yellow-billed cuckoos, mockingbirds, eastern kingbirds, field sparrows, chipping sparrows, and grasshopper sparrows: all birds that used to be common in Maryland. If each pair of these species requires thousands of caterpillars to successfully breed, imagine the number of caterpillars my yard would need to produce to support stable populations of all of these birds!

Which Plants Should We Use?

It is clear that if we need to make as many caterpillars in our yards as possible, we need to use plants that serve as hosts for the most caterpillar species. But which plants are those? Assembling this information is not a trivial task. There are some 2,112 native plant genera in the lower 48 states, and most of them contain species that serve as host plants for one or more species of caterpillars. Records of these host associations have been made over the past century by naturalists, ecologists, and particularly by lepidoptera taxonomists, and these are scattered in their writings throughout thousands of papers and books. Needless to say, finding and categorizing this information requires a combination of old-fashioned library work plus the handy search tools of the digital age.

With financial support from the US Forest Service, Kimberley Shropshire, my research assistant at the University of Delaware, created a mammoth database which has become the basis of a search tool developed by the National Wildlife Federation called Native Plant Finder.[17] Kimberley has ranked plant genera that occur in every county of the United States in terms of their ability to host caterpillars. Now, simply by entering your zip code, you can find out which woody and herbaceous plant genera native to your area are best at hosting caterpillars. This tool has removed one of the biggest obstacles to homeowner restorations. We no longer have to wonder what plants we should add to our landscape to enhance their ecological productivity.

Keystone Plants

Kimberley's work revealed a striking pattern: wherever one looks—be it in the north, south, east, or west; the plains, deserts, forests, or mountains—just a few plant genera are producing most of lepidoptera so important to food webs. We knew from our previous work in the mid-Atlantic states that not only were native plants far superior to introduced species in their ability to generate caterpillars, but native plants themselves varied by orders of magnitude in their production of caterpillars.[18] Some Maryland genera like oaks, cherries, and willows host hundreds of caterpillar species, while for others like Virginia sweetspire, there are no records at all of caterpillars using them. This is interesting itself, but when Kimberley assembled data for each county, we saw that this pattern held everywhere and we could quantify it: wherever we looked, about 5% of the local plant genera hosted 70–75% of the local lepidoptera species!

I like to call such hyperproductive plants "keystone plants" because they so closely fit the meaning of Robert Paine's classical terminology:[19] like the center stones in ancient Roman arches, keystone plants enable other species in the ecosystem to coexist. Remove the keystone and the arch falls down. Keystone plants function in the same way: they are unique components of local food webs that are essential to the participation of most other taxa in those food webs. Without keystone plants, the food web all but falls apart. And without some minimal number of keystone plants in a landscape, the diversity and abundance of the many insectivores—the birds and bats, for example, that depend on caterpillars and moths for food—are predicted to suffer.

The implications of this phenomenon for homeowners, land managers, restoration ecologists, and conservation biologists are enormous: to create the most productive landscapes possible—that is, landscapes in which the most plant matter is turned into edible insects—we have to include keystone plant species. This is a nuanced but incredibly important extension of our knowledge about how native plants contribute to ecosystem function. Before discovering the existence of keystone genera, we overestimated the degree to which most native plants contribute to food webs and assumed that if a plant was native it contributed a lot. We now know that a few native genera contribute so much more than most others that we cannot ignore them if we are to produce complex, stable food webs. A landscape without keystone plants will support 70–75% fewer caterpillar species than a landscape with keystone plants, even though it may contain 95% of the native plant genera in the area.[20]

Let It Be an Oak

Oaks are aptly placed in the genus *Quercus*, a name derived from the Celtic *quer* meaning "fine" and *cuez* meaning "tree," and oaks are indeed fine trees. With hundreds of species globally (taxonomists argue about the exact number, with estimates ranging from 400 to 600 species) and over 90 species in the United States, oaks occur in and often dominate all forest ecosystems in North America except the great coniferous forests of the north and the driest deserts of the Southwest. Ecologically, oaks are superior plants, and it would be easy to make a convincing case that they deliver more ecosystem services than any other tree genus. Many species are massive and sequester tons of carbon in their wood and roots and pump tons more into the soil they grow in. They are long-lived with some species achieving 900 years if you include periods of growth, stasis, and decline. Thus, the carbon they pull from the atmosphere is locked within their tissues for nearly a thousand years! In many ecosystems, oaks are also superior at stalling rainfall's rush to the sea. Their huge canopies break the force of pounding rain before it can compact soil, and their massive root systems prevent soil erosion and create underground channels that encourage infiltration instead of runoff. Lignin-rich oak leaves are slow to break down once they fall from the tree and create excellent, long-lived leaf litter habitat for hundreds of species of soil arthropods, nematodes, and other invertebrates. For me, though, all of these contributions to ecosystem function pale before the contribution oaks make to food webs.

Our early work showed that oaks in Delmarva supported hundreds of caterpillar species (511 to be exact), and at least 950 species nationwide, making them by far the best plants if we want to support food webs. If you think of plants as bird feeders, which is exactly what they are, then oaks are the best. To put this level of productivity in perspective, most other common trees in Maryland are slackers in comparison. Tulip poplar (*Lireodendron tulipifera*), for example, supports only 20 caterpillar species, black gum (*Nyssa sylvatica*) 36, sycamores (*Planatus occidentalis*) 46, persimmon (*Diospyros virginiana*) 46, and sweetgum (*Liquidambar styraciflua*) 35. Like oaks, native willows and cherries are also highly productive, but they do not surpass oaks in Delmarva. You don't need to understand precisely why oaks help food webs better than other plants; you just need to know that they do and that we should use them accordingly in our landscapes and restorations.

Completing the Life Cycle

Regardless of which trees, shrubs, or perennials we employ to increase the abundance and diversity of caterpillars on our properties, we have to use them in our landscapes in ways that enable the caterpillars they support to complete their life cycle. Caterpillars undergo what is termed "complete metamorphosis," a type of development that comprises four distinct stages: egg, larva, pupa, and adult. What's relevant here is that, for most caterpillar species, only two of these life stages, the egg and larval stages, are completed on the host plant. Most caterpillars crawl off of their host plant before molting to their pupal stage. Oaks on the Delmarva Peninsula, for example, serve as hosts for 511 species of caterpillars (see figure 11.2). A few of these, such as the polyphemus moth, spin their cocoon on the host tree itself after they have eaten their fill of oak leaves (see figure 11.3). But 480 species, some 94%, fall to the ground when the caterpillar is fully grown and either burrow into the soil to pupate underground, or spin a cocoon in the leaf litter under the tree.

FIG. 11.2. The spun glass slug caterpillar, one of the 557 species of caterpillars in the mid-Atlantic states that develops successfully on oaks. Photo courtesy of Douglas Tallamy.

The exodus most caterpillars make from their host plant before they pupate is not just something oak eaters undertake. Monarch caterpillars almost never form their chrysalis on milkweed; they crawl off to some other structure, often yards away from the milkweed plant they developed on, causing many monarch watchers to think a bird has gotten them. The pipevine swallowtails in our yard really take a hike when fully grown. I have found their chrysalides halfway up our oak trees, attached to the side of our house, and one was even hanging from one of our picture frames in our living room! Some of these caterpillars have crawled more than 25 yards from their pipevine hosts. Experts think the evolutionary motivation for such trekking is for caterpillars to put distance between themselves and their host plant before they enter the defenseless pupal stage.

Lepidopteran pupae are not only prey for hungry birds; they are also targets for numerous species of other predators as well as hymenopteran and dipteran parasitoids, many of which go to a host plant to search for food. The longer a caterpillar stays on its host, the greater the chances it will be

FIG. 11.3. This polyphemus moth cocoon is one of the few moth species that completes its life cycle on its host tree. Photo courtesy of Douglas Tallamy.

discovered and attacked by one of its enemies. By crawling some distance from its host plant before it pupates, the caterpillar has decentralized the effective search zone for its enemies. Instead of being able to efficiently search just a few square feet for its prey, the caterpillar has forced its enemies to search thousands of square feet, a low-return task that is usually not worth the time or energy involved.

This survival mechanism is very effective at reducing mortality in the pupal stage, but it forces us to think beyond the needs of caterpillars when we landscape. Not only do we have to provide food for developing cater-pillars, we also must provide the microhabitats their pupae require to survive. Your yard may contain an oak tree that can feed hundreds of caterpillars, but more often than not, that oak will be surrounded by mowed lawn growing in compacted soil. When your caterpillars drop from the tree, they will find no leaf litter in which to spin their cocoon, because each year you neaten up after the leaves fall. If they are a species that tun-nels into the soil, they will have to search far and wide to find soil loose enough to permit burrowing, and the longer they have to search, the greater the odds of being pulverized by your lawn service or squashed on your driveway or street. These challenges are even greater in urban envi-ronments, where trees are often surrounded by cement.

Fortunately, providing safe pupation sites in our landscapes is not an insurmountable problem. In fact, it can be a new and satisfying gardening goal. The needs of our caterpillars can be easily met when we replace some of our lawn with three-dimensional plantings. Annual or perennial beds, spring ephemeral showcases, ground covers, or shrub plantings of species appropriate for one of Maryland's physiographic regions, depending on where you live—any portion of your landscape that is not regularly tram-meled by feet and lawnmower wheels will quickly develop a thick O hori-zon of loose organic matter perfect for pupating lepidoptera, as long as you don't rake away your leaf litter during your spring and fall cleanups. To the surprise of many, some caterpillar species such as the beautiful woodnymph, the greater oak dagger moth, and Harris's three spot tunnel into soft wood to pupate; adding a decaying log to your garden (artfully placed of course) would allow such species to complete their development.

There is one final safety precaution we need to take to help our cater-pillars complete their life cycle. Once they emerge as adult moths from their pupae, they have to survive long enough to find a mate and for the females to locate their host plant and lay eggs. A few species like the white-marked tussock moth accomplish this in just a few hours, but most require several days to a few weeks. During this time, those species with mouthparts need to eat to maintain their energy, and most of the time they are eating nectar from nocturnal flowers. A recent study by Eva Knop at the University of Bern suggests that our propensity to light up the night sky is not helping moths in this regard.[21] Using night vision goggles, Eva counted insects that visit flowers in areas with no artificial lights, then added lights to those same areas. She found that when lights were on, moth visits declined 62%. Either the moths simply avoided spaces that were well lit, or they were fatally attracted to the lights as if the lights were Sirens.

Don't ask why insects are so drawn to lights; two centuries of research have not produced a satisfying answer to this puzzling question. The point is, insects do fly into light sources by the millions, and each night we light up nearly the entire world to their detriment. Lights reduce insect populations in several ways. A light can kill insects directly by repeated collisions with the bulb. Or the frenetic flight about the bulb can fatally exhaust them by burning up their energy reserves. Those insects that don't beat themselves to death or die of exhaustion are nevertheless waylaid from their normal activities of seeking host plants or mates. That is, lights waste precious time in the short adult lives of insects. Finally, an insect drawn to a light becomes an easy target for hunting bats or, if it's sitting near a light, for arthropod predators like daddy longlegs, carabid beetles, hanging scorpion flies, damsel bugs, ants, assassin bugs, mantids, and spiders. Those insects that make it to dawn are often picked off by birds that quickly learn that lights are an easy place to find breakfast. Whether you use lights at night so you can see your way around your property, or to discourage bad guys, please consider putting motion sensors on each one. That way, instead of illuminating your property all night every night, the lights only turn on when you really are out and about, or when the bad guy actually comes. That simple act will make an enormous difference for our insect populations, but particularly for night-flying moths and thus for caterpillars.

Restoring Native Bees

The second group we should be sure to subsidize in our landscapes is native bees, not because bees are important components of food webs (they are not; other than a number of specialized parasitoids, very few animals eat bees), but rather because, as pollinators, they maintain a diverse plant base for terrestrial food webs. Though many types of insects are credited with important roles in pollination, including moths, butterflies, beetles, wasps, flies, and ants, it is members of the Apoidea, our bees, that perform the lion's share of pollination duties (see figure 11.4).

When most people think of bees, an image of the domesticated European honeybee comes to mind. Honeybees were brought to North America with the earliest colonists because they could be managed and deployed where they were most needed and because they were generalist pollinators that were so good at pollinating the many old-world crop plants the colonists also brought with them. Before we imported honeybees, however, all of the animal pollination in North America (13% of our plants are

FIG. 11.4. Contrary to popular misconception, butterflies like this Canadian tiger swallowtail do very little pollination compared to bees. Photo courtesy of Douglas Tallamy.

FIG. 11.5. Native bees like bumblebees are collectively our most important group of pollinators. Photo courtesy of Douglas Tallamy.

wind pollinated) was accomplished by native pollinators, primarily the nearly 4,000 species of bees native to North America (see figure 11.5).

Like so many of our insects, honeybees are in trouble. Although 10–15% of bee colonies have always been lost during the winter months, dramatic and in some cases sudden declines in honeybee populations became apparent across the country and, indeed, globally between 2003 and 2007. A suite of ills, from mites to viruses and bacteria to abusive pollinating demands, have been blamed for these declines, now called collectively "colony collapse disorder." The threat to agriculture from the loss of honeybees was so obvious that even our politicians noticed and today, saving pollinators has become a politically correct mantra.

The press is full of articles telling us that we have to save bees because they pollinate one-third of our crops (it's actually about one-twelfth of our crops). True enough, but as a sole motivation for saving bees—indeed, all of our pollinators—this is shortsighted in the extreme. There is an even more compelling reason beyond protecting a few species of human crop plants to save pollinators from local or global extinction. In addition to pollinating one-twelfth of our crops, recall that animals are responsible for pollinating 87% of *all* plants and 90% of *all* angiosperms.[22] That's right, if pollinators were to disappear, 87–90% of the plants on planet Earth would also disappear. Not only would such a loss be a fatal blow to humans, it would take most other multicellular species with it as well. Saving pollinators from human environmental aggression goes well beyond maintaining a diversity of fruits and vegetables in our supermarkets; it is essential to life as we know it on planet Earth.

One positive result of colony collapse disorder in European honeybees is that it has drawn long-overdue attention to our North American bee species. Although only a handful of our native bee species have been studied, it is no surprise that most have been found to be in steep decline: 50% of midwestern native bee species have disappeared from their historic ranges in the last century.[23] Four species of bumblebees have declined 96% just in the last 20 years,[24] and 25% of our bumblebee species are at risk of extinction.[25] It is not a stretch to assume the thousands of species not yet examined are similarly challenged in today's world of RoundUp-ready corn and soybeans, lawns, and deadly roads. It is clear we must all act quickly to save our pollinators, but to do so effectively we have to understand who our pollinators are and what they need to thrive in our yards.

What Is a Pollinator?

It is logical to assume that all animals that go to flowers for pollen and/or nectar actually pollinate those flowers. The opposite is true; most animals that go to flowers do not end up pollinating the flowers, even if they successfully remove pollen and nectar from those flowers. It is more accurate to call animals going to flowers "flower visitors" and reserve the term *pollinator* for animals that actually transfer pollen from the male stamens to the female pistil. Butterflies, for example, get lots of credit for being great pollinators because they spend so much time nectaring at flowers. But this credit is not deserved; most butterflies take from flowers without giving much back in return. Generally, butterflies do not have a body shape conducive to transferring pollen. Even bees that have specialized adaptations for pollinating a particular flower genus may visit other flower genera without transferring any pollen. Because pollen and nectar are costly for flowers to produce, many flower genera have developed elaborate shapes such as extremely long corolla tubes, very narrow corollas, or closed petals that make access to their nectar difficult. The evolutionary idea in these cases is to prevent generalist pollinators from taking the pollen and allow only specialist pollinators access to the pollen because they are more likely to deliver it to another flower of the same species. Specialized interactions between flowers and their pollinators are largely responsible for the myriad sizes and shapes of flowers and bees that we find in nature.

One more thing: to make supporting pollinators at home part of mainstream culture there is an educational hurdle many people must leap. Unfortunately, the thought of thriving bee populations in the front and backyard is too often a nonstarter. After all, where there are bees, there are bee stings, or so many people think. Bees do sting (at least the females do. Males have no stingers, which are modified egg-laying devices), but only in self-defense or in defense of their hive. This is important point number one. Out of all 4,000 species of native bees, only bumblebees, a mere 46 species, have what we might call "hives," though they are tiny compared to honeybee hives (see figure 11.6). The rest of the species are solitary and never aggressively defend a home space. When people are stung by a bee, the perpetrator is nearly always a honeybee that they have stepped on with bare feet (that was my first sting as a toddler), or that is defending its hive. Important point number two is that, while foraging at flowers, bees are not aggressive at all. They are focused solely on gathering as much pollen and

FIG. 11.6. Bumblebees are one of the few groups of native bees that are social and create small nests of daughters. Photo courtesy of Douglas Tallamy.

nectar as possible. You can prove this to yourself by petting the next bee you see at a flower. The bee might fly off, but it won't sting you! The passive nature of foraging bees means that we can walk among flowers crawling with bees with no fear of being stung.

The most common misconception about the source of painful stings stems from sloppy taxonomy: people frequently mistake yellowjackets for bees. Yellowjackets are predatory wasps, not bees (and not pollinators). But like honeybees, they are social species that construct large hives in the ground or in trees, which they defend as aggressively as they can. I include bald-faced hornets in this group; even though they are black and white instead of yellow and black, bald-faced hornets are close relatives of yellowjackets and behave just like them. And unlike honeybees, yellowjackets and hornets can sting repeatedly, making any close encounter with them an unpleasant one. The good news is, there is no national movement to encourage people to have yellowjacket nests in their yards. The solitary bees we need to share our yards and parks with are harmless.

Making Bees Feel at Home

Like most creatures, native bees need the basics to exist: a place to live, food, and water. Most native bee species nest in the ground, within wood or pithy plant stems, or in any nook or cranny of the appropriate size. Bumblebees, our only native bees that are always social, favor shallow holes in the ground that are protected from rain. Abandoned mouse nests within a rock wall are ideal real estate for bumblebees because they are protected from rain and from digging predators such as possums, raccoons, and foxes that love to eat bumblebee larvae. If you don't happen to have an abandoned mouse nest or a rock wall, you can simulate one by burying a full roll of toilet paper (preferably unused) about three-quarters of its length in the ground (center hole facing up) in a site totally protected from rain or runoff. If you are motivated, you can build a small three-sided wooden house with an inch-wide hole drilled in one side to protect the roll from rain. Bumblebee queens fly around each spring evaluating every hole they find as a potential nest site. Chances are good that a queen will enter your makeshift nest box and chew her way into its center to set up a cozy house.

Ground-nesting bees, some 70% of our native bee species, are easy to accommodate, as long as you have soil that is loose enough for bees to excavate—that is, almost all non-compacted soil types except hard-packed clay. Ground-nesters prefer bare patches of dry soil with a slight southern slope, so if you have such spaces, avoid walking on them as that compacts the soil beyond use for the bees. We are not talking about huge areas. Two square feet of bare soil can provide housing for a number of reproducing female bees. Please, no lawn fertilizer near nests either: lawn products usually contain pesticides that do not make life easier for our bees.

FIG. 11.7. Many species of native bees, like this *Colletes* bee, nest harmlessly in the ground. Photo courtesy of Douglas Tallamy.

Species of native bees, particularly in the families Colletidae, Halictidae, Andrenidae, and some Apidae like long-horned bees and squash bees, will construct nests in the ground all season long, but such nests are most evident in early spring before they are obscured by vegetation. Look for holes in the ground surrounded by small mounds of excavated soil (see figure 11.7). Active nests usually have a bee leaving or entering the hole every few minutes. The holes lead to tunnels several inches deep with side shafts here and there, each containing a ball of pollen and a developing bee larva. If you have a site on your property that is ideal for ground-nesting bees, many individuals may nest within that small space. This can result in lots of bees frenetically coming and going, a spectacle that may be a bit scary at first; but these bees are busy rearing their young and will do their very best to ignore you. You can sit and watch them for hours with no ill effects!

Pithy stem nesters are a bit more particular in where they nest: they need pithy stems! Stems of many herbaceous plants like goldenrod, blackberries, giant ragweed, or native hydrangeas are essentially hollow except for a loose fibrous material that bees can easily remove. These cavities make perfect nesting sites. Mason bees, small carpenter bees, and small resin bees will tunnel into the stem, remove the pith from a section several inches long, and then construct a sequence of cells starting at the end of the cavity farthest from the entrance hole. Each cell is packed with pollen on which the bee lays a single egg. She then seals off that particular cell and starts to provision the neighboring cell. In this way the stem contains several developing larvae at once, each one a day or two younger than the previous larva.

When development is complete, the larva pupates within the cell. If it is early summer, the young adult will emerge from its cell by chewing a hole directly to the outside and start its own family. If the larva matures at the end of the season, the resulting prepupa will stay within its cell all winter, complete its development early in the spring, and emerge as an adult as soon as there are flowering plants in bloom.

Woody stem nesters like carpenter bees and some species of mason bees behave almost identically to pithy stem nesters except they build their nests in soft wood rather than soft stems. Soft wood could be in the form of a downed log or branch or, just as often, a dead branch still attached to a tree. Dead elderberry branches are good examples of suitable nest sites for these bees because they are so easily excavated. Finally, species such as *Osmia* mason bees, yellow-faced bees, and leafcutter bees often choose existing cavities for their nests. I can't tell you how many times I have found the snout of my watering can or even my outdoor water spigot plugged with leafcutter bee nests!

In terms of bee conservation there is a common theme here. Bees cannot nest or overwinter in our yards unless we provide what they need to do so. Most people do have open patches of ground, plants with pithy stems, easily excavated wood, and nooks and crannies somewhere on their property, but many of us work hard to eliminate these valuable resources. Our fall cleanup is particularly hard on bee populations; the senescing stems of our black-eyed Susans, penstemon, sunflowers, and all of the other perennials we are so anxious to cut back after they have bloomed are where our pithy stem nesters are hoping to nest the following summer. Similarly, that dead elderberry branch we feel compelled to prune off and the large elm branch that fell during a summer thunderstorm are now homes for bees that favor soft wood.

The social edict to neaten up is often in direct conflict with the needs of our native bees. There are opportunities for compromise, however. Maybe we can gently cut off our goldenrod stems near the ground, but rather than mulching them, tie them together like a decorative bundle of cornstalks and stand them up for the winter somewhere out of public view. The bees and katydid eggs within should be able to make it through the winter just as if they were left standing in your garden. Another solution is to learn to appreciate the winter structure of perennials. Coneflower seed heads look great with a cap of snow, milkweed pods are fun to look at all winter, and birds certainly appreciate the seed heads we have left behind.

In recent years we have discovered how easy it is to attract many species of stem- and wood-nesting bees, as well as species that nest in nooks and crannies, using commercial or homemade bee hotels. There is scarcely a conference or trade show these days that does not offer a variety of bee hotels for sale, and we humans love them. Hang one in a dry space in your yard at the end of winter and in short order the bees will be busy rearing their young. The number of bees you attract is often a simple function of the size of your bee hotel, and so each year we make our hotels a little bigger. Bee hotels are the perfect solution to increasing the nesting capacity of our yards. Or so we thought. What should have been obvious from the start is that by concentrating nesting opportunities in one place, we have made it very convenient for bee predators, parasitoids, and diseases to wipe out our bees![26] This doesn't mean bee hotels are useless, but it does mean we need to make them much smaller, comprising just a few cells each, and scatter many throughout our yards.

Meet the Needs of Our Specialists!

Now that we have provided housing for our bees we need to feed them; that is, we need to meet the nutritional needs of both adult bees and their developing larvae. Adult bees eat pollen and nectar while larval bees develop exclusively on pollen. That seems simple enough, but there are some essential particulars here that we need to pay attention to.

First, we have to think about timing. Bees, like most other multicellular organisms, must eat every day. Because the pollen and nectar they need comes from flowers, we need to have blooming plants in our landscape throughout the season. And bee communities are active most of the year in most parts of the country. In Maryland, there are native bee species on the wing from early March through the end of October.[27] The need for a continuous sequence of flowering plants in our landscapes is not a trivial challenge. It requires plant choices that are choreographed to enter your ecological stage one after another. Moreover, most native bee species do not forage over large areas as do honeybees. This means that to provide good habitat for native bees at home we need flowering plants at home. There is no problem in having more than one species of flowering plant blooming at once: in fact, that is desirable and gives bees nutritional options. But a landscape that goes through a two- or three-week period with no available blooms is deadly to bees.

Blooming phenology is only one thing to consider when landscaping for bees. We also have to recognize that many bee species require pollen from particular plant genera in order to reproduce. That is, like most of the caterpillars we discussed earlier, many bee species have become host plant specialists over evolutionary time. In fact, nearly 30% of the native bees in the mid-Atlantic region are host plant specialists.[28]

And it's no wonder that natural selection has favored bee specialization: plants differ from each other in so many ways. Some bees can reach their nutritional goals more easily if they develop the specialized adaptations that allow them to find, gather, transport, and digest the pollen of particular plants efficiently. Plants differ in when they flower, how long they flower, and the size, shape, and color of their flowers. They also differ in their pollen morphology and type of amino acids, lipids, proteins, starches, sterols, and chemical defenses. Some generalist bees are good at dealing with all of this variation and can do well on a variety of pollen species. But others become finely tuned to the characteristics of particular plant genera; if we don't include those genera in our plantings, those bees cannot survive in our yards.

How do we support both our specialists and our generalists? Sam Droege of the Patuxent Wildlife Research Center says we should meet the needs of our specialists and the generalists will follow. For example, in Maryland, there are 32 species of bees that can rear their larvae only on the pollen of native asters. Violets support 26 more species; goldenrod 11 species; native willows 16 species; 17 that need evening primrose (*Oenothera* spp); and 18 species that cannot reproduce without sunflower pollen.[29] If you are trying to help native bees but you plant butterfly bush, xenias, and impatiens, you will see generalist bees and you may bask in a false sense of accomplishment. But without goldenrods, asters, evening primrose, sunflowers, violets, and native willows, 120 species of specialist bees that might have been able to use your yard as a refuge will be absent. Sam has said many things during his career as a native bee authority, but in my view, this is the most important thing he has ever said: saving our bee specialists by planting what they have specialized on is the key to saving diverse bee communities around the country.

Notes

1 Edward O. Wilson, "The Little Things That Run the World: The Importance and Conservation of Invertebrates," *Conservation Biology* 1, no. 4 (May 1987): 344–46.

2 Jeff Ollerton, Rachael Winfree, and Sam Tarrant, "How Many Flowering Plants Are Pollinated by Animals?," *Oikos* 120, no. 3 (February 2011): 321–26.

3 Grateful acknowledgement is offered to the University of Delaware Press, in whose volume *The Delaware Naturalist Handbook* (McKay Jenkins and Susan Barton, eds., 2020) an earlier version of this essay first appeared.

4 Roz Weis, "National Insect Killing Week," *Decorah Journal*, July 16, 2009.

5 Gwen Pearson, "You're Worrying about the Wrong Bees," *Wired*, April 29, 2015.

6 Adrian Burton, "Crickets in Crisis," *Frontiers in Ecology and the Environment* 15, no. 3 (April 2017): 121.

7 Caspar A. Hallmann, Martin Sorg, Eelke Jongejans, Henk Siepel, Nick Hofland, Heinz Schwan, Werner Stenmans, et al., "More than 75 Percent Decline over

27 Years in total Flying Insect Biomass in Protected Areas," *PLoS ONE* 12 no., 10 (October 2017): e0185809.

8 Rodolfo Dirzo, Hillary S. Young, Mauro Galetti, Gerardo Ceballos, Nick J. B. Isaac, and Ben Collen, "Defaunation in the Anthropocene," *Science* 345 (July 2014): 401–6.

9 Tapio Eeva, Samuli Helle, Juha-Pekka Salminen, and Harri Hakkarainen, "Carotenoid Composition of Invertebrates Consumed by Two Insectivorous Bird Species," *Journal of Chemical Ecology* 36, no. 6 (June 2010): 608–13.

10 Roger Tory Peterson, *Eastern Birds* (Boston: Houghton Mifflin, 1980); Roger Tory Peterson, *A Field Guide to Western Birds* (Boston: Houghton Mifflin, 1990).

11 Ashley C. Kennedy, "Examining Breeding Bird Diets to Improve Avian Conservation Efforts" (PhD diss., University of Delaware, 2019).

12 Desirée L. Narango, Douglas W. Tallamy, and Peter P. Marra, "Nonnative Plants Reduce Population Growth of an Insectivorous Bird," *Proceedings of the National Academy of Sciences* 115, no. 45 (November 2018): 11549–554.

13 Robert M. Stewart, "Breeding Behavior and Life History of the Wilson's Warbler," *Wilson Bulletin* (March 1973): 21–30.

14 Stephen G. Martin, "Polygyny in the Bobolink: Habitat Quality and the Adaptive Complex" (PhD diss., Oregon State University, 1971).

15 Louise De Kiriline Lawrence, "A Comparative Life-History Study of Four Species of Woodpeckers," *Ornithological Monographs* no. 5 (1967): 1–156.

16 Richard Brewer, "Comparative Notes on the Life History of the Carolina Chickadee," *Wilson Bulletin* 73, no. 4 (December 1961) 348–73.

17 "Native Plant Finder," National Wildlife Federation, https://www.nwf.org /NativePlantFinder/Plants.

18 Douglas W. Tallamy and Kimberley Shropshire, "Ranking Lepidopteran Use of Native Versus Introduced Plants," *Conservation Biology* 23, no. 4 (July 2009): 941–47.

19 Robert T. Paine, "A Note on Trophic Complexity and Community Stability," *American Naturalist* 103, no. 929 (January–February 1969): 91–93.

20 Narango, Tallamy, and Marra, "Nonnative Plants."

21 Eva Knop, Leana Zoller, Remo Ryser, Christopher Gerpe, Maurin Hörler, and Colin Fontaine, "Artificial Light at Night as a New Threat to Pollination," *Nature* 548, no. 7666 (August 2017): 206–9.

22 Ollerton, Winfree, and Tarrant, "Flowering Plants," 321–26.

23 Laura A. Burkle, John C. Marlin, and Tiffany M. Knight, "Plant-Pollinator Interactions over 120 Years: Loss of Species, Co-occurrence, and Function," *Science* 339, no. 6127 (March 2013): 1611–15.

24 Sydney A. Cameron, Jeffrey D. Lozier, James P. Strange, Jonathan B. Koch, Nils Cordes, Leellen F. Solter, and Terry L. Griswold, "Patterns of Widespread Decline in North American Bumble Bees," *Proceedings of the National Academy of Sciences* 108, no. 2 (January 2011): 662–67.

25 Paul Williams and Sarina Jepsen, eds., *Bumblebee Specialist Group Report 2014* (Switzerland: International Union for Conservation of Nature, March 2015), https://www.iucn.org/ssc-groups/invertebrates/bumblebee-specialist-group.

26 J. Scott MacIvor and Laurence Packer, "'Bee Hotels' as Tools for Native Pollinator Conservation: A Premature Verdict?" *PloS One* 10, no. 3 (March 2015): e0122126.

27 Jarrod Fowler, "Specialist Bees of the Northeast: Host Plants and Habitat Conservation," *Northeastern Naturalist* 23, no. 2 (June 2016): 305–20.

28 Fowler, "Specialist Bees."

29 Fowler, "Specialist Bees."

CHAPTER 12

Fish

LETHA GRIMES

For fish, everything starts with water. Water is their highway, byway, communications medium, nursery, playground, school, bed, board, drink, toilet, and grave. Every one of a fish's vital functions—feeding, digestion, assimilation, growth, responses to stimuli, and reproduction—is dependent on water. And as is true for the rest of us, the quality of a fish's life depends on the quality of its water. Fish are especially sensitive to *dissolved oxygen*, *dissolved salts*, *light penetration*, *temperature*, *toxic substances*, *disease organisms*, and ample opportunities to *escape enemies*. Over time, and especially given the pressures of living in a densely populated region, virtually all of these elements are in a constant state of flux. To adapt to these pressures—and to this ever-changing, human-impacted environment—fish have evolved a variety of physical, physiological, and ecological protocols to increase their ability to survive and thrive. There are a number of things a Master Naturalist can learn about water—and watersheds—to help support health and diversity of our aquatic life.[1]

Topography of Maryland and Freshwater (and Brackish) Fish Distribution

The state of Maryland has four ancient river basins, or watersheds: the Monongahela (Maryland's only river basin draining to the Mississippi), the Potomac, the Susquehanna, and the Delaware. These ancient river basins—and the lakes, streams, and estuaries within them—have been altered by major geologic events, especially the advancing, retreating, and melting of glaciers and the rising and falling of sea levels. Over shorter time scales, streams are influenced by the frequency and magnitude of rainfall and the subsequent water and sediment flowing across the land surface and into defined channels. Beyond the changes left on the land itself, these events have also determined the *fish composition* in each of Maryland's river basins. Although many species are native to more than one river basin, several species are *endemic* and found in only one of them.

As we see more fully in the geology chapter in this volume, Maryland's current topography consists of three different regions:

The western *Appalachian Plateau*, or Mountain Region, consists of mountains and valleys, lakes and streams, including Backbone Mountain— the state's highest elevation point at 3,360 feet—and Deep Creek Lake, the state's largest man-made lake.

The Mountain Region's *cold-water streams* provide Maryland unique habitats not found anywhere else in the state. Approximately 2,750 miles of mostly *headwater streams*, the small streams at the highest end of a watershed, compose this unique habitat and are *high gradient, riffle dominant*, and *fast moving*; they are also typically cool (with maximum daily mean temperatures of less than 20°C) and oxygen-rich (generally with dissolved oxygen levels greater than 5 mg/L). These conditions make mountain streams ideal for brook trout (Maryland's only native trout) as well as non-native brown and rainbow trout.

Mountainous *limestone streams* are exclusively found in Washington County's Antietam and Beaver Creek watersheds. Fractures and cracks in limestone make *springs* and *seeps* common in these systems and make these streams physically and chemically distinct from non-limestone *freestone* streams. The exchange of water between the land surface and the limestone beneath it, or its *connectivity*, helps stabilize both the stream's pH and its water temperature. The Albert Powell State Trout Hatchery, located in Hagerstown, raises trout in water supplied from a 3,600 gallon-per-minute limestone spring, which provides a constant flow of water naturally cooled to 12°C (54°F), which is perfect for raising trout.

Maryland's *Piedmont Plateau* stretches from the western border of the Catoctin Mountains to the *fall line*, which runs roughly northeast parallel to the I-95 corridor from Washington, DC, through Baltimore to Havre de Grace. The fall line separates the hilly Piedmont Region from the lowland *Coastal Plain* Region, which is characterized by flat-country farmlands, urban developments, pastures, and quarries.

Maryland's 1,800 miles of Piedmont streams are among the most biologically productive systems in the state, mostly due to their varying topography and geology. The region's streams are typically *low to moderate in gradient* with *silt, sand*, and *gravel substrates*. Streams closer to the fall line have a higher gradient and a *cobble-boulder* substrate. The relatively gentle topography of this area has promoted extensive urban development, much of which has been historically focused near larger streams and rivers.

The easternmost and largest physiographic region in Maryland, the *Coastal Plain* begins at the fall line at its western edge and is subdivided into the Chesapeake Bay's Western Shore and Eastern Shore. The region is characterized by *lowlands* and some *marshes* with very few changes in elevation; the highest point on the Western Shore is just 400 feet above sea level. This region has also experienced a great deal of urban and suburban development, especially just east of the fall line.

Coastal Plain streams extend from the fall line eastward toward the Atlantic Ocean. These streams are typically *low gradient*; most have a peak elevation of less than 50 feet above sea level. Except for streams near the fall line, which have a higher gradient and harder substrates, most coastal streams contain only the *runs* and *pools* typical of slower-moving water and are characterized by *silt and sand substrates*.

The physiographic provinces of Maryland have contributed to a wealth of fish diversity in the state. The western provinces, rich with isolated mountain streams, have shaped fish diversity through evolutionary time, providing long-lasting homes to minnows, darters, catfishes, and perch. Farther east, the mouth of the Chesapeake Bay connects directly to the Atlantic Ocean, allowing many estuarine and marine fishes to enter the Bay each year where they feed and grow. Some of these fishes, such as striped bass (also called rockfish), which is Maryland's state fish, provide both sport and game fish for the region's many people who have come to love fishing. However, it is this love of sport fishing that has also led to the intentional human introduction of non-native species, such as brown trout and rainbow trout.

An Introduction to Fish

WHAT MAKES A FISH A FISH?

Fish are *poikilothermic* animals, equipped with *backbones, gills,* and *fins,* rather than limbs that are *pentadactyl* (five toed or fingered). Fish are the most diverse vertebrates in the world, with over 20,000 species. By contrast, birds are commonly estimated to have around 8,600 species; mammals, about 4,500 species; reptiles, 3,000; and amphibians, 2,500. There are about 350 different species of fish here in Maryland, about 100 of them living primarily in freshwater.

Phylogenetically, there are three major living classes of fish:

1. *Cephalaspidomorphi:* the hagfishes and lampreys that are *jawless* (agnatha) and have *pouched gills.*

2. *Chondrichthyes:* chimeras, sharks, rays, and skates that possess *true jaws* (gnathostomes), have gills with wall-like partitions between gill chambers (or a single chamber in chimeras), and have a cartilaginous skeleton that may be *calcified* but not *bony.*

3. *Osteichthyes:* fishes that possess true jaws (gnathostomes), a bony skeleton, and a *branchial chamber* that both protects the gills and serves as a facial support structure.[2]

BASIC ANATOMY

Physiologically, we can think of a fish as composed of ten systems of bodily organs that work together to make up the whole animal. These systems (1) cover the fish; (2) handle its food; (3) carry away wastes; (4) deliver nutrition to its tissues and organs; (5) interface its processes with its external environmental conditions; (6) provide for respiration; (7) provide movement; (8) protect against injury; (9) support the body in movement; and (10) work to perpetuate the fish as a species.

SHAPE

Every fish has a physical shape that has evolved to optimize its place in a particular ecological niche. For example, a body that is *torpedo shaped* (fusiform) and slightly *ovoid* in cross section reduces friction, increases energetic efficiency, and is frequently found in free-swimming *pelagic* (ocean-dwelling) species, such as tuna. But as any fisherman or scuba diver knows, there are a great many other body shapes that account for the tremendous diversity of habitats in which fish live. For example, cownose rays (they get their name from their square nose) have a kite shape and long whiplike tail and can grow to have a "wingspan" of 40 inches. Cownose rays are found along Maryland's Atlantic coast, and move into the shallow parts of the Chesapeake Bay to spawn. Another very distinguishable *spaded fish* is the Atlantic sailfish, found off the coast of Ocean City. This billfish has an extremely large dorsal fin, sometimes taller than the fish is long, which gives it the name. Thanks mainly to its torpedo shape, sailfish are thought to be the fastest fish in the ocean, reaching speeds of over 60 miles per hour.

RESPIRATION

One of the most important aspects of a fish's aquatic habitat is *dissolved oxygen.* Humans need a lot of oxygen, breathing in approximately 21%

atmospheric oxygen with every breath. Fish need far less—respiring oxygen in *parts per million*—but consistently oxygenated water is still vital for them to live. Fish access oxygen through *vascularized gills, lungs*, or *skin* that transport oxygen and unload it into tissues. Similarly, gills (or other respiratory structures) vent the waste carbon dioxide (CO_2) transported in the blood. In some *scaleless fish*, such as the American eels and some catfish species, this gas exchange takes place directly through the skin.

Several fish, known as *bimodal breathers*—like the bowfin, gar, and northern snakehead—have also evolved an adaption called *aerial respiration* to take oxygen directly from the air: the air is absorbed into a highly vascularized gas bladder that exchanges respiratory gases. Some fish species—like mummichogs and mosquito fish—are adapted to extract oxygen from the surface of water using *aquatic surface respiration*: highly vascularized gas bladders and—critically—upturned mouths that allow them to hover at the surface of the water column with their mouths oriented toward the surface of the water. Adult mummichogs can tolerate low oxygen levels (to less than 1 mg/L), by performing aquatic surface respiration and swallowing oxygen-rich water near the surface. By extracting oxygen directly from the air, these fish can even survive for a few hours in moist air outside of water.[3]

COLORATION

Coloration in fish does far more than attract potential mates.[4] A fish's coloring can serve several purposes, including *concealment, disguise*, and *advertisement*. More often than not, a fish's color is primarily adapted for camouflage, survival, and hunting. *Color resemblance* allows a fish to camouflage itself by blending its own hues and shades with its background habitat. The spotting on summer flounder, for example, enables it to blend with sandy bottoms, where it hides and waits for unsuspecting prey to swim into its attack zone.

Many fish exhibit darker backs and whiter bellies, called *countershading*. By blending their pale bellies with sunlight pouring in from the surface, this shading disguises fish from potential predators (or prey) swimming below. Conversely, by blending with the darker depths, their backs may go unnoticed by prey and predators swimming above and looking down. Sharks provide a widely recognized example of countershading, with most species having a darker back and lighter belly. While these color patterns may not be perfect at providing camouflage or invisibility, even a slight second of confusion may be all that is needed for a fish to swim away from a predator, or to attack its prey.

Coloration can also cue for communication with other members of the same species (*intraspecific signals*) and communication with other kinds of animals (*interspecific signals*). Especially during the spawning season, fish use coloration and patterns to recognize and interact with one another, and intraspecific signals can serve as both *social* (recognition, threat, and warning) and *sexual selection*. Sunfish are good examples: their coloration is heightened during the spawn, but murky water (in which fish lose the ability to recognize particular coloration patterns) can lead to hybridism.

Coloration may also offer cues to invite smaller fishes—such as cleaning wrasses—that browse on (and thus clean) parasites from a host fish's body surface. Interspecific signals can also help warn (or intimidate) potential predators and other assailants. Lionfish, which are native to Indonesia, present a good example of a species with bright orange and red warning coloration, screaming to other predators of the venom lurking within its spines.

Current Issues Affecting Fish

Freshwater fishes are among the four most imperiled fauna in North America. Perhaps the trouble lies in the fact that fish are a lot less visible than "charismatic megafauna" like polar bears and Bengal tigers. As a former boss once told me, it is hard to protect something you cannot see.

This pressure has increased dramatically in the last 50 years due mostly to human-caused environmental changes and habitat loss. Deforestation (which causes stream bank erosion), the presence of dams (which cause fluctuating water flows and interrupt spawning runs), and runoff from farms and suburban development all contribute to pollution and the silting of streams. Sediment from erosion and runoff not only reduce water quality but can also smother eggs on the bottom of the stream. In Maryland, several species are presumed extirpated or extinct, including the bridle shiner, the longnose sucker, the redside dace, the cheat minnow, and the trout-perch. Many more species are considered threatened. The Maryland darter (Maryland's very rare—only endemic—vertebrate species) was once found only in the swift currents of Swan Creek and Deer Creek; it was last found in the state in 1988. Pollution from agriculture and development runoff and water fluctuations caused by releases from the Conowingo Dam are considered the main causes of declining water conditions needed by the darter to survive.

Urbanization and impervious surfaces: As explained more fully in this volume's chapter on Maryland's environmental history, early European settlers cleared the region's forests to establish farms and homesteads and to burn wood for fuel. They also cleared wetlands to cultivate crops; straightened streams to drain their farm fields; and—especially during industrialization—turned creeks into "channels" to send sewage and other waste downstream. Over the last century—especially in western Maryland—streams began choking on the waste generated by coal mining. Today, even though many mines have been abandoned, creeks remain contaminated with legacy acidic water.

Meanwhile, across the Piedmont—and especially along the fall line—developers continue to cut down forests to build suburban and exurban homes. One unwanted result? Streams (and their many inhabitants) suffer from excessive flooding, temperature fluctuations, sedimentation, bank erosion, chemical changes, and a dramatic loss in biodiversity. The replacement of forests with "development" can greatly compromise aquatic life, especially because of the carpeting of the land with *impervious surfaces*—the many roads, parking lots, and rooflines that make up our human communities.

Take brook trout. Brookies need water temperatures to be cooler than 20°C (68°F) and require high dissolved oxygen level (greater than 5 mg/L). Just a 5% change in the amount of impervious surface in a local watershed can disrupt brook trout populations.[5]

Shad, a once bountiful Chesapeake Bay fish that was caught and used by both Native American and early European settlers, has suffered severe population demise from overfishing, industrial growth, and—especially— the construction of smaller mill dams in the 1800s and the much larger river hydroelectric dams in the mid-1900s. Dams are particularly onerous to shad because they are an anadromous fish, meaning they live in the ocean but migrate up in freshwater rivers and streams to spawn. Dams block this migration of fish to ancestral spawning grounds. The commercial shad fishery was closed in 1980, the Potomac River shad fishery in 1982, and recreational harvest is closed Bay-wide, although catch-and-release fishing is allowed in nontidal waters. Ballyhooed (and expensive) fish ladders on dams like the Conowingo in fact allow only a tiny fraction of fish to complete their reproductive life cycles.

Stormwater runoff and thermal pollution: The explosive expansion of impervious surfaces in Maryland's suburban landscape has also dramatically increased the *stormwater runoff* that follows major rainfalls. Surface runoff that is not interrupted by forest cover picks up volume, speed, contaminants, and heat. When all this water finally hits a stream, it can destroy streambanks; dump pollutants collected across all those farms, lawns, roads, and parking lots; and change the water's temperature in what is called *thermal pollution*.

Invasive species: Although Maryland is home to many *native* fish, most freshwater fishes are *non-native*, meaning that they were not naturally found in Maryland, but have been—intentionally or accidentally—introduced to the area. Non-native species may not cause harm or damage to ecosystems, but *invasive species* can—and do—cause both economic and environmental harm.

At one time, it was commonplace to introduce fish to new habitats. The first introductions recorded were carp introduced to Europe from Asia in the 13th century. In the United States during the 1800s, government agencies routinely introduced new species to feed and provide recreation for burgeoning societies. Not all these introductions came from other countries. For example, largemouth bass and smallmouth bass, which are native to the Ohio and Mississippi Rivers, were introduced to Maryland in the 1800s.[6] These introductions significantly affected Maryland's local fish communities, though initially the impact was not officially measured. Centuries later—with many competitive fishing tournaments held in tidal waters providing millions of dollars to Maryland's economy each year—these species have become valuable sportfish for anglers and local businesses: approximately half of Maryland freshwater anglers fished for bass in 2008 and 2011 according to the US Fish and Wildlife Service. But over time, as new species continue to be introduced—especially the northern snakehead, blue catfish, and flathead catfish—scientists have also chronicled potential dangers to the Chesapeake Bay watershed (see figure 12.1).

Once introduced, invasive species can flourish; many have no known predators, which allows them to vigorously reproduce. Northern snakehead, for example, can reach sexual maturity in just one year, and can reproduce up to five times a year. When their eggs hatch, the young fish form "fry balls" which are then aggressively protected by the parents, allowing for high survival rates of young snakehead.

FIG. 12.1. Branson Williams, Maryland's invasive fish program manager, holds an invasive blue catfish. Photo courtesy of the Maryland Department of Natural Resources.

Invasive species also compete with native species for food and habitat or alter and damage vegetation, which many fish depend on. Blue catfish, which can grow to over 100 pounds, feed opportunistically on readily available seasonal prey and can easily outcompete other native sport fish for food.

Intersex: Another eerie consequence of industrial pollution has been the disturbing evolution of fish considered *intersex*. As hormone-disrupting chemicals found in a great many consumer products—from birth control pills to flame retardants and synthetic fragrances—leak into our waterways, they can cause male fish to develop eggs in their testes. This problem was first discovered in the South Branch Potomac River in 2003.[7] A report released by the US Geological Survey notes that intersex conditions are widespread in the Potomac River basin, especially in male smallmouth bass.

Road salt: Excess road salt left on the roads from winter storm treatments can threaten aquatic life that is sensitive to salt levels. Salt concentrations can also infiltrate groundwater, which then flows into surface water.

What Has Been Done to Improve Maryland's Waterways and Fishes

As discussed more fully in this volume's chapter on Chesapeake Bay ecology, in 2010 the EPA established a comprehensive "pollution diet" for the Bay, limiting runoff pollution through a measurement known as a total maximum daily load (TMDL). The TMDL seeks to restore clean water in local streams and rivers by requiring the Bay region's six states and the District of Columbia to reduce the amount of nitrogen, phosphorus, and sediment running off their farms, lawns, and developments, a daunting task when our human population continues to expand. To help achieve these goals, farmers have initiated manure management systems, added *exclusion fencing* to keep livestock out of streams, and repaired or restored riparian vegetation buffers.

Jurisdictions are also making changes with *structural stormwater controls* as listed in the state's Stormwater Management Act of 2007. In an effort to mimic natural runoff and minimize land development's impact on water resources, these controls are designed to capture and treat runoff closer to—or at—the areas where rain first hits the ground. You can see these efforts at work in the construction of stormwater retention ponds installed along many of the state's highways.[8]

Indeed, the state's highway administration continues to design and install a variety of stormwater retention mechanisms. Depending on the area, there are many different types of control structures:

Grass swales: grass-lined channels.

Bioswales: linear constructed filters that treat and decrease the flow of stormwater runoff.

Infiltration trenches: excavated areas filled with stone to temporarily store runoff where it percolates into the surrounding soil. Pollutants are removed naturally before recharging the groundwater.

Stormwater wetlands or wet swales: areas that normally retain water or are in an area with a high groundwater table. They use wetland plants to filter runoff by *biological uptake.*

As a Master Naturalist, there are numerous ways to protect or improve your watershed and the animals that call it home:

- Get to know your watershed. Identify any issues affecting the watershed you live in. The Environmental Protection Agency monitors watersheds and current impairments.

- Conserve water when and where possible. Take shorter showers and use only cold water to wash laundry. Collect rainwater in rain barrels to water your garden. Minimize lawn watering.

- Consider planting a rain garden at your home. Work with locals in your community to expand this work into schools, parks, and other spaces.

- Plant trees and riparian vegetation (preferably native species) to help reduce runoff and erosion near streams.

- *Never* dispose of chemicals, oil, paint, or pharmaceuticals in drains or toilets. Check your county's household disposal/ recycling program for proper disposal protocol.

- Volunteer at local stream/river cleanups.

Notes

1 An excellent resource for more background on fish can be found in Karl F. Lagler, John E. Bardach, Robert R. Miller, and Dora R. May Passino, *Ichthyology*, 2nd ed. (New York: Wiley and Sons, 1977).

2 Lagler, Bardach, Miller, and Passino, *Ichthyology*.

3 "Mummichog: *Fundulus heteroclitus*," Chesapeake Bay Program, https://www .chesapeakebay.net/discover/field-guide/entry/mummichog.

4 Bob Fenner, "The Physiology and Behavior of Color in Fishes," Wet Web Media, http://www.wetwebmedia.com/AqSciSubWebIndex/coloration. htm#google_vignette.

5 Scott A. Stranko, Robert H. Hilderbrand, Raymond P. Morgan II, Mark W. Staley, Andrew J. Becker, Ann Roseberry-Lincoln, Elgin S. Perry, and Paul T. Jacobson, "Brook Trout Declines with Land Cover and Temperature Changes in Maryland," *North American Journal of Fisheries Management* 28 (2011): 1223–32, https://doi .org/10.1577/M07-032.1.

6 Albert M. Powell, *Historical Information of Maryland Commission of Fisheries: With Some Notes on Game* (Annapolis: Maryland Department of Natural Resources, 1967).

7 V. S. Blazer, L. R. Iwanowicz, D. D. Iwanowicz, D. R. Smith, J. A. Young, J. D. Hedrick, S. W. Foster, and S. J. Reeser, "Intersex (Testicular Oocytes) in Smallmouth Bass *Micropterus dolomieu* from the Potomac River and Selected Nearby Drainages," *Journal of Aquatic Animal Health* 19 (2007): 242–53.

8 Stormwater Management Act (Environment Article 4 §§201–215).

CHAPTER 13

Reptiles and Amphibians

RAYMOND V. BOSMANS

The scientific study of reptiles and amphibians—some of our most interesting, prehistoric, beautiful, and misunderstood animals—is known as *herpetology*. This group of animals (often referred to as "herps") invokes curiosity, appreciation, and fear in many people. Consider snakes: probably no other animal has been the subject of more myths, superstitions, and hatred than snakes. Master Naturalists can help change this perception.

The mid-Atlantic region has a diverse population of "herps." Many are common in both rural and suburban environments; others are in severe decline due to habitat destruction. Countless animals are killed simply trying to cross roads. This chapter sets out to give Master Naturalists a better understanding of reptiles and amphibians, to help them learn techniques to teach others, and to emphasize that we all should appreciate and learn to coexist with these amazing animals.

Reptiles and amphibians have remained (essentially) physically unchanged (though most are much smaller) since their prehistoric days on Earth. Herps are *cold-blooded* (or *ectothermic*), meaning their body temperature reflects—and is dependent on—the temperature of their environment.

For much of the year, herps can effectively regulate their temperature by moving in and out of the sun. *Basking* is an important part of life for turtles; they receive UVB rays from the sun to help metabolize vitamin D3, which (along with calcium) is a vital compound for shell growth. During winters here in the mid-Atlantic region, herps must go into a greatly slowed metabolic state called *brumation*, which permits them to survive very cold temperatures. A few species even freeze—and survive!—until the return of spring.

Reptiles and amphibians differ greatly in their life cycles. Reptiles typically hatch from eggs, although some snakes in our region bear live young. Baby reptiles look just like adults, although their coloring is often different when young. Reptiles lay their eggs in the soil, or under logs or piles of mulch. Eggshells can either be soft and flexible or hard, depending on the species. Parents do not tend to the eggs or protect their young.

By contrast, amphibians mate and lay their eggs in water during spring. Their young look different from their parents. Consider the *tadpole* of a frog or the *nymph* of a salamander: their young breathe with gills and feed on algae, but as they mature, they grow legs, lose their gills, develop lungs, and begin eating insects and worms. This is magic indeed!

Reptiles (turtles, snakes, lizards) have dry scaly skin, while amphibians (frogs, toads, salamanders) usually have smooth, soft, moist skin (though a toad's skin is rougher, drier, and wartier than a frog's). While frogs and toads are quite vocal—calling out in great numbers in the spring and summer—and a few tropical lizards (like geckos) vocalize to call to each other, reptiles in the mid-Atlantic do not make vocal calls.

Reptiles and amphibians came on the scene even before the dinosaurs: about 350 million years ago, during the *Carboniferous* period. The prehistoric environment was ideal for these animals, and they thrived and grew much larger than we see today. Among modern-day reptiles, the closest we have to a prehistoric-sized turtle in the Chesapeake Bay region is the leatherback sea turtle, which has a soft carapace and weighs up to 2,000 pounds!

Like all animals, reptiles and amphibians have their own ecological niches and roles to play: they feed on a long list of insects, rodents, plants, fish, and other herps, and are—in turn—a significant food source for other wildlife such as hawks, eagles, and skunks. Many herps are beneficial in helping manage pest insects, including those that damage our crops and negatively impact human health. Snakes, especially those that prey on rodents, are particularly helpful with human disease control. For example, areas that have a healthy population of black rat snakes have less Lyme disease. Why? Because Lyme disease spreads through deer ticks, which begin their lives feeding on mice, which rat snakes love to eat: fewer mice, less Lyme disease.

All herps, particularly frogs, toads, and salamanders, are highly sensitive to chemical pollution, especially pesticides, toxins, and fertilizers (amphibians are quick to die when their ponds and marshes become contaminated from chemical runoff, or from algae blooms caused by excessive lawn or farm fertilizers). Thus, the presence (or absence) of these animals in your community can be a good indicator of your area's environmental health.

Creating a Herp-Friendly Landscape

We all enjoy beautifully designed and maintained home landscapes. As Doug Tallamy and others point out elsewhere in this volume, it is not necessary to manicure and "control" your home landscape for it to be beautiful. The effort to physically (and chemically) sanitize the landscape takes a lot of time and money, and some of the things we do are harmful (or deadly) to wildlife. Here are some tips:

> **Don't mow your lawn at night:** Evening mowing kills countless numbers of animals that forage at night, especially toads, frogs, and snakes. At night, box turtles often hunker down in mulch at the edge of the lawn and get struck by mower blades.

> **Mow your lawn high:** Cool season grasses such as the fescues and bluegrasses perform best when mowed at two to three inches. This reduces weed encroachment, gives a much better-looking lawn, and reduces injury to turtles and snakes.

Use bio-rational pesticides: Many traditional pesticides are over-used and poison wildlife and non-pest insects. *Integrated pest management* is a "total system" approach that includes using natural pest control methods, such as selecting plants with fewer pest and disease problems and using less toxic chemical inputs, such as insecticidal soap, neem oil, horticultural oil, and diatomaceous earth (a powder that comes from the shells of aquatic organisms).

Leave some "wildness": We should no longer consider the sterile 1950s lawn our aesthetic ideal. Leave a place in your yard to pile leaves and branches collected in the fall. Instead of ecologically use-less turf grass, plant native species instead. You will be amazed at how many types of herps find refuge in these places—to say nothing of birds, butterflies, and other wildlife!

Create a water garden: The addition of a small pond will provide an excellent breeding area for toads and frogs. Build it and they will come! The frogs will stay the entire summer; toads will come to mate and lay their eggs and then disperse out into the landscape.

A Look at Common Reptiles and Amphibians of the Mid-Atlantic Region

ANATOMY OF TURTLES

A turtle, with its iconic hard outer shell, is one of the most recognizable animals in the world. It is thought that the first turtles were *soft*-shelled, but that over time their ribs fused together to form their characteristic hard shell. Today's soft-shelled turtles remain the closest thing we have to the first turtles.

The upper shell of a turtle is called the *carapace*, and the lower part is the *plastron*; both bones are integral parts of a turtle's skeleton, are never shed (a turtle's shell grows with it), and are covered with keratin *scutes*—a thin material very similar to that found in our fingernails. Scutes are colored and help with camouflage; in many aquatic species, scutes (unlike shells) are shed with new ones underneath. A turtle's hips, collar, and leg bones are all attached inside the shell; the spine runs down the middle of the carapace.

Turtles live either on land or in water. They have fully or partially webbed feet or (in the case of sea turtles) flippers. In our region we have box turtles, wood turtles, and spotted turtles, along with many types of aquatic turtles.

Tortoises live only on land, and are usually found in warmer climates. We do not have native tortoises in our region. Although pet tortoises escape and may be found here, most pet tortoise species cannot survive our winters outdoors.

Terrapins—the word, which comes from an Algonquian Indian word meaning "little turtles," refers to the diamondback terrapins that live in brackish water in the Chesapeake Bay's marshes and tributaries. If you call them all "turtles" (as the University of Maryland slogans often do) people will still know what you mean!

TERRESTRIAL TURTLES

Eastern Box Turtle (*Terrapene Carolina*)

The eastern box turtle is one of the most beautiful turtle species in the world; its nickname is "gem of the forest." No two box turtles are identical: their colorful patterns of orange, yellow, brown, and black are entirely unique and allow them to blend in well in the forest. Box turtles get their name because they have a *hinged plastron* that allows them to close tightly into a little "box." Only a few other turtle species in the world can do this.

Because of their beauty and gentle disposition, box turtles are very popular as pets, which—along with habitat destruction and fragmentation, road kills, and egg and hatchling predation by raccoons—has greatly contributed to their demise in the wild. Many states have adopted laws that regulate their collection as pets.

In the wild, box turtles do not usually reach sexual maturity until they are close to 20 years old; their average life span is 75–100 years. They typically mate in the spring, and because encounters between male and female are rare, females have adapted a way to store sperm for a few *years*! Eggs are laid June–July, with three to five eggs in a typical clutch. Females bury their eggs in flask-shaped nests in the soil in a sunny location. Incubation usually takes 90 days.

Male box turtles are more colorful than females and usually have bright red eyes. Some females may also have reddish eyes, though not quite as bright as the males. The plastron of males is concave while that of females is flat. Baby box turtles emerge in early fall, though sometimes they spend the winter in the nest and emerge in spring. Baby box turtles are dark brown and do not develop their distinctive coloration until they are about two inches in size.

Box turtles are omnivorous, feeding on a wide variety of worms, fallen tree fruit, strawberries, mushrooms, soft-bodied insects, and (occasionally) carrion.

North American Wood Turtle (*Glyptemes insculpta*)

The wood turtle is another iconic American species. It prefers habitats similar to that of the box turtle, but is more aquatic. Its brown, sculpted carapace (which can look very much like a piece of wood) can reach a length of around eight inches. Males have bright orange front legs and neck and a larger head than females. They mate underwater: eggs are also laid on land in June and July. Like box turtles—and for many of the same reasons—wood turtle population numbers have been greatly reduced, and their possession in captivity is strictly regulated. It is illegal to possess a wood turtle without a special permit from the state Department of Natural Resources.

LARGE FRESHWATER TURTLES (8–20 INCHES)

Our lakes, rivers, streams, and ponds have a diverse number of *semi-aquatic* turtles. Except for a few species, most populations are relatively stable, thanks in part to the restoration of marshes and stormwater management ponds. Another lucky fact: these species do not typically venture onto highways. Aquatic turtles cannot swallow their food on land. They must have their head underwater to swallow food.

Common Snapping Turtle (*Chelydra serpentine*)

Snapping turtles are the largest freshwater turtle in our region. Adults can max out with a 20-inch carapace and weigh as much as 70 pounds! Snappers have a prehistoric "dinosaur" appearance: their heads are large, and their tails are long, with dorsal ridges resembling those of an alligator. They are strong and can climb low walls and even chain-link fences to get where they want to go. It's common to find older ones covered with a thick growth of algae, which gives them the nickname "mossbacks."

Snappers get their name from their defensive behavior of striking and snapping with incredible speed and strength. Indeed, they can inflict a very painful bite! The safest way to pick up a snapper is by grasping it at the *back half of the carapace*. Be sure to keep your hands away from the front half of the carapace—they have very long necks, and may reach back and bite you! (Author's note: I have had a common snapper for almost 35 years. She lives outdoors year-round. At first she was a typical, defensive snapper, but after a few years of handling and caring for her, I have found her to be "puppy-dog tame." She comes to me, allowing me to pick her up with absolutely no biting. I use her for lectures all the time. She has learned to trust people!)

Snappers spend most of their time in the water, only coming out occasionally to bask in the sun or lay eggs. Females usually go great distances from their home pond to lay 24–75 eggs, which is when (often in June and July) people see them in their yards or crossing roads. Common snappers are omnivorous, feeding on frogs, worms, sick fish, ducklings, other turtles, snakes, carrion, and—as they get older—a lot of plants.

Like box turtles, snappers have a long lifespan, about 80–100 years; some field records report some living even longer. In regions of our country where people routinely eat turtles, it is the snapper that is hunted the most.

The alligator snapping turtle is another species found in the Midwest and the South. This massive turtle has a mature carapace length of 36 inches and often weighs some 200 pounds. Sadly, it is endangered because of over-harvesting for meat markets. Legislation is now in place in many states to help restore their numbers.

Northern Red-Bellied Turtle (*Pseudemys rubriventris*)

Next to the common snapping turtle, the red-bellied turtle is our second-largest freshwater species. A handsome, dark-colored animal, it has faint red or orange markings on its somewhat domed carapace. Its head and legs are black with thin yellow lines. The name originates from the juveniles, which have bright red plastrons, a color that fades with age to more of a salmon pink. Red-bellies typically grow to about 14–16 inches. Their young feed on aquatic insects, tadpoles, and worms, though in maturity they also switch to a mostly aquatic vegetation diet.

Red-bellied turtles are found in lakes, rivers, and canals throughout most of our region. Just a few years ago their numbers were in broad decline, but with dedicated conservation efforts they seem to be making a comeback in parts of their ranges. Although considered a freshwater turtle, they have been observed occasionally venturing into brackish water in the tidal areas around the Chesapeake Bay. Like painted turtles, red-bellies love to bask on logs in full sun.

Red-Eared Slider (*Trachemys scripta elegans*)

The red-eared slider (so named because of a red stripe behind each eye) is another large freshwater turtle found through the mid-Atlantic area. Though native to the southern states (and technically not native to our region), red-ears are very well established here largely because they are raised by the millions and shipped here (and around the world) for the pet trade. From the 1950s to the 1970s, a great many of these turtles died in captivity because very little information was published on the care of turtles. Some did survive, of course, and when they grew too large, their owners released them. Today, the red-eared slider is very successful in the wild and is even considered an invasive species in much of the United States and around the world.

In the 1960s and early 1970s, many children were getting very ill from a salmonella bacterium traced back to keeping these little turtles (children would handle them and then put their fingers—or the turtles themselves!—in their mouths). In the early 1970s a federal health law made it illegal to sell or possess turtles under four inches—because a four-inch turtle could not fit in a child's mouth! (The feds left it up to each state whether to enforce the law). Today a few states do allow the sale of baby turtles, but most do not. Adult red-ears are legal to sell and possess and, when properly cared for, can live for 50 years.

Female red-ears can grow up to 10 inches, males about 6–7 inches. Males have very long front claws that they use (believe it or not) to stroke the female's face during mating courtship. Red-ears usually lose their green coloration as they mature; their carapace becomes more of a greenish brown with a varied striped pattern, though their coloration and patterns can vary. Young turtles are carnivorous and switch to eating more aquatic plants as they mature.

SMALL TO MEDIUM FRESHWATER TURTLES (4–7 INCHES)

Eastern Mud Turtle (*Kinosternum subrubrum*)

The mud turtle is small (3–5 inches) and brown, with a semi-hinged plastron. Although their plastron has a hinge, it is not capable of closing up tightly like a box turtle's. Mud turtles live on the bottoms of ponds. They are not very strong swimmers, preferring to walk along the pond bottom looking for worms and aquatic insects. They usually do not venture out onto land, except during their egg-laying season in May and June. When first handled, mud turtles can be quite defensive, reaching back with their long necks to bite whoever is holding them. They can also release a foul-smelling musk—all in hopes that you (or a predator) will let them go. If you can hold out long enough, mud turtles will often drop these defensive actions when they realize there is no threat.

Common Musk Turtle (*Sternotherus odoratus*)

This small relative of the mud turtle is also aquatic; the two species often share the same habitat. Its carapace coloration is more grayish than brown, and it does not have a hinged plastron. Its plastron is similar to (but smaller than) that of the common snapping turtle.

Musk turtles have a pointed nose and a white/yellowish stripe running down the head and neck. Their disposition can also be quite defensive when first picked up and—as their name suggests—they will also release a foul scent from glands under their carapace. Mud and musk turtles may not be the best swimmers, but they can climb trees—and have been known to get into passing canoes!

Spotted Turtle (*Clemmys guttata*)

The spotted turtle measures just three to five inches across, but it is one of our most attractive species. Its carapace is typically black (though some are brownish black), with bright yellow spots that increase in number with age. The head and neck are black with brilliant orange markings. It inhabits ponds, slow-moving creeks, and marshes, but sadly, their populations are also declining due to habitat destruction and road kills. They are also highly sought after by hobbyists, and their collection from the wild is now restricted in most states. Fortunately, captive turtle breeders can offer healthy captive hatchlings in states where sale of turtles under four inches is permitted.

Bog Turtle (*Glyptemys muhlenbergii*)

The state reptile of New Jersey, the bog turtle is a small, close relative of the wood turtle and is critically endangered in our region (it is protected under the federal Endangered Species Act). Adults average about four and a half inches wide and have a dark brown carapace very similar in appearance to the wood turtle. Their bodies are dark, with distinctive orange blotches on each side of the head.

As their name implies, these turtles live in freshwater, peaty *bogs*, not swamps, ponds, or lakes. They like to hide out on tufts of grass in fields known as "fens," thick with predominately herbaceous vegetation; as woody plants begin to encroach, these turtles tend to leave. Recent efforts using goats to clear woody plants have helped to preserve bog turtle habitat.

Eastern Painted Turtle (*Chrysemys picta picta*)

The eastern painted turtle is perhaps the most beautiful turtle in our region, if not in the world! This common species successfully survives among human activity. Here again, with new regulations requiring housing and commercial development to install sediment and stormwater-retention ponds, these turtles have proven able to establish populations in these ponds. Although some road mortality occurs, these turtles typically do not travel long distances from their home pond.

Painted turtles have colorful carapaces; their background color can range from a deep olive green to almost black, with light colored edges on the scutes and bright red markings on their carapace margins. The plastron is a yellowish orange. The skin of their legs, neck, and head is black, striped with red and yellow. They are fast, excellent swimmers and can be found in rivers, lakes, ponds, and (sometimes) in streams. Like most aquatic turtles, they are highly carnivorous when young and become more herbivorous with age. Their mature size is six to seven inches for females and approximately five inches for males. Males have very long front claws that they use in courtship with females.

BRACKISH WATER TURTLES

Northern Diamondback Terrapin (*Malaclemys terrapin*)

The beautiful diamondback terrapin, Maryland's state reptile, is truly a treasure of the Chesapeake Bay's tidal marshes. These creatures have "sculptured" carapaces that range in color from black to brown; some are adorned with concentric patterns with yellow and light gray markings (terrapins are rarely identical). The original terrapins in the Chesapeake were mostly dark, but after the 19th century introduction of southern "concentric" turtles for the meat industry, there is now quite a bit of color variation. Their skin is smooth and soft and usually light gray, though some are dark gray and black, and others are almost white with various patterns of black. Female terrapins are larger than the males, reaching eight inches in length, three inches longer than typical males. Terrapins feed on aquatic snails, crustaceans, small crabs, and carrion found in the water.

Terrapin meat has historically been a significant seafood relished by many cultures; in Maryland, terrapins used to sell for as much $70 per dozen! Today, diamondbacks are protected in many states and may no longer be collected from the wild for the seafood market (those that are sold now must be "farm raised"). Terrapins face other threats, however, with many drowning in unattended crab traps. A special turtle excluder (also called a bycatch reduction device) is required on crab traps to keep terrapins out.

SEA TURTLES

There are five species of sea turtles found along our section of the Atlantic coast: the Atlantic hawksbill, green sea turtle, Kemp's ridley, loggerhead, and leatherback sea turtle. All are very large, from three to six feet in length (the leatherback is the largest). All are on the endangered species list. Sea turtles usually travel south to lay eggs but forage our waters in late summer and fall for fish, crabs, and jellyfish.

Lizards

People are often confused about the difference between salamanders and lizards. Consider: salamanders are amphibians, with soft skin (and no scales), usually found in moist places near water, where most mate and lay eggs (slimy and redback salamanders do their egg laying under moist logs).

By contrast, lizards (like other reptiles) have dry, scaly skin. They mate and lay their eggs on land, not in water, and have moveable eyelids and ear openings. Lizards are fast, salamanders are slow.

In the mid-Atlantic region there are four species of lizard: eastern fence lizard, common five-lined skink, broad headed skink, and little brown skink. All our lizard species are harmless to people and can help in managing insect pests.

EASTERN FENCE LIZARD (*SCELOPORUS UNDULATATUS*)

The eastern fence lizard is a common inhabitant of forests and farmland. They grow to about six inches, and have a silvery gray, scaly appearance with some fainter dark marking across their backs. Like all lizards, the fence lizard (also called fence swift) is quick to escape and hide. They typically hang out on wooden fences, logs, rocks, and old farm buildings. They feed exclusively on insects and lay eggs underneath logs and boards on the ground. Males have a shiny blue patch of scales on each side of the abdomen.

COMMON FIVE-LINED SKINK (*PLESTIODON FASCIATUS*)

The five-lined skink is an attractive, smooth, and shiny lizard that also moves quickly (skinks have small legs compared to the size of their body). They are black, with five yellow lines running down their backs. At maturity, they grow to about five to eight inches. Juveniles have bright blue tails. As they mature, the yellow lines gradually fade away, as does the blue on the tail. Five-lined skinks are found in similar habitats to the swifts, except they will often live in closer proximity to people, sometimes even venturing inside houses to look for insects.

BROAD-HEADED SKINK (*PLESTIODON LATICEPS*)

The broad-headed skink is the largest of the skink species, reaching a length up to 13 inches, and is mostly found in the southern and far-eastern regions of the mid-Atlantic region. Their heads are large and red when mature. They live in similar habitats to the other lizards mentioned here but are more common in the southeastern states.

LITTLE BROWN SKINK (*SCINELLA LATERALIS*)

Little brown skinks grow to about five to six inches long and are seldom seen because their color beautifully matches the dead leaves on the forest floor. Like other skinks, they are quick. Little brown skinks are typically found in the Coastal Plain region.

Amphibians

FROGS

Frogs are indeed a successful amphibian species, well adapted to living almost any place there is fresh water. They play an important role in managing insect populations, and many people enjoy having them in their yards and hearing them vocalize. Sadly, in the past several years, many frogs in our region have suffered from a deadly fungal disease that has reduced populations.[1]

American Green Frog (*Rana clamitans*)

The American green frog is a very common inhabitant of almost any body of fresh water, large or small. It grows to about three inches in length, and its coloration can vary from a brownish green to a bright green. Males have large ears (or *tympanum*) compared to those of females and have a yellow throat. Green frogs like to bask on the edge of a pond or on water lily pads. When startled by an intruder, they quickly plop into the water for protection.

All frogs are quite vocal, and this one is no exception. Males will make a sound that resembles a banjo being plucked, and will do this all day and night long. This is usually an announcement of their claim on a territory and a call for attracting females. Male green frogs will often fight each other if one gets too close to another's territory.

Green frogs mate in the late spring or early summer and lay large numbers of eggs on the water surface, usually at night. Within a few days, the eggs hatch and tiny tadpoles swim away to feed on algae on the pond's bottom. They remain as tadpoles until the next year, when they metamorphosize into frogs. As adults, green frogs feed on insects, worms, or just about anything else moving that fits into their mouth. At night, green frogs frequently leave the pond and forage for food on land, especially on rainy nights.

Wood Frog (*Lithobates sylvatica*)

Once scarce in much of our area, wood frogs seem to be making a nice comeback, in part because so many homeowners have created aquatic gardens in their backyards. The wood frog is more terrestrial than the green frog. They are smaller and tan to brown colored with a dark mask around the eyes. This is an attractive and distinctive looking frog indeed.

Wood frogs are among the first frogs to show up at ponds to mate in early spring, often doing so even when there is still some snow on the ground. Their call sounds a little like a duck quacking. In the cold, wood frogs (temporarily) turn very dark, almost black.

Spring Peepers (*Pseudacris crucifer*)

Spring peepers are small tree frogs measuring just a half-inch in length. They are a light tan, which can become darker as needed to help them camouflage. As enthusiastically audible as they are in springtime, peepers are so small and secretive that people seldom actually see them; as soon as you walk near, they suddenly stop their chorus until you leave. Spring peepers are the first species of frog to emerge and begin singing even as early as a winter "warm spell" in February. After mating season, they move out into the woodland and live in trees.

Tree Frogs (*Dryophytes* spp.)

There are several species of tree frogs in the mid-Atlantic region. Here we will focus on the most common: the gray and green tree frog species. These medium-sized frogs (two inches long) look more like toads than frogs because their skin is a little rougher. After mating in water, they spend the rest of the summer in trees, shrubs, and other structures, where they catch insects. Their metamorphosis is quick—just a few weeks pass from hatching to tadpoles to tiny "froglets."

Tree frogs produce a loud and somewhat shrill call at night, especially during a rainy and warm summer night. Tree frog skin can produce a mild toxin meant to protect them from being eaten by birds and snakes; it is recommended that you wash your hands after handling them.

American Bullfrog (*Lithobates catesbeianus*)

These are our real giant: some bullfrogs reach up to eight inches in body length. They have a loud, deep call (resembling a bellowing cow) that can be heard for great distances. Bullfrogs rule the pond! They will eat anything that they can catch, including other frogs, small snakes, goldfish, baby turtles, and even small birds and rodents. Bullfrogs are nomadic and often move from pond to pond. If you create a backyard fish or lily pond, you will surely have bullfrogs visit (see figure 13.1).

FIG. 13.1. Bullfrog. Photo courtesy of Raymond V. Bosmans.

Toads

Toads are the more terrestrial cousin to frogs. They are similarly shaped, but their back legs are small, and their skin is dry and "bumpy," often covered in what people call "warts" (don't worry, toads cannot give you warts). But beware: the large glands behind the eyes can exude an irritating toxin when they feel threatened by a predator. The irritant will cause a predator to release its grip.

There are several species of toads in our region: American, Fowler's, spadefoot, and eastern narrow-mouthed toad. The most common toads in our region are the American and Fowler's. Both reach a mature size of about three inches for females, smaller for males. Toads mate and lay eggs in ponds, marshes, and vernal pools in early to mid-spring, and sometimes do a second mating later in spring. Toads brummate on land under logs, rock piles, or mulch, with males emerging first in the spring to start moving to water. There they issue a high shrill call for females to come for mating. Females lay long, gelatinous strands of eggs among aquatic vegetation.

By summer, tiny baby toads (about one-eighth of an inch) develop from tadpoles and leave the water en masse (sometimes numbering in the hundreds) to start their new life on land. I have seen this hoard movement in my own yard; it looked like a dark mass of insects moving from a pond and across the lawn!

Toads can become comfortable living around people, and can seem quite tame. At my home, we have had a large female American toad who—every night—hopped up onto the front porch to eat insects attracted by our porch lights. Every time the front door was opened, she hopped into the house. She quickly became a member of the family!

In the garden, toads eat a variety of insects, as well as slugs. Toads can be encouraged to live in your yard by placing items like inverted clay flowerpots used as "toad houses" for daytime refuge.

Salamanders

Salamanders are lizard-like amphibians that inhabit moist woodlands. Like all other amphibians, salamanders start life in water as nymphs. All but two species develop from eggs laid in vernal pools. The exceptions, the white-spotted slimy and redback salamanders, deposit their eggs under moist logs and skip the nymph stage. The nymphs of all other salamanders have feathery gills that become lungs before they migrate to land. All salamanders have smooth moist skin and tiny legs, and prey primarily on insects and worms. Fun fact: the Allegheny Mountains have the world's greatest diversity of salamanders.

White-spotted slimy salamander (*Plethodon cylindraceus*): Grows to about five inches. A slender white-speckled black salamander common in moist woodlands. Frequently found underneath rotting logs. Feeds on small insects and worms. The redback salamander (*Plethodon cinerus*) is a close relative (frequently sharing the same habitat); it is similar in size and body color but has a red back.

Eastern spotted salamander (*Ambystoma maculatum*): A "mole" salamander that lives in burrows dug by other animals, it grows up to nine inches, with large yellowish-orange spots. Common throughout the eastern United States but becoming less common in the mid-Atlantic. It lays eggs in vernal pools. Spotted salamanders secrete a white fluid that is toxic to predators.

Northern red salamander (*Pseudotriton ruber*): Inhabits forests, wetlands, and creeks. Grows to seven inches. Its basic color is reddish brown to bright red, covered with many tiny black spots. Breeds in water. Larvae eat insects and worms; adults eat almost anything alive that will fit into their mouth.

Eastern newt (*Notophthalmus viridescens*): An aquatic salamander, with skin that is rougher than other species. Grows to four inches; green in color, with small black spots and red spots on the sides of the back. Yellow belly, with numerous black spots. During mating season, the tails of males become broad. When young newts leave the water, they turn red and live on the land for up to three years in what is called the red eft stage. They then return to spend the rest of their life in water, take on traditional newt coloration, and feed on small worms and aquatic insects. Young efts (and adult newts) also secrete a skin toxin to ward off predators. Popular as aquarium pets, especially in Europe; in captivity they learn to eat pelleted food and can live up to 15 years.

Eastern hellbender (*Cryptobranchus alleganiensis*): At 12 to 30 inches, the largest salamander in North America. Lives along the bottom of fast-moving rivers and streams. Lives 15 years in the wild and up to 30 in captivity. Now a threatened species, with habitats protected by law. Rarely found except in pristine forest streams.

Snakes

Snakes elicit emotions from uneasiness to deep phobia, and too often this fear results in the needless killing of a snake. Snakes are in fact unique members of mid-Atlantic wildlife and must not be harmed. In some states, the killing of snakes is illegal and punishable by fines.

Biologists consider snakes the most recently evolved of all reptiles. They were once equipped with small legs but have evolved to live without them and indeed have adapted to live in almost all parts of the world except Greenland and the polar regions.

Some common questions:

How do snakes move without legs? Snakes get around remarkably well without legs. They can climb, swim, and catch live prey. Their movement is accomplished by the strong muscular contractions of their many ribs. These muscles are attached to the *belly scutes* that grip the surface much like the treads of a bulldozer. The slow, relaxed movement of a snake is slow and straight—some people call it a "cat-erpillar crawl"—but when frightened, a snake will move its body in an "S" shape, pushing against the ground much like an ice skater's blades pushing across ice.

Why do snakes shed their skin? Snake scales are made of *keratin* and are covered by a thin clear skin. As they grow, or their skin becomes worn, snakes will hide (and not eat) for a few days as they shed old, dull skin and emerge with a new, brilliantly colored skin underneath (young snakes may shed every couple of weeks, older ones only a few times a season).

Why does a snake flick its tongue? A snake's tongue is a forked sensory structure used to explore its environment and find mates or food. It "tastes" molecules in the air using a specialized organ in the roof of its mouth called a *Jacobson organ*. A snake's tongue is harmless and—contrary to superstitions—does not sting.

What do snakes eat? Snakes have small teeth that are curved inward to help grasp prey. Their bite, except for venomous species, is harmless to humans. All snakes eat live food. Larger snakes, including constrictors, prey on rodents; some feed on other snakes, birds, and eggs. Smaller species feed on worms, fish, soft-bodied insects, and amphibians. Most snakes eat once or twice a week. Giant snakes such as pythons and boa constrictors (none of which live in the mid-Atlantic!) can go a year without feeding.

Are snakes mean? No! Snakes prefer to remain still, hoping predators don't notice them. Under perceived threat, they will slip away to seek cover; most snakes are as surprised and frightened as you are! Remember: smaller things generally get eaten by bigger things. To a snake, a human is a real threat! That said, if they feel cornered, most snakes will defend themselves by striking, hissing, and—if needed—biting.

Some wild snakes are more docile than others. The common black rat snake, also known as the eastern rat snake, is quite tolerant of human activity. I have rescued large rat snakes off roads, and after gentle handling they all calmed down within a few minutes. Water snakes and black racers are another story—they simply do not like being handled!

Common snake species in the mid-Atlantic region:

Small Snakes (10–16 inches)
Eastern worm snake (*Carphophis amoenus*): Harmless. Smooth, shiny, reddish-brown like an earthworm; pink belly, a very small head, and tiny black eyes. Very secretive, seldom seen in the open; often found underground or under logs and rocks in suburban as well as rural areas. Feeds on earthworms and salamanders. Lays eggs.

Dekay's brown snake (*Storeria dekayi*): Harmless. Dull grayish brown, with black lines or dots down the back and sides. A white diamond-like pattern is visible when the snake flattens its body or is engorged with food. Very common even in the city. Feeds on earthworms. A close relative (but not as common) is the **red-bellied snake** (*Storeria occipitomaculata*), which is similar in color but with a red belly. Bears live young.

Ring-neck snake (*Diadophis punctatus*): Harmless. Very common. Slender, slate gray or black, with a yellow or orange ring around the neck and a yellow belly. Frequently found under flat rocks, logs, and mulch. Lays eggs.

Smooth earth snake (*Virginia valeriae*): Harmless. Very seldom seen. Glossy brown without patterns. Lives mostly underground and in forest debris. Lays eggs.

Medium Snakes (24–40 inches)

Eastern garter snake (*Thamnophis sirtalis*): Harmless. Very common. (Often mistakenly called a "garden snake," but this is incorrect; named "garter" because of its resemblance to garters once worn to hold up socks.) Found in suburbs and rural areas. Brownish green, with light yellow stripes down the back and sides. Some have prominent diamond-shaped dark spots. Coloring varies depending on locale. White diamond pattern visible when the snake flattens its body or is engorged with food. Feeds on earthworms, salamanders, frogs, small mice, and fish. Most will release a foul musk (and empty bowels) when handled. Unpleasant indeed! Bears live young.

Ribbon snake (*Thamnophis saurita*): Harmless. A close relative of the garter. Similar in appearance, but more slender, with more prominent stripes. Unlikely to be found in a landscape unless there is water nearby. Feeds on frogs and fish and likes to hang out on vegetation along the edge of a pond. Bears live young.

Rough green snake (*Opheodrys aestivus*): Harmless. Uncommon. A very beautiful slender green snake with a white belly. Arboreal—and often hard to see because of how well it blends with foliage. Named for its *keeled* scales. Feeds on insects and spiders. Found mostly in rural areas. Lays eggs.

Northern water snake (*Nerodia sipedon*): Harmless. One of the region's most common snakes. Found in almost all bodies of fresh water and sometimes in brackish. Has a stout body, with dull dark reddish-brown bands across the back over a lighter brown background (pattern is more apparent in young snakes). Pattern fades over time, turning brown, though banding is more visible when wet. Feeds on frogs, fish, and small rodents.

Often confused with the venomous cottonmouth (or water moccasin), but we do not have these snakes in this region. Water snakes have a reputation for being ill-tempered and aggressive. They are not aggressive but defensive. Their environment has many predators that eat them, so can you blame them? Bears live young.

Queen snake (*Regina septemvittata*): Harmless. Water dweller. Unlike the northern water snake, it is slender and docile, seldom biting when handled. Dull brown with heavily keeled scales and three black stripes down the back. Belly is a yellowish-ivory color that extends up the edge of the belly and is visible from the side. At home in ponds, lakes, or fast-running streams. Feeds on newly molted crayfish. Often lays on tree branches and shrubs overhanging water; some are said to have dropped from trees into canoes. Bears live young.

Brown or mole king snake (*Lampropeltis calligaster rhombomaculata*): Harmless. Rarely seen; spends most daytime hours under logs or underground. A shiny medium brown, with a series of small dark reddish-brown bands across the back. Lives mostly near rural farmland, seldom found in suburbs. Lays eggs.

Eastern hog-nosed snake (*Heterodon platirhinos*): Harmless. Very stout; can vary in coloration. Most are a bright orange, with broad markings against a golden yellow, dark brown or black background; some are all black. It has a turned-up "hog" nose that it uses to dig in soft soil looking for food. Its diet is almost exclusively toads. Although totally harmless to humans, it does have a mild venom that "sedates" toads while they are being swallowed.

Hog-nosed snakes are famous for their *bluffing displays* meant to scare away predators. When under stress they will flare their heads and necks like a cobra, hiss loudly, and strike—but will never bite. When these theatrics fail, they will turn over on their back and act dead, with their tongues hanging out! (Author's note: I've seen these displays in the wild. When playing dead, if I turned them right side up, they quickly flip over to their back. They have never bitten me. When kept in captivity, they stop all these efforts at deception.) Lays eggs.

Scarlet king snake (*Lampropeltis alapsoides*): Harmless. Very rarely found. Smallest member of the king snake genus. Much more common in the southeastern United States, but some are occasionally found in the southeastern regions of the mid-Atlantic. Very beautiful, with a brilliant red/black/white stripe pattern—a harmless mimic of the venomous coral snake of the South. Remember the old saying "red next to black: friend of Jack; red next to yellow: kill a fellow." Feeds on frogs, toads, and small rodents. Lays eggs.

Eastern milk or red king snake (*Lampropeltis Triangulum*): Harmless. Another beautiful snake, very similar in coloration and pattern to—but longer than—the scarlet king snake. Common in both rural and suburban areas. Young have a pattern of brick-red rectangular spots on a white background. As they mature their color becomes duller. Feeds on rodents and other snakes. Lays eggs.

Large Snakes (40–72 inches)

Eastern chain king snake (*Lampropeltis getula*): Harmless. Shiny black, with a white or yellow argyle pattern. Like all king snakes, it inhabits woodlands and open fields. More common in rural areas, but occasionally found in suburban areas. Capable of quickly catching and subduing its prey by constriction. Feeds on rodents, frogs, and snakes, including venomous ones (it is immune to their venom). Lays eggs.

Black rat, aka eastern rat, snake (*Pantherophis alleghaniensis*): Harmless. Large, beautiful, black that tames nicely. (Author's note: this is my all-time favorite mid-Atlantic snake. I still have a couple that I raised from babies that are now 21 years old and almost seven feet long! They continue to be great "snake ambassadors" for the countless classes I taught over the years). Heavy bodied, usually showing thin white lines that are remnants of their juvenile pattern. All rat snakes are constrictors and excellent climbers. Can easily climb a stone or brick wall, a barn, or a tree. Juveniles are not black but have an attractive pattern of dark rectangles over a light gray background; they gradually become completely black when about 20 inches long. Lays eggs.

Corn or red rat snake (*Pantherophis guttatus*): Harmless. A relative of the black rat snake, though not as common, it is brilliantly colored, with large orange rectangular markings bordered in black against a light brown or beige background. More common in the mid-Atlantic's southern regions. Ubiquitous in the world's pet trade. Very docile, they thrive and breed easily in captivity. Sometimes called the "designer snake" because of the many color morphs created by breeders. Lays eggs.

Eastern black racer (*Coluber constrictor*): Harmless. A slender, entirely black snake; unlike the eastern rat snake, it completely loses any white marking from its juvenile stage (juveniles look similar to those of the rat snakes). A rat snake's body in cross section is shaped like a loaf of bread: flat on the bottom and rounded on top. The black racer's body is completely round, like a garden hose. Racers are fast moving and defensive when handled. Unlike rat snakes, they do not constrict their prey. They feed on frogs, other snakes, and rodents. Lays eggs.

VENOMOUS SNAKES

Rattlesnakes and copperheads (timber rattler, genus *Crotalus*, and copperhead, *Agkistrodon*): Dangerous. People confuse "poisonous" with "venomous." *Venom* is a toxin that is *injected*, while a *poison* is *ingested*. So, there are no "poisonous" snakes in our region. Venomous snakes use their venom to subdue their prey; they prefer to be left alone and only bite as the last resort to protect themselves.

Of our two venomous snakes, the rattlesnake is the more dangerous. Both rattlesnakes and copperheads are *pit vipers*. The *pit* is a small, heat-sensing hole between the eyes and the nostrils that helps the snake to find warm-blooded prey.

Both rattlesnakes and copperheads can be found in similar rural and mountainous areas. They are secretive and not often encountered by people.

Snakes will vigorously vibrate their tail whenever frightened. Because of the rattle on their tails, rattlesnakes make a loud rattling or buzzing sound, warning strangers to stay away! Whenever working or hiking in areas where venomous snakes occur, learn to identify them, and be mindful of where you place your hands and feet.

Rattlesnakes are quite distinctive, but copperheads can easily be confused with harmless snakes that resemble them. People often confuse harmless water snakes, milk snakes, and corn snakes with copperheads and kill them. This is a shame (and is illegal in most states). In fact, of the few venomous snake bites that do happen, many occur when a person is killing the snake.

Snake myths and superstitions:

Myth: When a snake's head is severed it will not die until sunset. False. Like any other animal, a snake dies when its head is cut off. However, a dead snake's body may briefly squirm from muscular contractions.

Myth: Snakes hypnotize their prey. False. Snakes may quietly stare at their prey as they try to sneak up and catch them.

Myth: Milk snakes drink milk from cow's udders. False. Milk snakes are often found around barns looking for prey, but they are rodent feeders.

Myth: A snake uses its tongue and tail as a stinger. False. The touch of the tongue or tail is completely harmless.

Myth: Black snakes and copperheads can crossbreed, producing a venomous black snake. False. This is not biologically possible. These snakes may overwinter together in the same places, but they are not related and cannot produce viable offspring. Copperheads are live bearers and the black snake lays eggs.

Our region has many unique and beautiful animals that all play an important role in our ecosystem. Whether it be a rural or suburban environment, many animals have learned to adapt as best as they can to man's activities. Reptiles and amphibians are frequently encountered by homeowners and gardeners. Many people do not know much about them and need the help of a Master Naturalist to help them to identify, understand, and appreciate them. Your learning and eventual future advising of people will be exciting and appreciated!

Note

1 Ben C. Scheele, Frank Pasmans, Lee F. Skerratt, Lee Berger, An Martel, Wouter Beukema, Aldemar A. Acevedo, et al., "Amphibian Fungal Panzootic Causes Catastrophic and Ongoing Loss of Biodiversity," *Science* 363, no. 6434, March 29, 2019, http://science.sciencemag.org/content/363/6434/1459. See also Matthew Wright, "Fungal Disease Threatens Hundreds of Amphibian Species Worldwide," University of Maryland College of Computer, Mathematical, and Biological Sciences, March 28, 2018, https://biology.umd.edu/news/fungal-disease -threatens-hundreds-amphibian-species-worldwide.

CHAPTER 14

Mammals

LUKE MACAULAY

When people are asked to name their favorite animals, they usually start by saying "dogs" or "cats." The same is true when you ask about their favorite wildlife species. Almost invariably, they will name deer, chipmunks, bears, bobcats, fox (or maybe whales or dolphins!). What do all of these animals have in common?

They are all mammals.

For human beings, mammals have always been particularly engaging. Perhaps this is because of our affection for their fur, their eyes, or the ways they move. Or maybe it's because they are (in fact) our closest animal relatives. Mammals deliver their babies live, as we do, and their mothers feed their young with milk, signaling commonalities deep in our ancestral heritage as living creatures. Because of this connection, Master Naturalists are often interested in what they can do to help mammalian wildlife.

Maryland has approximately 64 extant terrestrial mammals, and 4 that once lived here but have since been extirpated. The exact number of species can vary due to factors like habitat changes, ongoing conservation efforts, and new discoveries. This chapter's list includes two large-bodied

TABLE 14.1. Mammal Species

CATEGORY	DETAILS	SPECIES COUNT
Rodents	7 squirrels and chipmunks, 5 mice, 4 voles, 1 lemming, 1 beaver, 2 jumping mice, 1 muskrat, 1 porcupine	22
Rabbits	3	3
Burrowing insectivores	8 shrews, 3 moles	11
Bats	10	10
Mustelids (weasels)	7	7
Hoofed mammals	1 native, 2 extirpated, 2 exotic	5
Canids (dogs)	4 (1 extirpated)	4
Felids (cats)	2 (1 extirpated)	2
Opossum	1	1
Skunks	2	2
Bear	1	1
Raccoon	1	1

exotic mammals—sika deer and wild horses—and mammals no longer found in the state (wolves, bison, elk) but excludes other exotic mammals associated with human habitation such as the black rat, Norway rat, and house mouse. Table 14.1 shows the number of mammal species in the 12 groups discussed in this chapter.

Evolutionary *phylogeny* is an organizational "tree" showing the evolutionary relationships among a group of organisms by illustrating when they diverged from a common ancestor. Phylogeny helps illustrate how these mammal species relate to each other, based on DNA data that is used to analyze similarities given mutation rates over time.

In the following sections we will break down the unique characteristics of these mammals by category and show how they are related to each other.

Rodents

With 22 species in Maryland, rodents constitute Maryland's most common mammal family and make the largest single contribution to mammal diversity in the state. Rodents are also the most diverse order in the mammalian class worldwide, accounting for about 40% of all mammal species.

The word *rodent* comes from the Latin *rodere*, meaning "to gnaw." Rodents share a common trait of continuously growing top and bottom incisor (front) teeth, which must constantly be ground down—or the teeth can grow so long they can prevent the animal from eating—or even puncture the rodent's own skull.

Although many species (typically exotic mice and rats) have a tendency to occupy (and sometimes damage) human homes, rodents also play a key role in our ecosystems, serving as a primary food source for various predators. Others serve vital roles in seed dispersal, forest regeneration, and as part of the broader food web.

SQUIRRELS AND CHIPMUNKS: SEVEN SPECIES

In Maryland, squirrels and chipmunks encompass seven distinct species, each with unique characteristics and behaviors. Their presence and behavior offer insight into the health and dynamics of local habitats.

The **eastern chipmunk (*Tamias striatus*)** is a small, ground-dwelling species known for its cheek pouches and striped fur. They are most often found west of the "fall line" in Maryland and inhabit wooded areas and parks, storing food in burrows for winter.

The **groundhog, or woodchuck (*Marmota monax*)**, is a larger rodent known for its burrowing habits and its capacity to damage homes, gardens, and crops. In North American folklore, they are perhaps most famous for their role in (allegedly) predicting the arrival of spring.

The **eastern gray squirrel (*Sciurus carolinensis*)** is a common sight in both urban and rural settings. These adaptable squirrels are known for their gray fur and bushy tails as they skillfully navigate tree canopies and ground areas.

The **eastern fox squirrel** (*Sciurus niger*) and its subspecies, the Delmarva fox squirrel (*Sciurus niger cinereus*), are larger than the gray squirrel. The Delmarva fox squirrel is notable for its conservation success story, having been brought back from the brink of extinction due to habitat loss. The US Fish and Wildlife Service listed this squirrel as endangered in 1967, then embarked on a relocation and habitat conservation effort that resulted in the establishment of 11 new populations on the Delmarva. The squirrel was delisted in 2015.

The **red squirrel** (*Tamiasciurus hudsonicus*) is smaller, with distinct reddish fur, and is found in the foothill and mountainous regions of western Maryland. They primarily inhabit coniferous forests and are known for their loud, chattering calls.

Lastly, the **southern flying squirrel** (*Glaucomys volans*) is a nocturnal species with a unique ability to glide between trees using a membrane stretching from its wrist to its ankle. These squirrels are smaller and have a softer, grayish-brown fur. Far more common than many people realize, flying squirrels are found in wooded suburban settings but rarely seen unless caught at bird feeders at night (they are attracted to peanuts and sunflower seeds).

"NEW WORLD" MICE: FIVE SPECIES

In Maryland, the natural history of mice includes several interesting species, each adapted to specific habitats and ecological roles.

The **marsh rice rat** (*Oryzomys palustris*) is a semi-aquatic species found in wetlands and marshes throughout Maryland. It is a strong swimmer, able to swim over 10 feet underwater, and it eats a diet that includes aquatic insects, seeds, and plants. This species plays a crucial role in wetland ecosystems, contributing to the nutrient cycle and serving as prey for larger predators, including red foxes.

Maryland is the northernmost extent of the range of the **eastern harvest mouse** (*Reithrodontomys humulis*), though this species is currently considered extirpated from the state. This small rodent prefers grassy or brushy habitats, often in lowland environments, and is notable for building intricate nests, often constructed aboveground in dense vegetation.

Because their diet primarily consists of seeds and insects, they play a significant role in both seed dispersal and pest control.

The **deer mouse (*Peromyscus maniculatus*)** is a versatile species that inhabits a variety of environments, from forests to grasslands. They are recognized for their agility and adaptability and their ability to survive in harsh conditions. Deer mice are also important in forest ecosystems for their role in seed and fungal spore dispersal.

The **white-footed deer mouse (*Peromyscus leucopus*)** closely resembles the deer mouse but is distinguished by its white feet and underbelly. This species is widespread and adaptable and often comes into contact with human environments. They are omnivorous, feeding on a variety of plant and animal matter, and play a key role in the food web as both predator and prey.

Lastly, the **Allegheny woodrat (*Neotoma magister*)** is a larger, nocturnal rodent known for its preference for rocky habitats such as cliffs and caves. They are unique for their habit of collecting and storing various objects like bottle caps, feathers, and bones in their dens. The Allegheny woodrat is considered a species of conservation concern due to habitat loss and population decline.

VOLES AND LEMMING: FIVE SPECIES

Maryland hosts four voles and one lemming species (the southern bog lemming is closely related to voles and earns its place with voles as a result):

- Meadow vole (*Microtus pennsylvanicus*): The meadow vole is commonly found in open fields, meadows, and grasslands and can be recognized by its dense, soft fur and small size. These voles are prolific breeders, contributing significantly to the diet of many local predators. They can also cause damage to lawns by creating narrow runways or paths through the grass and can girdle young trees in orchards, gardens, or landscaping.

- Pine/woodland vole (*Microtus pinetorum*): Compared to its relatives, the woodland vole has a more underground lifestyle in its native deciduous forests and orchards. They are smaller and have a shorter tail than the meadow vole. Their burrowing activities are significant for soil aeration but can sometimes conflict with human agricultural practices when it eats the bark at the base of young trees, girdling and killing them. Their diet consists mostly of roots, tubers, and some above-ground plant parts.

- Southern rock vole (*Microtus chrotorrhinus carolinensis*): A subspecies of the rock vole, this vole is found in rocky, mountainous areas. It is less common than other vole species in Maryland. Known from just three Maryland sites, this endangered mammal occurs in wet, northern hardwood-hemlock forests with moss-covered boulders.

- Southern red-backed vole (*Clethrionomys gapperi*): Easily identifiable by the reddish stripe down its back. Its range includes the northeast United States and Canada; in Maryland, it is found in the western piedmont region. These voles prefer forested areas, particularly with coniferous or mixed woodlands. They are mainly nocturnal and are known for their diet of seeds, fruits, fungi, and insects.

- Southern bog lemming (*Synaptomys cooperi*): As its name suggests, this lemming is found in wet, boggy areas. A small, stout-body species, this animal has a shorter tail than voles and small ears, and it is more closely related to voles than to their lemming cousins found in the arctic.[2] They are primarily nocturnal and are known for their solitary and secretive nature. Their burrowing and tunneling activities are also important for soil aeration and nutrient mixing. Their diet consists mainly of grasses, sedges, and other herbaceous plants. In winter, they may also consume bark and roots.

FIG. 14.1. American beaver. Photo by Mike Budd. Courtesy of the US Fish and Wildlife Service, https://digitalmedia.fws.gov/digital/collection/natdiglib/id/34075/rec/3.

AMERICAN BEAVER

The **American beaver (*Castor canadensis*, figure 14.1)** is North America's largest rodent and is well known for its ability to modify landscapes through dam building. Indeed, beavers are ecological engineers: by building dams across streams using tree branches, vegetation, rocks, and mud, they create wetlands, which provide habitats for a variety of other species. Unfortunately, these activities can also flood roads, agricultural lands, and other human-developed areas.

Beavers are well adapted (and notorious!) for cutting down trees and shrubs, both for food and to use for building materials. This behavior can impact forest composition and can sometimes conflict with human land use, particularly in managed woodlands and urban areas where their girdling can damage and even kill specimen trees that people enjoy. Beavers primarily eat the cambium, the soft tissue that grows just under the bark of a tree, as well as aquatic plants and roots. This diet changes seasonally and influences their impact on different plant species.

Despite their potential for causing damage in human communities, the wetlands that beavers create support a range of wildlife, including fish, waterfowl, and amphibians. The ponds created by their dams are akin to having giant sponges on the landscape, which can help in (1) filtering pollutants from water and (2) holding water that can help reduce the effects of droughts and floods.

For centuries, and across their wide North American range, beavers were over-trapped for their fur to supply European markets for hats and clothing. This led to significant regional and continental population declines. Conservation efforts have successfully helped their numbers rebound, but this has also led to conflicts in areas where human and beaver habitats overlap. Effective management often involves balancing the ecological benefits beavers provide with the challenges they can pose to human land use.

JUMPING MICE: TWO SPECIES

The **meadow jumping mouse (*Zapus hudsonius*)** and the **woodland jumping mouse (*Napaeozapus insignis*)** are notable for long tails and long hind legs, which enable them to make large leaps. As their name suggests, the meadow jumping mouse prefers moist meadows and grasslands near streams or wetlands, while the woodland jumping mouse prefers woodlands near water sources.

MUSKRAT

The **muskrat (*Ondatra zibethicus*)** is a medium-sized semi-aquatic rodent native to North America, including Maryland. The muskrat is more closely related to the southern bog lemming and voles than other rodents, although it is much larger, typically weighing from one and a half to four pounds. They have stocky bodies, short legs, and a vertically flattened, scaly tail that—along with their webbed feet—make them excellent swimmers. Lodges built out of vegetation and mud in bodies of water serve as protective homes and nesting sites.

FIG. 14.2. Porcupine. Photo by Cristina Stahl. Courtesy of the US Fish and Wildlife Service, https://digitalmedia.fws.gov/digital/collection/natdiglib/id/33510/rec/3.

PORCUPINE

Found primarily in western Maryland, the **porcupine (*Erethizon dorsatum*, figure 14.2)** is—after the beaver—Maryland's second-largest rodent. Its coat of sharp quills, which are actually modified hairs, serves as a defense mechanism against predators. Porcupines have a plump body, short legs, and a small head with a blunt nose.

Porcupines are found in a variety of habitats, including forests, deserts, and grasslands, but they prefer wooded areas. They are mostly nocturnal and spend much of their time in trees. They are good climbers and often climb trees looking for food; they are primarily herbivorous, feeding on leaves, herbs, twigs, bark, and green plants. In winter, when plant food is scarce, they can come into conflict with humans by chewing on wooden objects like tool handles, cabin siding, and tree bark.

Porcupines have a relatively low reproductive rate. Females typically give birth to a single offspring after a gestation period of about seven months. Porcupine populations were listed as state endangered until 1980, when they were downgraded to a species in need of conservation, but have since made a comeback in Maryland and were delisted entirely in 2014.

Rabbits

Maryland hosts three species of rabbits: the snowshoe hare, the eastern cottontail, and the Appalachian cottontail. Like rodents, with whom they were once classified, rabbits also have continuously growing teeth, but they have four front incisor teeth instead of two.

The two cottontail species are nearly indistinguishable in the field, with the Appalachian cottontail being slightly smaller, sometimes having a black spot between its ears, and lacking a white spot on its forehead that is often found on eastern cottontails. Genetic analysis can distinguish them, and so can looking at their skulls: Appalachian cottontails have an irregular line where the nasal bones attach to the skull. On eastern cottontails this line is smooth.

As their name suggests, Appalachian cottontails are often found in the mountainous regions of the state; they prefer persistent natural high elevation shrublands known as *heath balds* and high-elevation red spruce forests.[3]

Eastern cottontails are found throughout the eastern United States and prefer early successional habitat characterized by open fields combined with nearby brushy and shrubby escape cover.

The snowshoe hare was once found in the western, mountainous areas of the state but is presumed to be extirpated from Maryland. Snowshoe hares prefer *young forests* with *dense understories*, so as forest harvests have declined and forests have matured in Maryland, many species that rely on young forest—and thick shrubby regenerating forest habitat—have experienced significant declines.

Insectivorous Mammals:
Mole, Shrews, and Bats

Maryland hosts three major categories of insectivorous mammals: shrews (8 species), moles (3 species), and bats (10 species).

SHREWS

Shrews and moles are closely related, residing in the order *Eulipotyphla*, which means "truly fat and blind." Shrews have smaller bodies than moles, with visible—though tiny—eyes (a mole's eyes are completely covered by fur), and front feet that are smaller, not as wide, and specialized for digging. Many shrews use tunnels dug by other animals instead of digging their own, and they spend a lot of time under leaf litter or under rocks. Maryland has eight species of shrew. The following five species are most commonly found in the more mountainous west:

- Masked shrew, *Sorex cinereus*

- Southern pygmy shrew, *Sorex hoyi*

- Smoky shrew, *Sorex fumeus*

- Long-tailed shrew, *Sorex dispar*

- Southern water shrew, *Sorex palustri*

Other shrews are found across the entire state:

- Southeastern shrew, *Sorex longirostris*

- Northern short-tailed shrew, *Blarina brevicauda*

- Least shrew, *Cryptotis parva*

The long-tailed shrew is limited to consistently moist (or *mesic*) forests containing large areas of loose rock outcroppings (talus), meaning it is found in relatively limited isolated locations.

Shrews have an incredibly fast metabolism—far higher than you might expect, given their body size. The southern pygmy shrew's heart beats *1,200 beats per minute* (20 times per second)! To fuel this rapid metabolism, they must consume up to three times their body weight each day and can only survive a few hours without eating. Likely connected to this fast metabolism, they have very short lifespans, typically living only one to two years.

The southern water shrew is listed as endangered in Maryland. Feeding primarily on aquatic insect larvae (e.g., mayflies, caddisflies, stoneflies), this small (~15 cm), uniquely adapted, semi-aquatic mammal is believed to be restricted to only six pristine, high-elevation headwater streams in the central and southern Appalachians in Garrett County, Maryland.

MOLES

Maryland has three species of moles. The hairy-tailed mole is found in a diversity of habitats in the mountainous parts of western Maryland. The eastern mole can be found throughout Maryland and in many environments where soils are not too wet or rocky. The star-nosed mole is also found throughout much of the state and prefers wet areas, often eating aquatic insects and their larvae. The star-nosed mole is immediately identifiable by its eponymous proboscis, which is surrounded by 22 fleshy rays with 100,000 nerve endings—an extremely efficient nervous system that makes it the most touch-sensitive animal in the world.[4]

Because moles are known to dig underground burrows, some people mistakenly believe them to be the culprits that eat garden plant roots. Although they do indeed create "molehills" and raised tunnels in lawns and gardens, they are actually insectivores—they focus on eating worms, insects, and grubs—and do not cause major problems to plants (and actually benefit gardens by aerating soil). The offending root-eating mammal is usually a vole. An easy way to remember the difference: *Moles are Meat-eaters and Voles are Vegetarians.*

THE OTHER INSECTIVORES: BATS

All 10 bat species in Maryland fall into the family *vesper* (meaning "evening"). Four are in the genus *myotis* (meaning "mouse-eared"). They are all insectivores and famous for their ability to consume huge quantities of insects using ultrasonic sounds (or *echolocation*) to find and chase down insects.

Although it is true that bats can contract (and spread) rabies, in Maryland they are rarely a vector for rabies infections; the culprit in Maryland is usually raccoons. However, given the potentially deadly consequences of rabies, if you are ever exposed to a bat that could have scratched your skin, you should immediately seek medical attention.

Of the 10 species of bat found in Maryland, 7 (at the top of the following list) hibernate in large groupings in caves or mines during winter. (The red bat, hoary bat, and evening bats are all believed to hibernate in trees— in hollows, under bark, or in coniferous trees.)

1.   Little brown bat, *Myotis lucifugus*

2.   Big brown bat, *Eptesicus fuscus*

3.   Indiana bat, *Myotis sodalis*

4.   Northern long-eared bat, *Myotis septentrionalis*

5.   Tricolored bat, *Perimyotis subflavus*

6.   Eastern small-footed bat, *Myotis leibii*

7.   Silver-haired bat, *Lasionycteris noctivagans*

8.   Red bat, *Lasiurus borealis*

9.   Hoary bat, *Lasiurus cinereus*

10.   Evening bat, *Nycticeius humeralis*

The six species appearing at the top of this list hibernate in larger colonies and have been confirmed to contract *white-nose syndrome*, caused by a fungus originally from Europe that forms on their noses during winter hibernation. While it doesn't kill the bats immediately, the disease leads to increased activity (and a subsequent loss of energy and fat reserves) during hibernation that eventually leads starving bats to seek food before winter is over (when food insects are rare). Tricolored bats, little brown bats, and northern long-eared bats have been particularly affected by white-nose syndrome. It is estimated that over 5 million bats have died as a result of this disease. There has been no diagnostic sign of white-nose syndrome on red bats, silver-haired bats, hoary bats, or evening bats.[5]

Although the disease has spread to bat colonies across the United States, people can help by improving habitat for bats by (for example) planting native wildflower species to attract insects and by leaving dead trees standing, limiting the use of insecticides, and—for those who might visit caves where bats congregate—decontaminating gear that could be contaminated from fungal spores. Properly constructing and maintaining bat houses can be a somewhat complicated process, so be sure to conduct best practices for design and placement.[6]

Mustelids

Mustelids (belonging to the family *Mustelidae*) are a diverse group of carnivorous mammals known for their slender bodies, short legs, and musky odor produced by their anal glands. They are set apart from other mammals by their agility, strength, and adaptability, which make them proficient hunters and survivors in a variety of environments. Most mustelid reproduction involves *embryonic diapause*, during which a fertilized embryo does not implant in the uterus wall immediately but can remain dormant for up to a year. This allows the young to be born under favorable environmental conditions.

In Maryland, the mustelid family is represented by seven fascinating species, each with its own niche and adaptations: the **American marten (*Martes americana*)** is tree dwelling, with a preference for mature, dense forests. It primarily consumes voles but will also sometimes take squirrels and other larger prey. Martens are considered extirpated from Maryland, though Pennsylvania is planning a reintroduction program in coming years. The **fisher (*Pekania pennanti*)** is another forest-dweller. Extirpated (mostly due to habitat loss) from the eastern United States in the early 1900s, the fisher was successfully reintroduced to nearby West Virginia in 1969 and has recovered throughout most of its historic range in western Maryland. It is known for its versatility and ability to hunt prey ranging from small mammals to porcupines.

The **American ermine** (*Mustela richardsonii*), also known as the short-tailed weasel, is a small, fierce predator that adapts to the winter by changing its fur to a camouflage white. They live in a diversity of habitats, including woodlands near rivers, marshes, shrubby fencerows, and open areas adjacent to forests or shrub borders.

The **least weasel** (*Mustela nivalis*), the smallest of the bunch, is a tiny but voracious predator that lives in a wide variety of habitats but prefers forests and woodlands with rocky slopes. It is secretive, often going unseen, hunting rodents with remarkable efficiency.

The **long-tailed weasel** (*Neogale frenata*), with its distinctive long tail and energetic demeanor, is a common but secretive resident of Maryland's diverse habitats. Often found near water is the **mink** (*Neovison vison*), equally at home in the water and on land, typically feeding on a diet rich in fish and crustaceans. The **northern river otter** (*Lontra canadensis*) is a charismatic mustelid and a symbol of healthy waterways that is often seen playing and sliding in rivers and streams. River otters are found throughout creeks, rivers, and other water bodies in Maryland, and their populations have been increasing in recent decades thanks to relocation efforts. They are now considered nearly fully recovered to their former historical range in Maryland and the United States.

Hoofed Mammals

Maryland hosts native white-tailed deer and exotic sika deer and wild horses. The state once had American bison and elk, but both were extirpated, in 1775 and 1874, respectively. Wild horses are still found in a few locations, most famously in Chincoteague National Wildlife Refuge.

White-tailed deer (*Odocoileus virginianus*), an integral part of Maryland's natural history, are found throughout the state. Their range extends from the dense forests of western Maryland to the agricultural landscapes of the central region and the coastal marshes of the Eastern Shore. Before European colonization, these animals were abundant in the region but were believed to be found in lower densities than today due to the presence of higher predator populations and subsistence hunting by Indigenous people. The arrival of European settlers in the 17th century led to increased hunting pressure, which culminated with *market hunting* that nearly extirpated white-tailed deer from the state by 1900. Hunting regulations that stopped the harvesting of female deer allowed for population recovery, but by 1980 populations had increased to such levels that deer began to be considered a nuisance, impacting agricultural crops and landscaping and causing a big jump in deer-vehicle collisions.

Recognizing the importance of maintaining a balance between deer populations, human interests, and ecosystem health, the Maryland Department of Natural Resources (DNR) has implemented a comprehensive deer management strategy. Central to this approach is *regulated hunting*, which serves as the primary tool for controlling deer numbers. Hunting seasons are carefully scheduled, and *bag limits* are set based on ongoing population assessments and ecological research. This not only helps prevent overpopulation (and associated issues such as habitat degradation and increased disease transmission) but also helps mitigate vehicular collisions and other human-deer conflicts.[7]

Additionally, the DNR issues *crop damage permits*, allowing landowners and farmers to remove deer outside of regular hunting seasons, providing immediate relief to those whose livelihoods may be impacted by deer overabundance. The DNR also promotes and supports the use of *nonlethal management techniques*, such as fencing and repellents, particularly in suburban and urban areas where hunting may not be feasible. Administering contraceptives to control populations has proven ineffective, due to the labor cost of applying yearly doses to every member of the breeding population. Darting deer—typically from just 20 yards away—is also a challenge and gets harder as deer learn to avoid shooters.

SIKA DEER

Sika deer, native to East Asia, were introduced to Maryland in the early 1900s by Clement Henry on James Island. Their current range in Maryland has expanded and is now primarily concentrated on the Eastern Shore, with the largest populations found in Dorchester County. The Blackwater

National Wildlife Refuge is a well-known habitat for these deer. Sika deer are smaller than native white-tailed deer and have a different set of behaviors and habitat preferences. They tend to favor marshy areas and dense underbrush for cover. Management strategies, including regulated hunting seasons, are in place to control their population and limit their spread.

ELK

Extirpated in 1874, elk (*cervus canadensis*), also known by their Indigenous name *wapiti*, were once widespread across much of North America, including the East Coast. Elk are a *keystone species*, meaning they play a crucial role in shaping the ecosystems they inhabit. As herbivores, they influence plant communities through grazing and browsing, which can affect the structure and composition of forests and grasslands. Along with American bison, these large herbivores likely influenced the historical landscape in Maryland, maintaining open meadows and grassland habitats as they have done in the American West.

Although there are no current plans for reintroducing elk to Maryland—in part due to their size and the lack of expansive rural areas to support a population—they have been successfully reintroduced to parts of Kentucky, Pennsylvania, West Virginia, Virginia, North Carolina, and Tennessee.

AMERICAN BISON

The largest terrestrial animal in North America and an iconic symbol of the American West, American bison (*bison bison*) once roamed widely in vast herds across the continent's grassland. Before European colonization, bison also roamed Maryland's diverse landscapes. Just as they did in other parts of the bison's range, European settlers pushed bison to extinction through unregulated hunting and habitat loss.

WILD HORSES

The wild horses of Maryland, specifically the Assateague horses, present an exotic twist to the state's mammalian narrative. Although they evolved elsewhere—their closest relatives are European horses—they have become a symbol of wildness and freedom on Maryland's coastal landscape. As-sateague horses are more accurately described as *feral*, having reverted to a wild state from domestication. Their ancestry can be traced back to do-mesticated horses, believed to have been introduced to the area in the 17th century, possibly by shipwreck or by early settlers seeking to avoid livestock taxes.

Assateague's horses have become integral to the dynamics of the barrier island ecosystem. Although they consume large quantities of grasses, which can lead to overgrazing and can alter the natural plant communities, their grazing has also served as a surrogate for extirpated historical grazers, including bison and elk. This grazing helps maintain habitat for other na-tive species by opening areas within marshes and can prevent the over-growth of certain plant species. Although their population has been limited to a target of 80–100 animals, their impact remains a subject of ongoing study and some contention.

Canids

Coyote (*Canis latrans*): Now common across North America, coyotes are versatile and adaptable animals with a varied diet, including rodents, rab-bits, fruits, and—occasionally—livestock or pets. Coyotes are known for their distinct howl, or yipping sound, which is used for communication among individuals and packs. Their presence in Maryland reflects their ex-pansion eastward over the past century.

Gray wolf (*Canis lupus*): Once present in Maryland, gray wolves have long since been extirpated in the state due to habitat loss, hunting, and trapping. They are the largest member of the Canidae family, primarily preying on large mammals like deer and elk. Gray wolves are social ani-mals, living and hunting in packs. Conservation efforts have been focused on their reintroduction and recovery in several parts of western North America. The nearest populations to Maryland are found in northern Michigan.

Red fox (*Vulpes vulpes*): The red fox has a reddish coat, bushy tail, and pointed ears. It is highly adaptable and can thrive in a variety of habitats, including forests, grasslands, and urban areas. Red foxes have a diverse diet, feeding on rodents, rabbits, birds, insects, and fruit. Solitary hunters known for their cunning and intelligence, they are distinguished from gray foxes by their red coat and a white-tipped tail. In late winter and early spring, female red foxes (and occasionally males) emit an unsettling screeching call associated with mating behaviors called the *vixen scream*.

Gray fox (*Urocyon cinereoargenteus*): Unique among canids, gray foxes can climb trees, a skill they use for foraging and escaping predators. They inhabit a variety of habitats but prefer areas with dense brush or woods. Gray foxes have a varied diet, similar to that of red foxes, including small mammals, birds, and plants. The gray fox is distinguishable from the red fox by its gray fur and the black stripe running down its tail.

Felids

The **mountain lion (*Puma concolor*)** and **bobcat (*Lynx rufus*)** are the two wild cat species found in Maryland. The mountain lion, also known as the cougar, puma, or panther, was once widespread across the eastern United States, including Maryland, but has been declared extirpated from many eastern states, including Maryland, due to eradication efforts in the 1800s. Sightings of these majestic animals are rare and often unconfirmed, leading to debate about their current presence in the region. They are capable of long dispersals from populations farther west. One mountain lion is believed to have traveled up to 1,500 miles from its home in South Dakota, when it was struck by a vehicle in Connecticut. It is not believed that breeding populations exist east of the Mississippi River.

The bobcat is significantly smaller than the mountain lion and is identifiable by its short "bobbed" tail, tufted ears, and spotted patterning. These cats are adaptable and can be found in a variety of habitats, including forests, swamps, and even suburban areas. Bobcats are solitary and primarily nocturnal, with a diet consisting mainly of rabbits, rodents, birds, and (occasionally) deer. They continue to have a strong presence in western Maryland, are commonly sighted in the Piedmont region, and are occasionally seen on the Eastern Shore.

Virginia Opossum

The Virginia opossum, or possum, is North America's only extant marsupial, making it the most genetically distinct mammal in Maryland, having broken off from the other mammals in the state nearly 160 million years ago. It is so unique that the next major evolutionary break occurred nearly 80 million years later, when rodents and rabbits broke away from all the other remaining mammals in Maryland about 83 million years ago.

Possums are nocturnal, omnivorous, and habitat generalists and can be found in most parts of the state. Habitat improvements that could support possums include creating brush piles, leaving old dead trees (snags) and downed logs for denning sites, and supporting native food plants such as crabapples, persimmons, and pawpaws.

Skunks

The **striped skunk (*Mephitis mephitis*)** is the most common skunk found in Maryland, and there is some question about whether **spotted skunks (*Spilogale putorius*)** are still found in the state. The striped skunk is easily recognized by its black fur and the distinctive white stripes running from its head down its back. They are omnivorous, feeding on insects, small mammals, fruits, and plants. Skunks are known for their *defensive spray*, a potent-smelling mist used to deter predators. Like deer, they prefer habitats like open fields and mixed woodlands and for this reason are often found near human habitation.

The eastern spotted skunk is distinguishable by its black fur with white spots and short, broken stripes. This species is more secretive and less likely to be encountered by humans. They inhabit wooded areas, farmlands, and brushy fields. Similar to the striped skunk, they are omnivorous and have the ability to spray a foul-smelling liquid for defense. Unlike its cousin, the eastern spotted skunk is known for its distinctive handstand behavior performed before spraying.

Black Bear

The black bear is Maryland's largest land mammal. They are primarily found in western Maryland, particularly in the Appalachian and Allegheny Mountains. They have black fur, though some may be brown or cinnamon-colored, and a relatively short tail. Black bears are omnivorous, eating a varied diet that includes berries, nuts, insects, small mammals, and carrion. They are also known to forage in human-populated areas, occasionally leading to conflicts. While generally shy and avoidant of humans, they can become a nuisance if they associate human areas with food. Black bear populations have increased in Maryland over the last several decades, with sufficient recovery by 2004 to allow for a short hunting season to help manage the population and reduce conflicts with humans.

Given their range of movement, black bear populations are difficult to estimate, but the state's DNR estimates that the population is healthy and consists of over 2,000 bears in the state. The hunting season has resulted in an average annual harvest of 85 black bears; an additional 60 bears are killed annually in vehicle collisions. Black bears have continued to move eastward to more populated areas of Maryland, with regular sightings and complaints occurring in central and southern Maryland, suggesting the population is continuing to grow and expand.

Raccoon

Raccoons are nocturnal and highly adaptable, and they can live in a variety of habitats, including forests, marshes, and urban areas. They are easily recognized by their black facial mask and ringed tail. Raccoons are omnivorous and opportunistic feeders, known for their dexterous front paws and intelligence. Their paws are highly sensitive—matching the sensitivity of human hands—which (among other talents) helps them get into feed and trash containers, including in urban and suburban areas. They are *nest predators*, often consuming the eggs of ground-nesting birds, snapping turtles, and other animals.

Habitat Management for Mammals

There are four major land covers that make up the majority of Maryland: agricultural areas, including crops, hayfields, and pastures; forested areas; wetlands; and suburban/urban areas (micro-habitats such as rocky outcrops also host important habitat for certain mammals). Here are some general habitat management guidelines for each of these land covers. They are considered best practices for supporting wildlife, including mammals.

Since species have divergent habitat needs, the same location cannot always serve as habitat for all species. Consideration should thus be given to the status of the surrounding landscape, what kind of habitat one would like to provide, and for which species. In some cases, an isolated patch of habitat may not attract a specific species of interest due to *dispersal constraints* from existing populations, but even small patches can serve as important *refugia* (places in generally degraded landscapes where animals can survive) and corridors for wildlife expansion in the future.

As it is for birds and insect populations, a diversity of native vegetation is generally a beneficial component to managing for mammalian wildlife. Improved varieties of certain legumes, while not native, can also provide forage for deer and (because they can fix atmospheric nitrogen into the ground) enhance soil fertility.

AGRICULTURAL AREAS (ROW CROPS, HAYFIELDS, PASTURES)

Row Crops

Planting buffer zones of native vegetation around crop fields can provide shelter and foraging areas for mammals and other wildlife. Incorporating brushy, native hedgerows between agricultural fields can create habitat and travel corridors for wildlife. Land managers can use a variety of techniques (disking, prescribed fire, herbicides, and mowing) to prevent woody encroachment in buffers and to enhance plant diversity.

Disking: Wintertime disking of field borders can help promote broadleaf plants and wildflowers (called *forbs*). Disking can be conducted on a rotation to maintain vegetation cover as a refugia for wildlife. For example, one can disk one-third of an area each year (or every other year) between December and February to set back vegetation and increase forb cover and landscape biodiversity.

Herbicides: Selective and targeted herbicide application can be used to kill woody species encroaching on herbaceous buffers. Always carefully read the label of any herbicides and follow local regulations for human and ecological safety and to avoid any unintended killing of nearby plants. In certain weather conditions, some herbicides can *volatilize* into the air and accumulate on vegetation, causing the unintentional killing of plants. Some herbicides work primarily through *absorption* through plant leaves with minimal entry through roots, while others will work on leaves and can enter into the soil and kill nearby plants through their roots. The *half-life* of chemical compounds is an important consideration when using herbicides to better understand the length of time the herbicide will remain in the environment.

Mowing: Mowing tends to promote exotic grass cover that is less beneficial for many wildlife species. Mow sparingly, mostly to create trails and a diverse mosaic of vegetation height. For example, create some figure eight patterns in a field while leaving other areas of thick cover where deer, rabbits, voles, foxes, and myriad other species (especially insects and birds) can forage and hide. Remember that mowing between April 15 and July 15 will destroy the nests of several bird species. Late summer mowing also reduces the thermal cover needed by wildlife in wintertime, so to enhance wildlife habitat, leave vegetation to overwinter and try to conduct the bulk of mowing between March 15 and April 15.

Hayfields

Mowing hayfields in spring and early summer can be detrimental to ground-nesting bird populations. If economic return is not a priority of hayfield management, ground-nesting birds will benefit by delaying mowing as late as possible, to July 1 (or even July 15) to reduce nest destruction. Leave uncut grass buffers around field edges, and consider leaving one or more uncut strips in a field to provide some cover, habitat, and wildlife corridors after cutting.

Pastures

Again, if economic return is not a priority of a pasture and grazing operation, use stocking rates that maintain a patchy vegetative cover, including
areas of shrubby cover and grass height of 12 inches or more. Introduce
hedgerows, and allow shrubby islands to grow to enhance habitat diversity.

FORESTED AREAS

Maintaining diversity: Natural disturbance through fire and herbivory by
large mammals historically created a diverse forest structure in Maryland,
with a wide array of age classes that many wildlife species needed to survive.
Today, with most large mammals extirpated, young regenerating forests in
particular make up a very small proportion of Maryland's landscape, and
the populations of many wildlife species reliant on this habitat have declined. Mammals tend to be more resilient to a diversity of habitats, but
these impacts are most apparent in the state's bird populations.

Work with a licensed forester to create or maintain a mix of tree ages
and species to support different mammalian needs. In many parts of
Maryland, historical timber harvests have led to many tree stands of similar age class—between 60–100 years of age. Diversifying forest age and
structure can create more niches for a variety of wildlife in our forests.

Forest harvests and thinning: If your property is larger than five
acres, the Maryland Forest Service will (for a nominal fee) help you develop
a forest management plan both to create enhanced forest diversity and reduce your property taxes. Properly planned forest harvests, thinning, and
even girdling trees can enhance understory growth, as can intentionally
leaving fallen logs and standing dead trees (or *snags*) that provide important dens, cover, and forage for insects, birds, and mammals alike.

Edge feathering: Habitat diversity can be achieved or enhanced by
creating "feathered" edges, where forests meet agricultural fields or other
land covers. This can involve cutting trees (or killing less-preferred trees)
to allow increased light to reach the forest floor and a diversity of understory species, wildflowers, and other plants to be able to grow.

Invasive species: Invasive species can fundamentally change the vegetation composition in an area and can occur anywhere on the landscape—including unmanaged forests, pollinator gardens, field buffers, and harvested or thinned forests. The impact and presence of invasive species is highly site dependent, and whether considering a management action or planning to leave an area unmanaged, consideration should be given to how to manage invasive species and maintain a healthy native ecosystem.

WETLANDS

Water quality: Always test your soil before adding fertilizers, and read and follow labels on pesticides and herbicides so that you apply the proper amount of chemicals in the proper time. In general, avoid applying chemicals when rain is in the forecast to reduce runoff into nearby water bodies, which can significantly impact water quality and be especially harmful to aquatic life.

Buffer zones: Use native vegetation to generate buffer zones around wetlands. This will help create filters to reduce the runoff impacts from nearby land disturbance.

SUBURBAN AND URBAN AREAS

Green spaces: Develop and maintain green spaces, parks, and urban forests that can serve as mini-refuges for wildlife. Whenever possible, expand areas planted with native vegetation.

Wildlife-friendly gardening: Use native plants in home landscaping to provide natural food sources and habitat. Follow the label guidelines on pesticides and herbicides used to minimize impacts on non-target species. Leave areas unmowed to grow into native wildflower meadows for pollinators and refugia for small mammals.

Maryland's diverse landscapes, stretching from the Atlantic coast to the Appalachian Mountains, create a unique sanctuary for a remarkable array of mammal species. As Master Naturalists, we are in a singular position to share our knowledge and enhance stewardship of the natural world, ensuring the continued existence of mammals and biodiversity in Maryland's dynamic environment. Whether wandering along the Chesapeake Bay, traversing the Appalachian foothills, or observing urban wildlife, we are invited to marvel at the diversity and resilience of Maryland's mammalian inhabitants. Through enhanced awareness and conservation efforts, we can contribute to sustaining the rich tapestry of life that these creatures represent in our natural heritage.

Notes

1 Nathan S. Upham, Jacob A. Esselstyn, and Walter Jetz, "Inferring the Mammal Tree: Species-Level Sets of Phylogenies for Questions in Ecology, Evolution, and Conservation," *PLOS Biology* 17, no. 12 (2019), https://doi.org/10.1371/journal.pbio.3000494.

2 D. Andrew Saunders, *Adirondack Mammals* (Syracuse: State University of New York, College of Environmental Science and Forestry, 1988).

3 Joseph J. Apodaca, Corrine A. Diggins, and Liesl Erb, *Distribution, Habitat Preferences, and Landscape Genetics of Appalachian Cottontail (Sylvilagus obscurus) in Western North Carolina* (Raleigh: North Carolina Wildlife Resources Commission, 2020).

4 Kenneth C. Catania and Fiona E. Remple, "Tactile Foveation in the Star-Nosed Mole," *Brain, Behavior, and Evolution* 63, no. 1 (December 2003): 1–12, https://doi.org/10.1159/000073755.

5 Elyse C. Mallinger, Katy R. Goodwin, Alan Kirschbaum, Yunyi Shen, Erin H. Gillam, and Erik R. Olson, "Species-Specific Responses to White-Nose Syndrome in the Great Lakes Region," *Ecology and Evolution* 13, no. 7 (July 2023): e10267, https://doi.org/10.1002/ece3.10267.

6 Reed D. Crawford and Joy M. O'Keefe, "Avoiding a Conservation Pitfall: Considering the Risks of Unsuitably Hot Bat Boxes," *Conservation Science and Practice* 3, no. 6 (2021), https://doi.org/10.1111/csp2.412.

7 *Maryland White-Tailed Deer Management Plan 2020–2034* (Annapolis: Maryland Department of Natural Resources, 2020).

CHAPTER 15

Citizen Science

BOB HIRSHON

For dozens of avid native bee fans, Thursdays and Saturdays are Bee Lab Days. Each week, up to 50 volunteers arrive at the United States Geological Survey Interagency Bee Lab in Laurel to help answer a longstanding science question: what sorts of bees visit what sorts of flowers?

Bee Lab director and USGS entomologist Sam Droege (see figure 15.1) explains that while some bee species happily visit and pollinate multiple flower species, a significant percentage are *specialists*. In fact, as Doug Tallamy explains in his chapter on insects, many flowers and bees have co-evolved to rely on one another for their survival. Considering there are about 400 species of bees native to Maryland (and over 2,500 species of native flowering plants), figuring out who's pollinating whom is a gargantuan task, beyond the abilities of even the dedicated graduate students at Droege's lab.

Luckily, a small army of *citizen scientists* are willing and able to help. *Citizen science* is an evolving term referring to the collection and analysis of natural history data by members of the general public, often in partnership with professional scientists.[1]

FIG. 15.1. Wildlife biologist Sam Droege and a volunteer examine a bee at the USGS Native Bee Inventory and Monitoring Lab. Photo by Bob Hirshon.

Bee Lab volunteers not only tend and monitor beds of native flowers and report on the bees that visit them, but also contribute seeds that they collect, help prepare them for sowing, plant them, and transfer seedlings to pots and pots to garden plots. In the process, they learn about Maryland ecology, pollinator interactions, native plant horticulture, and of course, bee identification and natural history. In sum, they both contribute to and benefit from ongoing research of vital importance to Maryland environmental science and stewardship.

Statewide, there are hundreds of citizen science projects, some as involved and consuming as the Bee Lab research. Others require intense effort limited to a short time period, like the Salamander Crossing brigades, that count and chaperone tiger salamanders and other amphibians across busy roads during the first warm rains of spring. Still others require nothing more than snapping occasional photos of organisms with a mobile phone and sharing them through an app. There are even projects consisting of data-gathering apps that reside on mobile phones, working automatically, without any attention from the volunteer at all, including CrowdMag, which maps terrestrial magnetism. Like the Bee Lab volunteers, Maryland Master Naturalists can both contribute to and benefit from these projects, which span a wide range of disciplines.

Meteorology and Phenology

In the United States, one of the earliest and most ambitious attempts at what we would now call citizen science was undertaken by Thomas Jefferson. While aspects of his biography are clearly problematic (see McKay Jenkins's chapter on environmental justice for more on this), Jefferson was an avid naturalist and made significant contributions to natural history, especially in the mid-Atlantic region. At his Monticello estate, he compiled detailed meteorological records, including daily measurements of lowest and highest temperature, cloud cover, and precipitation, along with what would now be termed *phenology observations*: the cycles of natural phenomena, such as precise dates for the first appearance of the migratory birds; the leafing and flowering of various trees; and the first calls of common birds, insects, and amphibians.

But observations from a single location were of little use when it came to predicting weather, or looking for meaningful trends in flora and fauna. So, Jefferson devised a plan to supply a thermometer to one weather watcher in each county in Virginia, who would take and report daily readings of temperature highs and lows, along with wind speed and direction. Once implemented successfully, Jefferson hoped to expand beyond Virginia, to create a nationwide network of dedicated weather watchers.

While the Revolutionary War interrupted Jefferson's efforts, his dream is now a reality in the form of citizen science meteorological projects that provide high-resolution weather data from tens of thousands of locations. The oldest of these projects is the National Weather Service (NWS) Cooperative Program, which started in 1890 and now has more than 8,700 volunteers reporting on meteorological observations from farms, urban and suburban areas, national parks, seashores, and mountains.

The NWS SkyWarn Storm Spotter program launched in the 1970s to get rapid reports of severe local thunderstorms. Today there are over 350,000 participants, whose timely observations, combined with doppler radar and satellite data, help NWS meteorologists track fast-moving storm cells and predict tornado activity.

Volunteers can download the free app for the NWS mPING (Meteorological Phenomena Identification Near the Ground) to help report local weather conditions using drop-down menus. Volunteers select rain, snow, hail, wind, or various other conditions, choose the intensity of the event on another menu, and submit the report. Observations can be made any time, and as frequently as one per minute. The reports provide NWS with extremely fine resolution data not available from their other programs.

Sponsored by the National Oceanic and Atmospheric Administration (NOAA) and the National Science Foundation, the CoCoRaHS (Community Collaborative Rain, Hail, and Snow Network) program consists of people coast to coast providing regular observations of precipitation that they collect with rain gauges and hail pads. The network consists of about 8,000 observing stations run by individuals and school classrooms. The data is mapped daily on the CoCoRaHS website for use by the NWS, meteorologists, hydrologists, emergency managers, ranchers, farmers, insurance adjusters, and many others.

A few other comprehensive phenology data collection projects:

- *National Phenology Network/Nature's Notebook:* A national program to collect phenology information like flowering and leafing-out dates for plants and first and last sightings for migratory birds and other animals. For Maryland, there are currently seven active data gathering campaigns. The site has protocols for reporting on 600 Maryland plant species and 234 Maryland animal species. Campaigns include the Redbud Phenology Project, monitoring eastern redbud flowering and leaf out; Nectar Connectors, looking at pollinator activity; and Pest Patrol, tracking 13 species of pests that damage trees,

10 of which are currently active in Maryland. You can also start your own local phenology project through Nature's Notebook. In Maryland, there are currently six Active Local Phenology groups listed.

- *Budburst:* Currently running "Milkweeds and Monarchs" project to determine whether monarch butterflies prefer laying eggs on flowering or not-yet-flowering milkweed stems.

- *Journey North:* Active since before the internet, it currently tracks the migrations of monarch butterflies and seven types of birds, all native to Maryland, and also conducts other topical phenological research programs.

Monitoring the Environment

Just south of Blackwater National Wildlife Refuge (and north of Toddville), there's a marshy forest graveyard where the bleached trunks of once-towering pine trees dot the bleak landscape. It's a *ghost forest,* defined as a once-forested coastal area where saltwater intrusion caused by sea level rise has killed the trees and allowed the spread of marsh grasses through the understory. In the citizen science project Ghosts of the Coast, volunteers report sightings of ghost forests and document them with photographs using a mobile phone app.

This is just one of dozens of citizen science projects dedicated to monitoring the environment—an occupation especially relevant to Maryland, home of environmental science champion Rachel Carson, and a landscape acutely at risk from the effects of climate change, intensive agriculture, and the legacy of coal mining.

Dozens of local and regional groups collect stream and river water quality data throughout Maryland, including Trout Unlimited, the Izaak Walton League, the Riverkeepers, Bluewater Baltimore, Nature Forward, Maryland Department of the Environment, and many more. This local data collection helps launch and monitor cleanup efforts and contributes to governmental zoning and planning. The Chesapeake Monitoring Cooperative, a partnership between six leading organizations, supports 135 such local and regional efforts and presents its data in the Chesapeake Data Explorer, an interactive, online map showing a variety of water quality parameters. By combining and visualizing locally collected data from seven Chesapeake watersheds, the tool provides a big picture of water quality trends.

Debris Tracker, developed at the University of Georgia with support from NOAA, also pools data from many local organizations, including many in Maryland, around the occurrence of plastic litter. The project relies on the Debris Tracker mobile app, with which volunteers report plastic litter wherever they see it.

Another debris project, the Visible Trash Survey from the Alice Ferguson Foundation, asks volunteers to monitor a 200-foot-long track within the Potomac River watershed and conduct a detailed survey tallying up trash in 50 categories. Another, more specific litter tracking project is the Maryland Coastal Bay Nurdle Patrol. *Nurdles* are tiny plastic pellets used as the raw material for making virtually all plastic products. They're transported to factories via rail and truck, and regularly spill into the environment on the way, ending up in streams and on beaches. Nurdle Patrol members conduct 10-minute surveys on beaches and report on nurdle density. This work helps researchers sleuth out major sources of environmental nurdles and try to reduce them.

There are, of course, other environmental threats beyond the degradation of air, water, and soil. Globe at Night is a citizen science program in which participants report on light pollution. Artificial light not only makes it difficult to view the night sky, it also disrupts the sleep and migration patterns of animals, birds, and insects; interferes with mating and other animal behaviors; and can adversely affect plant phenology. Volunteers choose a location and determine how many stars are visible in a target constellation. For a more accurate assessment, they can also use a device called a Sky Quality Meter, which the project is making available for loan from local libraries.

Documenting Biodiversity

Believe it or not, Silver Spring, Maryland, is one of the world's hot spots for insect diversity. After all, nearly 5,000 insect species have been collected and described there, including many Maryland state records, some species never before seen in the United States, and at least one species—a small dagger fly—entirely new to science.

This is not the result of an extraordinary ecosystem; it's the work of an extraordinary entomologist named Gary Hevel, who began documenting the insect diversity of his backyard just over 20 years ago. "I had had the opportunity to collect insects around the world for the Smithsonian," explains Hevel. "And then I was influenced by a book, *A Lot of Insects* by Frank Lutz, an entomologist from the American Museum of Natural History. After four years, he had found 1,406 insect species in his backyard. So I accepted the challenge, and easily bested that number in four years." Hevel says that over the past few years, he's been collecting bugs faster than he can identify them: he's currently still working his way through insects collected in 2020.[2]

Hevel's work clearly testifies to the importance of citizen scientist contributions to the study of insect diversity; he is especially fond of using an app called iNaturalist, a crowdsourced species identification system through which members can post photos of organisms to be identified by experts. The platform also houses targeted projects to identify particular species or larger groups of organisms.

In addition to iNaturalist, there are a variety of other programs available to help young bug lovers, as well as programs to help citizen scientists interested in documenting biodiversity in everything from mites to whales. In October 2023, the Maryland Biodiversity Project registered its 21,000th species (a planthopper native to the American Southwest, very far from home, discovered by Derek Hudgens in Baltimore). This ambitious nonprofit project aims to catalog all living things in the state.

The first formally named *bioblitz* was held at Kenilworth Aquatic Gardens in 1996. Since then, bioblitzes have spread across the globe, and in many cases have become annual events. One such is the annual City Nature Challenge, which happens each April, and in our area pits teams in Baltimore against those in Washington, DC, to see who can photograph the most species in a single weekend. While bioblitzes focus primarily on outreach and education, they can also contribute to research by documenting occurrences of new species to an area or, just as importantly, document declines of once-abundant species.

In addition to these broad biodiversity efforts, there are also projects aimed at particular categories of organisms. HerpMapper reports on reptiles, Frogwatch USA and Frogwatch Montgomery map amphibian diversity and distribution, Firefly Watch tracks firefly species, Ask a Bumblebee looks at bee diversity, TreeSnap documents tree distribution, and FungiQuest covers mushrooms. It should be noted that some species tracked by these projects are also tracked by *poachers*, who can exploit reports of rare species and attempt to remove the organisms and sell them to wild animal collectors or to exotic cuisine and traditional medicine purveyors. Discretion is required when reporting on these target species: rather than report them on iNaturalist, which shares location data with all users, consider using projects like HerpMapper, which shields precise locations from all users except for formally vetted researchers.

Ornithology

Until the early 1900s, family and friends in the United States and Canada would gather during the Christmas holiday season, divide into two teams, and compete in what were known as "side hunts." The idea was to shoot as many birds and small animals as possible. The winning team wasn't the one who harvested the most food; it was the one that piled up the most kills.

In 1900, the ornithologist Frank Chapman, publisher of *Bird-Lore* magazine (later to become *Audubon Magazine*) suggested the world might be better served if folks *counted* birds instead of killing them. And this was the unlikely origin of the Christmas Bird Count (CBC), administered by the Audubon Society and today one of the most popular citizen science events in the world. For the first CBC, 27 birders reported spotting 89 species;

in 2022, at the 122nd annual CBC, over 75,000 participants tallied 2,554 species!

The CBC runs from December 14 to January 5 and is conducted by groups who lay claim to non-overlapping study circles 15 miles in diameter. In recent years, there have been 25 circles in Maryland. You can get more information at the Maryland Ornithological Society website, or go directly to the Audubon CBC site, where you can find the circle nearest you and sign up.

The CBC is just one of six annual bird counts in Maryland. There are also the Mid-Winter Count, the May Spring Migration Count, the Fall Migration Count, the Breeding Bird Atlas, and Hawk Watches.

The number of breeding birds in the state is of particular interest. While over 450 species have been observed in Maryland, fewer than half of them—about 200—build their nests and raise young here. Monitoring the success of these nesting birds is vitally important to assessing the health of bird populations.

The Maryland Breeding Bird Atlas is a five-year tally of nest sites located within more than 1,300 three-mile-square blocks, covering Maryland and the District of Columbia. Project contributors try to find as many breeding birds as possible and report them using the eBird mobile app (there's a dedicated Atlas portal within eBird, with a section for MD/DC observations). There have been two Maryland Breeding Bird Atlases completed so far, for the periods 1983–1987 and 2002–2006. As of this writing, the third (covering 2020–2025) is underway. The Maryland Ornithological Society website has details on how you can contribute.

Marsh birds are indicator species for the health of wetland ecosystems. The Secretive Marsh Bird Monitoring Project, operated by Chesapeake Bay National Estuarine Research Reserve, is our region's partner in the National Marsh Bird Monitoring Program. Survey sites are Jug Bay, Otter Point Creek, and Monie Bay. Participants wake up early and spend several hours on the water at a designated survey point, documenting species presence, total number, call type, and other data.

For less-adventurous bird science enthusiasts, Audubon Maryland/DC operates a whole suite of projects, including the Great Backyard Birdcount,

Project FeederWatch, Nest Box Monitoring, and Hummingbirds at Home. The Raptor Breeding Monitor project at Meadowside Nature Center asks volunteers to commit to walking their trails twice per month from March to June to monitor raptors nesting at Rock Creek Regional Park in Derwood, Maryland.

Eastern bluebirds are threatened by habitat loss and competition from non-native house sparrows and starlings. The Maryland Bluebird Society encourages Master Naturalists to build and monitor bluebird nest boxes, provides detailed instructions and reporting forms, and also runs an annual Bluebird Boot Camp each May. Individual nature centers and farms, like Fox Haven Farm, conduct their own bluebird nest box monitoring citizen science projects and seek volunteers to help.

The Cornell University Lab of Ornithology is a pioneer and national center for citizen science. The Ornithology Lab runs *eBird*, one of the world's largest biodiversity citizen science projects (with over 100 million bird sightings contributed annually!) and has created such programs as the Great Backyard Bird Count, Project Feederwatch, NestWatch, and a fun project (aimed at curious but inexperienced bird enthusiasts) called Celebrate Urban Birds (CUBs). The CUBs program has partnered with 12,000 community-based organizations, distributed more than 500,000 educational kits, and has awarded birding mini-grants. Participants report on 16 bird species that they are likely to find in typical urban and suburban environments. They aim to introduce ornithology to a wider, more diverse audience and work actively with local communities to co-create scientific investigations that serve the interests of the community members.

Tracking the Spread of Invasive Species

Upon arriving in Maryland—after decades spent in Massachusetts—I was immediately impressed by the lush, almost junglelike foliage draped across Sligo Creek Parkway and Rock Creek Park. But I soon learned that these exotic-looking vines and shrubs were (indeed) exotic, and also invasive, and their very robust vitality was suffocating the parks' native flora.

While most threats to our native biodiversity, like habitat loss and climate change, are slow moving, often invisible, and (frustratingly) out of our control, invasive species are right in front of us and tantalizingly susceptible to immediate intervention. As a result, they can excite murderous tendencies in even the most reserved and mild-mannered naturalist. Aptly named *Weed Warrior* programs harness those tendencies in a positive direction, training, arming, and guiding volunteers on sorties to seek out and destroy patches of (among many other things) English ivy, creeping euonymus, wineberry, and porcelain berry. I've seen soft-spoken librarians caked in mud grappling with a deep-rooted wisteria until they uttered a guttural cry of victory and raised the offending vine triumphantly overhead. It can be frightening—but it's righteous work!

While weed warriors' primary goal is restoration of native habitat, rather than data collection, they still provide valuable information on the location and spread of invasive species, especially if they (or their group leaders) are enrolled in the Maryland Department of Natural Resources Statewide Eyes program. This project dovetails with the Mid-Atlantic Early Detection Network, which provides a mobile app to report on invasive species incursions.

In addition to state-based invasive species programs, National Capital PRISM (Partnerships for Regional Invasive Species Management) runs its own weed warrior programs under the auspices of the National Park Service (NPS) for volunteers on NPS lands like Rock Creek Park. They also run the popular Invader Detectives program, which focuses on species detection rather than removal.

Of course, plants are only part of Maryland's invasive species problem. The Maryland DNR lists 58 non-plant invasive species, including feral hogs, nutria, mute swans, northern snakeheads, blue catfish, European green crabs, several crayfish, a half dozen mollusks, five species of beetle, fire ants, and chytrid fungus. Also present in Maryland (but not currently on the list) are spotted lanternflies, jumping earthworms, and two of the most dangerous invasive organisms on Earth: *Aedes albopictus* and *Aedes aegypti*, mosquitoes that are vectors for yellow fever, zika, chikungunya, and dengue.

Mosquito Habitat Mapper is part of the NASA-supported GLOBE Observer citizen science program. Volunteers download the GLOBE Observer mobile app and select the Mosquito Habitat Mapper tool which allows them to photograph instances of standing water and report on whether or not they contain mosquito larvae. Optionally, they can also sample and count the larvae and identify the species. Researchers combine these ground-based observations with satellite-based data to track mosquito infestations and predict outbreaks.

The project has collected nearly 30,000 data points over the last few years. "One of the most significant things that has come out of this is that the photographs that you have been providing to us by using our app have been used by scientists to develop what we call computer vision models," says project lead scientist Rusty Low. "And so we now have an AI classification of mosquitoes. This is really going to be revolutionary."[3]

To monitor and report sightings of invasive animals that are neither dangerous nor widespread enough to warrant a dedicated NASA project, a great many citizen scientists rely on iNaturalist. With a large community of professional scientists who can make authoritative identifications of organisms, and an easy-to-use mobile app for capturing and geolocating photos, iNaturalist is perfectly suited for keeping tabs on introduced and invasive species. iNaturalist members can also create projects that ask volunteers to submit sightings of targeted species in specific regions of interest, a feature that researchers can then use to collect data.

Monitoring and Protecting Threatened or Endangered Species

When the first warm rains of spring dampen the cool earth, frogs and salamanders sleeping underground open their eyes and wriggle their way to the surface. Then, with the singlemindedness of college undergraduates flocking to Fort Lauderdale during spring break, they set off for the stream or pool in which they were born for a mating frenzy. By emerging and traveling en masse, they can overwhelm the appetites of predators that await them; there simply aren't enough hungry birds and foxes to make a dent in their numbers. But their mass emergence does little to protect them from another killer: speeding cars that crush them by the thousands on stream- and pond-adjacent roads.

Fortunately, in many parts of Maryland, brigades of citizen scientists also emerge from their cozy quarters on these nights. Alerted by weather-watching project coordinators, they gather at well-known crossing points with flashlights and traffic cones to slow traffic and shepherd the amphibians across the roadways. Some neighborhoods allow street closures and detours during these times.

In addition to an active salamander brigade program, Jug Bay Wetlands Sanctuary in Lothian, Maryland, also hosts one of the longest-running eastern box turtle research projects in the country. Every year they affix transmitters to four turtles and track them each week until they begin their annual hibernation.

"We have been learning an incredible amount of information about how far the home ranges of box turtles are, how long they live, and how often they're able to successfully mate and have successful nests as well," says Liana Vitali, stewardship and citizen science coordinator at Jug Bay.[4] In addition, every box turtle found in the 300-acre sanctuary gets a *notch code*, a unique pattern of small, V-shaped notches on the outer edge of their carapace, or upper shell. "Everyone from visitors to school groups to volunteers, whenever they encounter a box turtle here at the sanctuary can report that notch code to us and their location and we can keep track of all these encounters and understand the health and the vibrancy of the population that we have here at the sanctuary," Vitali says.

While citizen scientists at Jug Bay work in a vibrant group of over 200 active volunteers, other citizen science pursuits are far more solitary. Fourteen-year-old Roshan Vignarajah has been active in entomology since he was 11, and in recent years has focused his attention on fireflies. Even though fireflies are some of the most charismatic and, well, flashiest of insects, they are poorly known. "There are 2,400 species *described*," he says. "No one knows how many species there *are*."[5] How many species of fireflies are in Maryland? Somewhere between 45 and 100.

One reason for the lack of specificity is that firefly species are so difficult to distinguish. Many species are morphologically similar, with only tiny differences in their coloration, head shields, wing cases, antennas, and mouth parts. It is their unique *flashing patterns* that reveal their identity to potential mates—and to keenly observant humans.

If there is uncertainty about these insects and their status in Maryland, it's not due to any lack of effort on the part of Roshan and his entomologically inclined friend Aaron. Aaron discovered the state's first-reported population of the critically endangered species *Photuris bethaniansis*, and Roshan conducted research forays to the site to study the population. Recently, Roshan documented the range and population size of the difficult-to-identify *Photuris eliza* on the Eastern Shore. Currently, he and Aaron are co-authoring scientific papers for the *Maryland Entomologist* and other journals, on the range extension of *Photuris cinctipennis* (a rare firefly found only in Maryland) and the description of a potentially undescribed species they discovered on the Eastern Shore.

Roshan benefits from having parents who are heroically supportive, driving him to remote wetlands in the dead of night, sometimes hours from home, and then waiting patiently while he conducts his surveys. He also consults with professional entomologist Christopher Heckscher at the University of Delaware and amateur entomologist Hess Muse. The data that he and Aaron collect go to the citizen science projects Firefly Watch, Firefly Atlas, and iNaturalist.

The monarch butterfly, *Danaus plexippus*, is another charismatic, familiar—and vulnerable—insect that has become the subject of numerous statewide and national projects. In partnership with the national *Monarch Watch* organization, the Audubon Society of Central Maryland conducts annual butterfly tagging events. There are also state park tagging events at Gunpowder Falls, Deep Creek Lake, Fort Frederick, and Point Lookout.

Citizen scientists who are also anglers can help monitor fish of concern. The Maryland Department of Natural Resources runs a variety of projects, including the Volunteer Angler Survey, where participants report their observations of rockfish, shad, blue crabs, and other target species. DNR researchers also mark target fish species with color-coded plastic tags to help them learn about the species' movement, mortality, habitat use, and growth rates. They then encourage anglers to report any tagged fish they catch and to record the location and size of the fish. The state DNR also runs programs that train anglers on how to tag fish themselves.

Once ubiquitous in the Chesapeake Bay region, horseshoe crabs are now of increasing concern, and the DNR monitors their population through the Horseshoe Crab Sightings project. Volunteers help track the crabs with a mobile app that collects and maps sightings by volunteers. Similarly, the University of Maryland operates the Dolphin Watch project, through which volunteers report dolphin sightings. There are also many national programs that benefit from mobile app–facilitated observations of threatened aquatic species, including manatees, sharks, and whales.

The Smithsonian Environmental Research Center (SERC), based in Edgewood, Maryland, studies a parasite that is depleting populations of the ecologically important mud crab. You can help them monitor these parasites through the SERC Chesapeake Bay Parasite Project.

It's worth repeating here that whenever citizen scientists work to identify vulnerable, threatened, or endangered species, they should always consider using methods (like HerpMapper for reptiles and amphibians) that limit broader public access to their projects' exact locations. Poachers involved in the pet trade, or seeking specimens for exotic foods and traditional medicines, have been known to use rare species mapping projects to find and poach plants and animals with high market value.

Citizen Science in Space

Some of the earliest citizen scientists studied the heavens. In 1847, naturalist and astronomer Maria Mitchell gained international fame for discovering and calculating the orbit of the comet C/1847 T1, which came to be known as Miss Mitchell's Comet. In 1850, she became the first female fellow of the American Association for the Advancement of Science (AAAS), and—despite having no college education—became professor of astronomy at Vassar College, director of the Vassar College Observatory, and editor of the astronomy column of *Scientific American*.

Citizen scientists continue to make important astronomical contributions, including a recent discovery of a near-Earth asteroid, TW-2023, by volunteers at the Daily Minor Planet project. NASA operates over 40 citizen science projects, some looking at Earth-based phenomena like cloud cover and the aurora borealis and australis. Others face outward, looking for exoplanets and Kuiper Belt objects and studying the Martian atmosphere and solar activity.

Community Science and Environmental Justice

While the term *community science* is often used as a synonym for citizen science, it has traditionally referred to the use of scientific methods to address matters of concern to a neighborhood or other local group. Community members generally lead these projects, with or without support from a professional scientist. Community science projects may grow out of local concerns regarding environmental contamination, health disparities, lack of access to basic resources, education opportunities, or through a shared interest in local natural history.

As McKay Jenkins explores more fully in the volume's chapter on environmental justice, city zoning plans often site factories, highways, waste storage, and other sources of pollutants in communities where people of color and those with low incomes work and play. Municipalities also tend to invest less in the protection of natural resources in those areas.

For Master Naturalists, both community science and environmental justice projects are great opportunities to engage with segments of the population outside of our insular community. But the relationship can also be challenging. When citizen scientists enter communities talking about the need for evidence and the collection of data, community members may "hear" something more like "we think you're either ignorant or lying." It is far more effective to collaborate with community members on equal footing, to reach mutually beneficial conclusions through consensus and trust.

In years past, corporate and governmental groups have also been known to bring in scientists and other experts to challenge community claims of pollution or injustice, leading community groups to distrust scientists. And lastly, given a poor track record of explaining their work to community members, professional and citizen scientists can appear self-interested and even callous, more interested in data than people.

To counter these perceptions, it's important for citizen scientists to be transparent, humble, collaborative, and nonjudgmental. It may be true that community groups can learn natural history and restoration practices from citizen scientists. But it is equally true that citizen scientists can learn about historic and systemic environmental justice issues from the people they are seeking to support. Collaboration works both ways. Working together on community science or environmental justice projects builds trust, spreads good will, and—in the end—proves far more effective and enlightening for all involved.

As Coreen Weilminster explains in this volume's climate communication chapter, Master Naturalists can learn more effective ways to talk about science to a wide variety of audiences. After all, no matter how welcoming and friendly group members initially are, it won't take long for visiting outsiders to feel like they've stepped into a strange, inexplicable world. "Who's Aldo Leopold? Isn't taxonomy all about taxes? One more time: tell me why we're killing all the nicest plants in this park? A bird has more than one song? Why does everyone love that worm but hate that other worm? Does everyone else here know Latin?"

Much of our energy as naturalists goes toward protecting the natural world from threats like climate change, habitat loss, and pollution. Those efforts can't be successful without the support of the larger community. So, how do we maintain our close community of like-minded souls while also broadening our appeal?

A useful first step is acknowledging that all people—everywhere and in all times—have a fundamental right to a healthy environment. All of us—whether we live in cities, suburbs, or rural areas—have our own relative degrees of ecological ignorance and cultural blind spots. Like all effective educators, Master Naturalists must seek to build on points of common interest. For example, everyone wants clean air and safe spaces for their children to play, so enhancing public green spaces might be a place to start. Local flooding inconveniences many people, so communities can be encouraged to protect and expand rain-absorbing forest patches and to build rain gardens. Additionally, there is a strong and widespread perception that science, technology, engineering, and mathematics (STEM) education helps children succeed, so parents may support extracurricular activities that are science focused, and these can include citizen science projects.

Most of all, it's important to engage in a positive and compassionate way with people who may have different feelings about (and experiences in) the natural world. In her book *Black Faces, White Spaces: Reimagining the Relationship of African Americans to the Great Outdoors*, Carolyn Finney writes that to some people, the "great outdoors" may be a place of peace and freedom and an escape from the noisy bustle of the city. But for others—especially those whose ancestors lived in fear—a forest can be a lonely, frightening place. Finney quotes from an interview she had with veteran National Park Service executive Bill Gwaltney, whose father told him, "There's a lot of trees in those woods, and rope is cheap."[6]

Unique Opportunities for Master Naturalists in Citizen Science

There are many opportunities for Master Naturalists to contribute to citizen science projects in capacities above and beyond those offered by regular volunteers. Reliable data provides the foundation for every successful citizen science project, and trained Master Naturalists have the basic science knowledge and skill set to deliver it. Even if the project is in a field unfamiliar to you, your input can be useful. In addition, it's relatively easy for Master Naturalists to multitask their data collection activities: an ornithological excursion can also be the perfect setting to note phenological phenomena for Nature's Notebook, or to collect and submit observations of non-avian species to iNaturalist. Any hike can yield magnetic field data for CrowdMag or litter information for Debris Tracker. Volunteering in this way also widens your knowledge of natural history beyond your primary discipline.

Another way to participate is to volunteer to help the scientists who oversee citizen science projects. Many if not all of these projects need support from volunteers with specialized skills and knowledge, such as those acquired through Master Naturalist training. For iNaturalist, there is a need for people who are able to inspect photos and confirm or reject the identifications offered at the site. For bioblitzes, bird counts, and other excursions, project leaders often appreciate having trained assistants who can help guide volunteers, answer questions, and otherwise manage the event. If there's a citizen science project that interests you, and that corresponds with your content knowledge and skill set, consider contacting the lead scientist to see if there's anything you can do to assist them.

Finally, if you have a research question that might be amenable to crowdsourced data or other volunteer contributions, why not create your own project? There are a variety of platforms and tools to help you do it, and most offer toolkits and other help to get you started. These include:

- *iNaturalist* offers one of the easiest on-ramps and is especially useful if you want to collect photos with expert species identification help. Friends of Sligo Creek have a very successful project here.

- *Zooniverse* began as an astronomy-focused platform but now offers a toolkit that allows you to create all manner of natural history projects, with a high level of staff support.

- *ArcGIS* and *FieldScope* are great choices if you need sophisticated mapping tools that allow you to view and analyze data in a number of dimensions.

- *Anecdata* and *CitSci.org* are useful citizen science hosting platforms. Confused about which one to use? There is a one-hour webinar available on YouTube in which representatives from all six of these organizations explain clearly what they offer.[7]

- Another type of data collection tool you can create is a *Chrono-Log Photo Station*. This involves setting up a mobile phone stand where anyone can place their phone, take a picture, and share it on your project site. Over time, you can collect enough photos to document incremental changes at your location.

Ensuring Data Quality

For years, citizen science faced the criticism that data provided by nonprofessionals was unreliable, useless, or worse, detrimental. The field has largely overcome this criticism through improvements in data quality control techniques and by better training of prospective citizen scientists.

Some projects minimize the chances for error through design: the tasks performed by volunteers are so constrained that it's almost impossible to make mistakes. For example, for some *camera trap* studies, volunteers identify anything that is clearly not an animal, like a falling branch or a rolling tumbleweed. So far, people are still better at this task than computers, and volunteers cut down significantly on the workload of the researchers, who can focus on only those images that contain animals.

For other projects, volunteers complete tutorials and gain formal certification to participate. Identifying a bee down to the species level may require a high level of knowledge, but almost anyone can learn to recognize major bee families and genera after just a few hours of training. Motivated volunteers can gain expert-level skill at distinguishing between two nearly identical species of cattail or firefly just by dedicating themselves to this very narrow area of expertise. In addition, through citizen science volunteer work, even people with no background in formal science sometimes discover that they are gifted at tasks very useful in science, like discerning between visually similar items, hearing nearly inaudible differences in sounds, or spotting subtle patterns in data that even experts overlook.

Many projects improve their data quality by relying on big numbers: observations from individual citizen scientists may not be dependable, but data from dozens of people, properly analyzed to eliminate outliers, may rival the accuracy of a single expert scientist. There is a large and growing field of *data quality assessment and control* devoted to finding the optimal number of people required to assure reliable data for a given task.

Maryland Resource Centers

Maryland is home to dozens of nature centers, wildlife sanctuaries, research centers, and other sites that host hundreds of citizen science programs. The Maryland Department of Natural Resources runs and/or participates in projects including the Maryland Breeding Bird Atlas, Statewide Eyes (invasive species management), the Volunteer Angler Survey, Wildlife Crime Stoppers, and the Horseshoe Crab Sightings project.

Jug Bay Wetlands Sanctuary offers citizen science volunteer projects on box turtles, butterfly phenology, songbird populations, and much more. Smithsonian Environmental Research Center (SERC) has the Chesapeake Water Watch, Chesapeake Bay Otter Alliance, Environmental Archaeology, and, seasonally, the Salt Marsh Census, Invader ID, Project Owlnet, Bluebird Monitoring, and many more.

Audubon Maryland/DC runs the Important Bird Areas (IBA) Champions project to help spot birds and monitor populations in your area, the Christmas Bird Count, Project FeederWatch, Nest Box Monitoring, Hummingbirds at Home, and the Great Backyard Bird Count. Nature Forward, at Woodend Sanctuary in Chevy Chase (formerly the Audubon Naturalist Society), hosts the Water Quality Monitoring Program, the Taking Nature Black, and Naturally Latinos annual conferences, and also runs Creek Critters, a project and mobile app to survey aquatic invertebrates.

Libraries are a large and growing resource for citizen science programs. The Citizen Science at Your Library program, run by the national citizen science group SciStarter, provides resources and grants to offer citizen science kits for loan to families, school groups and out-of-school programs, many of them in Maryland. Visit scistarter.org/library to see a list of current libraries or to learn how to add a library to the network.

Tips for Taking High-Quality Species Photos

Many well-meaning naturalists often contribute low-value photographs to iNaturalist, eBird, and other citizen science programs. Attempting to identify the organisms depicted in these images wastes an inordinate amount of expert time. Damon Tighe—a wildlife photographer and veteran iNaturalist contributor with over 80,000 observations, 13,500 species, and nearly 50,000 identifications—delivers presentations on how to take wildlife photographs that will be most useful to researchers who rely on eBird, iNaturalist, and similar image-based data sources.

Tighe reminds citizen scientists to be respectful of the environment and the organisms found there. Follow park or refuge guidelines, and avoid dangerous situations. Also, before shooting in any area, make sure you know what toxic or even deadly things could be in your environment, including plants that you shouldn't touch and animals that you shouldn't approach. There may not be many, but it's important to know about any toxic plants, venomous reptiles, and stinging/biting insects and spiders you may encounter.

As general rules, try to shoot your photographs in shadow, not sunlight. Organisms are easier to identify if their photos show texture and fine detail, and bright light will flatten out these elements. If necessary, use your body to create the shadow. If you are shooting with a mobile phone and bright light is unavoidable, touch your phone screen where you want to focus and drag your finger up or down to darken the exposure to reveal more texture in the object.

Purchase inexpensive clip-on macro lenses for your mobile phone. Tighe says that you need to fill the frame with your object, and for many organisms, that's possible only with the use of a macro lens. "The majority of life is smaller than us; it operates on scales that we are not at." The lenses are very inexpensive, generally under five dollars, and work surprisingly well.

Tips for Taking High-Quality Species Photos (Continued)

One photograph is usually not enough. "Use multiple photographs to distill out some of the essential details that will allow an expert to identify the genus or species," Tighe says. You'll need a shot of the entire organism, the organism and its surroundings or habitat, and close-ups of any defining features.

For plants, try to get a whole organism shot first, showing the plant in its entirety. Then get a habitat shot: step back to show the organism and its surrounding environment. Then a shot of the leaf top and another of the leaf underside. If the background is busy, hold the leaf so that your hand is the background so that the edges of the leaf are visible. Finally, take shots of a flower or fruit, if present.

For fungi, the whole mushroom is the first shot, and the habitat is second. For the whole mushroom, show the underside and the entire stem (or *stipe*), including the base. It's okay to pull the mushroom to show all of it. The third shot is a close-up of the gill attachment, which is the area of the underside of the mushroom where the gills meet the stem. Additional close-ups focusing on the mushroom cap and the stipe could be useful.

For organisms that move, start with a "safety shot": a photo that's not your preferred shot, but the best you can take before moving in close. That way, if the animal gets away, at least you have something. This is especially important with birds and flying insects. For snakes and lizards, take shots of the whole body, the head and the tail, as well as habitat. The location of the scales of the head is a key indicator of species.

For spiders, try to get a shot of their face to show the number and position of the eyes. Jumping spiders will approach the reflective lens of a camera, perhaps because they think the reflection is a rival spider. That will allow you to get a shot of the entire spider, filling the frame, facing the camera. You will also

Tips for Taking High-Quality Species Photos (Continued)

need a side body shot, to show any eyes on the side of the head. Finally, a rear shot showing the abdomen will reveal descriptive patterning.

For beetles, get a whole organism shot from above, a whole organism side shot, and a close-up of the mouth parts if possible. Plant-climbing beetles will often fall off the plant when approached, so put your hand or a paper plate underneath the beetle so it doesn't fall into the undergrowth. For flies, snap a full organism top shot showing wing veins, a side shot with haltiers, which are the small gyroscopic structures sticking out from under the wings, and a head shot if possible. For butterflies, the wing underside is most important; for moths, the top side is most important. Getting both top side and underside is ideal.

For land snails, take a whole organism shot from the back showing shell spiral, another shot from the other side of the shell, and a side angle that shows the width and tilt of the shell. For bivalve mollusks, shoot the whole shell from the top and the inside, plus an edge or profile shot. For fish, you need just a shot of the whole organism with all fins visible and the habitat. For birds, the head shot is most important, but whole body dorsal and ventral views are also very useful.

Thanks to these efforts, citizen science projects now routinely lead to studies published in prestigious peer-reviewed journals like *Science*, *Nature*, and *Cell*; data acquired through the portals eBird and iNaturalist have led to over 100 publications for each. According to a study published in the journal *PLOS One*, fully half of what is known about migration and climate change has been based on data from citizen science. And a study published in the journal *BioScience* found that about one-third of butterfly field studies have involved citizen science. Amateur naturalists discovered a novel meteorological phenomenon associated with the aurora borealis, have reported numerous near-Earth asteroids, and identified exoplanets orbiting distant stars. Clearly, data gathered by citizen scientists is making a significant and growing contribution to scientific research.

As Master Naturalists, we engage deeply with nature. Beyond merely enjoying nature recreationally, we are in *communion*, with a deep appreciation of its importance, complexity, and fragility. This relationship enriches our lives and makes us keenly aware of the challenges ecological systems now face from habitat destruction, climate change, invasive species, loss of biodiversity, and pollution. Beyond merely bearing witness and raising alarms, we can now work directly with scientists who have devoted their lives to diagnosing and remediating these threats. If they are Earth's doctors, we are the army of Earth health professionals that supports them. We can monitor Earth's vital signs and spot early warnings of disease. We can take the pulse of the planet. We can, and must, be citizen scientists.

Notes

1 We use the term *citizen science* here because as of this writing, it is still the most commonly used term referring to this practice. While in this case the word *citizen* is used in a general sense, referring to a member of a group with shared purpose and responsibilities, there has been concern that the word might sound exclusionary to people who aren't US citizens. The Citizen Science Association has already changed its name to the Association for Advancing Participatory Sciences, and many other groups have replaced "citizen science" with "community science." However, the term *community science* has already been claimed by groups of people, often people of color, working together in neighborhoods, generally on matters pertaining to environmental justice, and many of them are unhappy about that term being usurped. For a well-reasoned discussion of this topic, see M. V. Eitzel, Jessica L. Cappadonna, Chris Santos-Lang, Ruth Ellen Duerr, Arika Virapongse, Sarah Elizabeth West, Christopher Conrad Maximillian Kyba, et al., "Citizen Science Terminology Matters: Exploring Key Terms," *Citizen Science: Theory and Practice* 2, no. 1 (June 2017).

2 Email correspondence with Gary Hevel, November 15, 2023.

3 Video interview with Rusty Lowe conducted for the *SciStarter* podcast, recorded on May 24, 2023, in Tempe, Arizona, at the C*Science Annual Conference.

4 Online interview, November 3, 2023.

5 Online video interview for *SciStarter* podcast, November 3, 2023.

6 Carolyn Finney, *Black Faces, White Spaces: Reimagining the Relationship of African Americans to the Great Outdoors* (Chapel Hill: University of North Carolina Press, 2014), 118.

7 "SciStarter LIVE #26: Platforms and Resources for Citizen Science Project Leaders," posted February 21, 2023, by SciStarter, YouTube, https://youtu.be /q6SUtz4mrPg?si=eXdGi-eITahNdh85.

CHAPTER 16

Climate Communication

COREEN WEILMINSTER

More than 60 years ago, Rachel Carson published an article in the *New Yorker* magazine about a mythical town in the heart of America, "where all life seemed to be in harmony with its surroundings." Dense with glorious flowering plants and forests ablaze with color, the mythical town was famous for its rich array of fish and animals and—especially—"for the abundance and variety of its bird life." But then, ominously, a "strange blight" of toxic, man-made chemicals spread across the land, and the people in the town awoke to a "silent spring."[1]

Carson's mythic opening to what became the 20th century's most famous piece of science journalism alerted the reading public to the dangers of industrial toxins. Her writing led to congressional hearings and a major overhaul of federal environmental legislation and prompted the passage of the Toxic Substances Control Act of 1976, the law that regulates chemicals and requires safety testing of any new chemicals. Rarely has there been such a vivid example of the power of scientific storytelling.

Today we face the rise of an even more ominous threat. In September 2018, I drove to City Dock in historic downtown Annapolis to pick up my teenaged daughter from her evening shift at one of the restaurants on Ego Ally. Our family had gotten into the habit of checking the tides before we dropped off or picked up from those work shifts. But on this particular night, the tide levels were pretty high. I texted my daughter to let her know I was on my way, but that she would need to meet me a block or two away, since the parking lot was likely flooded. The street was indeed closed to traffic, and some of the businesses had already started sandbagging in anticipation of the high tide flooding. Rising water: this is now a regular part of owning a business in the heart of our state capital during our era of climate change (see figure 16.1).

Maryland has 3,190 miles of shoreline, a fact that makes our coastal residents particularly vulnerable to the consequences of sea level rise and coastal flooding. In addition to the chronic flooding events my family routinely navigated while my daughter worked downtown, storm surges, erosion, loss of forest and wetlands, and saltwater intrusion compromising drinking water and agricultural land are forcing coastal communities across Maryland and the greater Chesapeake Bay to adapt to accelerated high tide flooding.

Toxic chemicals in the 20th century and human-caused climate change in the 21st are morally reprehensible environmental issues that have proven difficult to adequately cement in our collective consciousness. In both cases, collecting data simply has not been enough: just as Rachel Carson was savaged by the chemical industry for her work, so have today's climate scientists found themselves the targets of industrial and political attacks for suggesting our lives may be threatened by our own habits.

And so, we look to another means of communication to complement the work of scientists and to help us understand how humans—though the cause of the problem—must also be the solution.

Storytelling

Humans have been sharing stories about environmental changes since we first drew symbols on cave walls: whether depicting animal migrations or the celestial patterns of the sun, moon, and stars, stories captured evidence

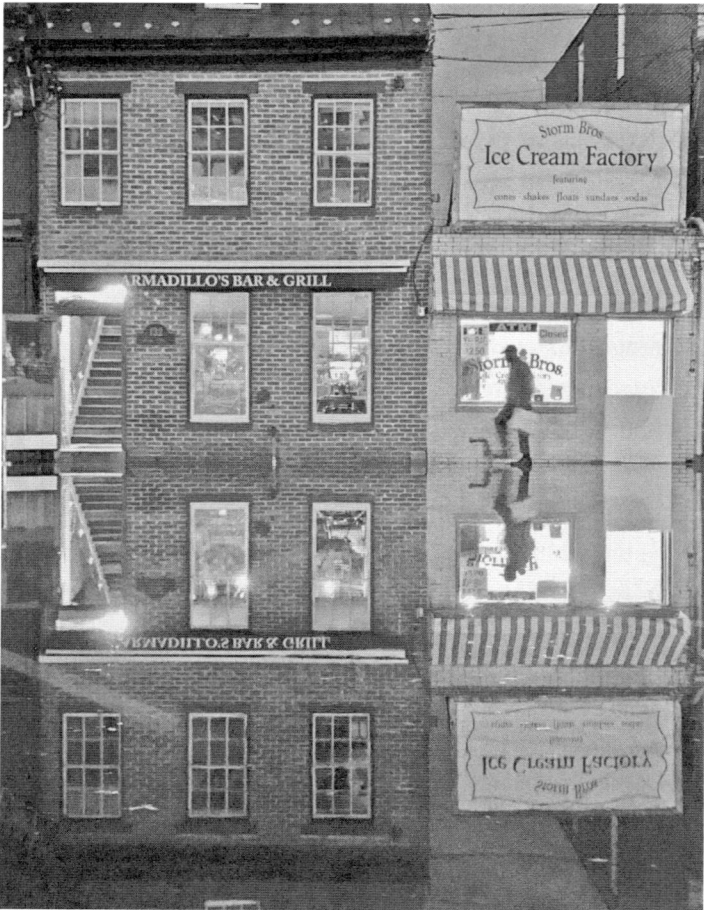

FIG. 16.1. City Dock flooding in Annapolis. Photo by Coreen Weilminster.

of natural changes but also left us insights into our planetary (and species) evolution. Indeed, our ability to represent the world in symbols has long been linked to the beginnings of language and the development of modern human intelligence. Partially as a result of this early art and penchant for recordkeeping, humans have evolved a brain that is wired for storytelling, the recognition of patterns, and the creation of art.

Stories have always helped us make meaning of our place in the world. Decades of research and experience show that stories can shift how people

CLIMATE COMMUNICATION 351

think and feel. They can make important actions feel right, normal, and inevitable. And as we know through Rachel Carson's work, they can even change policies that affect the whole world.

Few environmental issues have been as complicated for scientists as climate change, and how we discuss and interpret the issue can mean the difference between action and ignorance or indifference. But there's a rub: climate change research is based in complex and complicated scientific modeling, so trying to make sense of the issue—and to *communicate* this sense—can be exceedingly difficult, with gaps in our knowledge filled in with noise from rhetorical political positioning. Maryland's Master Naturalist training has added *climate communication* to its syllabus for a number of years. Master Naturalists have the capacity to reach unique audiences and to address natural history through a lens of our changing climate. Indeed, famed climate scientist Katharine Hayhoe believes that the most important individual action one can take to fight climate change is simply to *talk about it*. As a Master Naturalist, your voice is your superpower, and how we talk about it matters.

How Do We Talk about Climate Change?

Master Naturalists often seek to connect with people on an information-based, rational level, which is great for knowledge building, but it may not be enough to encourage action around that knowledge, to actually compel *behavior change*. While *intellectual understanding* is essential to increasing awareness, *emotional connection* adds the powerful pull to *act on that awareness*. Storytelling in science provides a means of inspiring others in a way that appeals to both our logical *and* emotional sides. Remember: people learn from stories from the moment they are old enough to listen. We all know what happened to "the boy who cried wolf," and that no matter how grand a journey we are on, or how far the road may take us, "there's no place like home."

So how do we talk about climate change? How do we bring it into a presentation about natural history, giving it the space and power to generate productive conversations about solutions? Climate change is clearly one of the defining issues of the 21st century, and the situation isn't pretty. Consider: the science of climate—mixing meteorology, climatology, earth science, chemistry, and physics—is complicated. The average adult American has a sixth-grade level of science understanding. And industrial polluters have long used sophisticated political and media operations to bend public opinion away from policy changes that most scientists consider critical to sustaining global environmental and human health.

Making sense of the issue so that we can see our role in its causes and solutions is imperative. As a Maryland Master Naturalist, you are in a unique position to educate, inspire, and motivate others to act on climate. What we say about climate change—and how we say it—affects how people think, feel, and act. You don't have to be a climate scientist to become a skilled and productive climate communicator. Values-based community-level solutions are the real "meat and potatoes" of your communication toolkit, and having reliable scientific resources can set you up for a successful discourse.

This chapter will guide you through the basics of *climate science, regional impacts*, and *strategic framing for productive conversations*. These tools are rooted in social science, basic natural science, and tested communication strategies.

Know Your Audience

GLOBAL WARMING'S "SIX AMERICAS"

Would you ever write a love letter "to whom it may concern"? No. Understanding your audience is essential to good communication. With climate communication, there are tools for that. The Yale Program on Climate Communication research has identified what they call "Global Warming's Six Americas": six distinct audiences within the American public that respond differently to climate information by feeling either *Alarmed, Concerned, Cautious, Disengaged, Doubtful*, or *Dismissive* (see figure 16.2). The six audiences were first identified in 2008 using a national survey of American adults; the survey measured the public's climate change beliefs, attitudes, risk perceptions, motivations, values, policy preferences, behaviors, and

Alarmed   **Concerned**   **Cautious**

Disengaged   **Doubtful**   **Dismissive**

FIG. 16.2. Global Warming's Six Americas, by Michael Sloan. Used courtesy of the Yale Program on Climate Change Communication and the Center for Climate Change Communication at George Mason University.

underlying barriers to action. Of the six, just three types of people—those who are Alarmed, Concerned, or Cautious—are likely to provide opportunities for productive conversations; fortunately, some 72% of Americans fall within these three groups.[2] This is a reason to take heart: there is real opportunity for active engagement in our discussions about climate solutions.

YALE CLIMATE OPINION MAPS

Public opinion about global warming influences how we (and our leaders) make decisions both on policies to reduce greenhouse gas emissions (*mitigation*) and on ways to prepare for its impacts (*adaptation*). The trouble is, American opinions on these issues vary widely depending on where (and how) people live.[3] Scientists at the Yale Program on Climate Communication

Climate Opinion Maps Key Takeaways

- US residents still underestimate the risks of climate change to their families and communities.

- People also underestimate how pervasive the desire for more action is across sectors and actors.

- There is widespread support for policies that reduce carbon emissions and improve sustainability.

have also mapped their survey findings *geographically*.[4] These maps show how people's climate change beliefs, risk perceptions, and policy opinions vary at the state, congressional district, metro area, and county levels. These maps provide an interesting (and useful) look at where Americans agree and disagree, which can shape how (and where) a Master Naturalist might engage in discussion.

Climate communicators can use the Yale Climate Opinion Maps (updated every one to two years) in myriad ways. For example, the maps show where to find points of greatest consensus, which can be a great place to begin climate change conversations. Take policy support: in Maryland, an estimated 82% of adults support funding research into renewable energy sources. Another 82% support providing tax rebates for energy-efficient vehicles or solar panels, and 83% believe schools should teach about global warming. (The maps also highlight topic areas that may be more fraught. Risk perceptions may be one of those areas: only 52% of adults in Maryland believe that global warming will harm them personally.)

More useful (if less encouraging) information: only 33% of adults in Maryland hear about global warming in the media at least once a week, and just 40% discuss it at least occasionally. This data is critical for Master Naturalists seeking to bring climate into their interpretation of natural history. And it brings us to the next point in understanding how to effectively communicate about climate change.

You Don't Have to Be a Climate Scientist, But a Little Science Helps

Productive conversations about climate change are often doomed before they begin. Even basic terms like *climate change* and *global warming* are used interchangeably but in fact have distinct meanings. Similarly, the terms *weather* and *climate* are easily confused, though they refer to events with different scales of time and space. The Yale Climate Opinion Maps confirm that some of the breakdowns in conversations about climate change begin at the definition of "what causes it." Arguments like "Earth's climate has been changing forever" and "Carbon dioxide is a good thing: we breathe it out and plants breathe it in" are true, but they have little to do with the fact that human activities have impacted the delicate balance of Earth's atmospheric composition. Let's explore some scientific fundamentals.

GOLDILOCKS

Earth is what is known as "a Goldilocks planet": it's not too hot, not too cold, it's *just right* for life to flourish. The factors that make it this way include the planet's mass, its distance from the sun, and its atmosphere. As the third planet from the sun in our solar system, Earth exists within "the habitable zone," which NASA defines as "the distance from a star that allows liquid water to persist on its surface—as long as that planet has a suitable atmosphere." In our solar system, Venus, Earth, and Mars are found within that range. But only Earth has a "suitable atmosphere" to support life.[5]

Without water, there is no life. What renders Earth a "water planet" (and the others too hot or too cold to be water planets) has a lot to do with the abundance of heat-trapping gases in each planet's atmosphere. Every planet (and some of the smaller moons) in our solar system has an outer layer of atmospheric gas that functions as a kind of "heat-trapping blanket." A "suitable" atmosphere in the habitable zone is one that doesn't trap too much or too little of the sun's heat energy. The most potent heat-trapping gases responsible for Earth's "suitable" atmosphere include carbon dioxide, methane, nitrous oxide, and water vapor.

Throughout Earth's history, this atmospheric blanket has been trapping heat from the sun that would otherwise escape into space. Natural sources of these gases come from Earth's own cycles: the carbon cycle, the water cycle, the nitrogen cycle, and the methane cycle. With these "natural" greenhouse effects, about 30% of the sun's solar energy bounces off of Earth's surface and goes back out to space. Human activity also generates rampant amounts of these heat-trapping gases. Burning fossil fuels (like coal, oil, and natural gas) increases the concentration of these gases, amplifying the planet's natural warming effect (thickening that atmospheric blanket).

CLIMATE CHANGE VERSUS GLOBAL WARMING

So, what is the difference between *global warming* and *climate change*?

Global warming refers to the long-term heating of Earth's surface—the kind that is currently caused by human activities, especially the burning of fossil fuels. This warming has been observed since the pre-industrial period (between 1850 and 1900), and its documented rise since the early 20th century—and especially since the late 1970s—has been stark. Since 1880, the planet's average surface temperature has risen about 1°C (about 2°F). The majority of the warming has occurred since 1975, at a rate of roughly 0.15 to 0.20°C per decade.[6]

Climate change refers to the much broader range of changes (including global warming) that are happening to our planet. The changes are measurable and clearly indicate that Earth is warming. They include rising sea levels; shrinking mountain glaciers; accelerating ice melt in Greenland, Antarctica, and the Arctic; changes in precipitation (remember, Earth is a water planet, and water vapor is one of the greenhouse gases influenced by excess warming); and shifts in flower/plant blooming times.

To be sure: Earth's climate has warmed and cooled many times throughout its long geologic existence. But today's warming is happening much faster than it has at any point on that long timeline; the past nine years alone have been the warmest since modern recordkeeping began.[7] A couple of degrees may not seem like much, but we now clearly understand that this climate destabilization can have big impacts on the health of all living things and many Earth systems.

WHAT IS NORMAL? UNDERSTANDING SHIFTING BASELINES

The concept of *shifting baselines* describes a form of *generational amnesia* that can profoundly influence how we understand (or underestimate) and manage (or mismanage) natural resources, especially given the changes amplified by a changing climate. Shifting baselines occur when what we experience as "normal" varies radically from what our ancestors considered "normal." For example, you may have been alive long enough to recall rivers or ponds that routinely froze over in winter. Today, a young child may have never seen such a thing. What was "normal" for you—frozen lakes—is considered "abnormal" for a younger person.

Consider that for almost 50 years, our climate has (historically speaking) been "abnormal." As a matter of fact, if you were born after 1976, you have technically never experienced a "normal" year. In the years since then, the annual global temperature has been above the long-term average temperature. So, our *baseline* for what constitutes a normal climate has *shifted*. And as Earth's climate changes, we are seeing extreme weather disruptions across the planet: record-breaking heat, thousand-year rain events, deadly floods, more extreme and intense wildfire seasons, and a more active hurricane season with more intense storms—all becoming more frequent. According to the 2021 Intergovernmental Panel on Climate Change (IPCC) Sixth Assessment Report, the human-caused rise in greenhouse gases "has increased the frequency and intensity of extreme weather events."[8]

WEATHER VERSUS CLIMATE

Despite these shifting baselines, there are still simple ways to distinguish climate from weather. *Weather* refers to atmospheric conditions that occur *locally over short periods of time*—from minutes to hours or days. Rain, snow, cloud cover, sun intensity, winds, hurricanes, tornadoes, and thunderstorms—these are all familiar components of weather.

Climate, on the other hand, refers to the *long-term* (usually at least 30 years) regional (or global) average of temperature, humidity, and rainfall patterns over seasons, years, or decades.

You might be wondering, "If weather and climate are different, how can we attribute extreme weather to changes in Earth's climate?" It's sort of like a baseball player on steroids: if a slugger is going to break home run records, he's going to have to hit a lot of home runs. If he starts "juicing" on steroids, the likelihood of his hitting a ball over the fence increases. So too with our "amped-up" climate: the increase of greenhouse gases in the atmosphere leads to warmer global temperatures. These not only break global temperature records, they also have other impacts, like increases in precipitation intensity and more intense droughts. It's a statistics game, with ever more records, greater severity, and greater intensity.

And just as a slugger taking steroids might hit more home runs but still (occasionally) strike out, a climate warmed by human-generated fossil fuel emissions can produce a string of record highs *even as record lows are still possible*. It's really not about the occasional outlier. It is about extremes.

HOW DO WE KNOW IT'S HAPPENING?

Scientists have been observing Earth and its systems for a long time. Using multiple instruments to collect lots of different types of information, their observations have provided compelling evidence that the planet is warming and the climate is becoming more unstable. Both the National Oceanic and Atmospheric Administration (NOAA) and the National Aeronautics and Space Administration (NASA) point to several long-term observations as indicators of a warming world: global temperatures are rising, as are air temperatures near Earth's surface (the troposphere). So are land temperatures, the amount of atmospheric water vapor (a "feedback" greenhouse gas), and temperatures over the ocean's surface. As it absorbs excess atmospheric heat, the ocean is also getting warmer both at surface levels and at depth. Arctic sea ice is declining, with ice sheets shrinking at Earth's poles (considered to be the planet's "air-conditioning units"). Glaciers are retreating;

snow cover is decreasing; sea levels are rising. Extreme weather events are increasing in frequency. Because our water planet's oceans absorb excess carbon dioxide, not just excess heat, our oceans are becoming more chemically acidic, which (given that acid is a corrosive) is bad news for organisms that use calcium carbonate to make their shells.

Other physical and biological changes confirm the climate is warming, including the *rate* of retreat in glaciers, changes in *phenology* (e.g., the timing of the leafing out of plants and the arrival of spring migrant birds), the growing intensity of rainfall events, and the shifting range of some plant and animal species.

SO, HOW DO WE KNOW IT'S US?

Since systematic scientific assessments began in the 1970s, "the influence of human activity on the warming of the climate system has evolved from theory to established fact," according to the IPCC.[9] But how do *the scientists* know? Looking into Earth's past shows the history of planetary cooling and warming. Scientists measuring carbon dioxide (CO_2) and other greenhouse gases (both in the atmosphere and in gases trapped in ice) have found that *atmospheric* CO_2 is increasing, with levels higher than anything the Earth has seen in nearly *a million years*. Ice cores drawn from Greenland and Antarctica (and even tropical mountain glaciers) show that Earth's climate responds to changes in atmospheric greenhouse gas levels.

But something even more compelling within those ice cores (and the trapped legacy CO_2 samples contained in them) indicates how historic atmospheric changes differ from current ones. Ice core samples are *layered*. Each year's precipitation lays down a layer of ice that can be counted like tree rings. Samples taken in Greenland and Antarctica give us the longest look back into Earth's climate history: Greenland samples go back about 125,000 years; Antarctic samples (from the Vostok ice core) go back over 400,000 years.

There is strong correlation between the two sites. As the ice hardens each year, tiny bubbles of air are trapped. Each bubble is a time capsule, a tiny sample of air from the past, from which scientists can isolate carbon dioxide

(CO_2) and examine it at a molecular level. Because CO_2 has a ratio of carbon isotopes (carbon 12, carbon 13, and carbon 14) that acts like a fingerprint, scientists can measure the isotopes in the CO_2. CO_2 from natural sources has more carbon 14; CO_2 from burning fossil fuels does not contain carbon 14. Ice core CO_2 samples have an abundance of carbon 14. Current atmospheric CO_2 samples contain significantly less carbon 14, which indicates clear evidence of the presence of gases released from fossil fuel burning.

Maryland Is Seeing Impacts from Climate Change Every Day

As Maryland Master Naturalists, it's critical that we understand—and communicate about—the impacts scientists are tracking in our own state. The Chesapeake Bay and its waterways support Maryland's economies (fisheries, tourism, agriculture), offer recreation (beaches, scenic views), and provide essential habitats and vital ecosystem services (like water filtration, bird flyways, and nursery grounds for countless species). For all these reasons, scientists, decision-makers, and many average Marylanders seek to understand how changes in land use and climate will affect the ecosystems—and the connected economy—of the Chesapeake Bay.

In 2015, managers, researchers, and educators from NOAA, the University of Maryland Center for Environmental Science, the Chesapeake Bay National Estuarine Research Reserves (CBNERR) of Maryland and Virginia, and Chesapeake Environmental Communications (CEC) got together to investigate this topic. Through an analysis of long-term climate data from 1895 to 2014 from the National Weather Service and CBNERR locations, scientists examined 26 different indices seeking physical evidence of climate change. There were four key takeaways from this Climate Sensitivity Study, as represented in a set of graphs prepared by CEC (now Green Fin Studio) in 2015.

GROWING SEASON LENGTH

Growing season length (GSL) is the number of days in a calendar year that plant growth can occur. The study found that the length of Maryland's growing season has increased by *more than 30 days* over the last century.

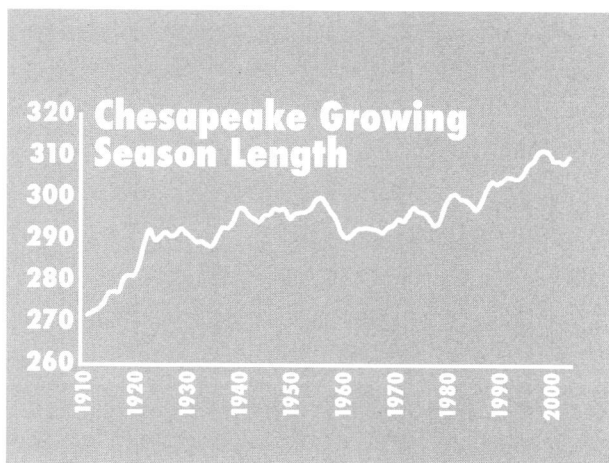

FIG. 16.3. The length of the growing season has increased by more than 30 days over the last 100 years. From K. A. St. Laurent, V. J. Coles, and R. R. Hood, "Climate Extremes and Variability Surrounding Chesapeake Bay: Past, Present, and Future," *Journal of the American Water Resources Association* 58, no. 6 (2022): 826–54. Courtesy of Green Fin Studios.

The *onset* of growing season is occurring earlier and is ending later than in the past. And changes in the Chesapeake have been much bigger than those observed over the whole East Coast. This means that trees, shrubs, and crops have *an additional month each year* when temperatures are warm enough for them to grow. The GSL index for the Chesapeake Bay is telling us that the region is getting warmer (see figure 16.3).

FROST DAYS

Frost days is a *cold event index*. Specifically, it is the number of days in a calendar year when daily minimum temperature is less than 0° C (32° F). There has been a clear *decrease* in the number of frost days each year: we have *30 fewer frost days* per year than we did a century ago (see figure 16.4). Among other things, this may mean that agricultural pests and vectors of disease (like ticks and mosquitoes) do not die off in winter. Warmer winter water also increases the likelihood of expanding potentially harmful vibrio, a bacteria that is dangerous for humans.

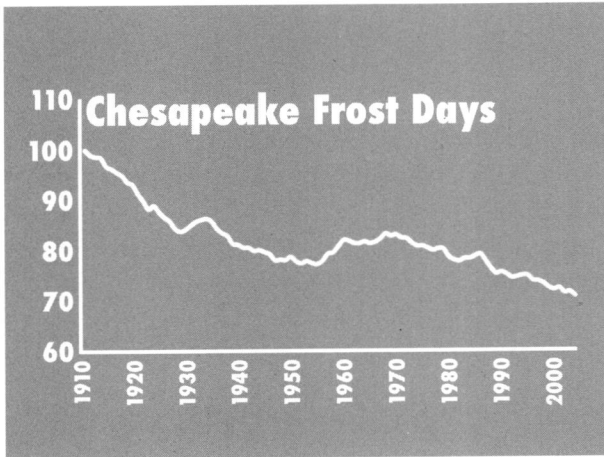

FIG. 16.4. Frost days. The Chesapeake Bay experiences 30 fewer frost days than it did 100 years ago. From St. Laurent, Coles, and Hood, "Climate Extremes and Variability." Courtesy of Green Fin Studios.

TROPICAL NIGHTS INDEX

The term *tropical nights* refers to the number of days in a year when the minimum temperature does not go below 68° F (20° C). Comparing the early 1900s to the early 2000s, there are about *30 more tropical nights annually* throughout the Chesapeake region. And four out of five years with the largest number of tropical nights occurred within the last two decades (2005, 2010, 2012, and 2014; see figure 16.5). There are 30 more warm summer nights (>68 °F) per year now than 100 years ago, which may decrease crop yields: more frequent warm nights create ideal conditions for mildew and fungus to grow and can negatively impact sensitive aquatic organisms in the Bay, such as submerged aquatic vegetation.

Warm nights are also a major cause of stress to both ecosystem health and human health. When temperatures don't drop much overnight, ecosystems (and people) aren't able to recover from the extreme heat of the daytime. Chronic stress can lead to increased risk of illness and disease. This can be especially risky for vulnerable populations like the very young and elderly, individuals with chronic illness, people who live in treeless "urban heat islands," and outdoor workers.

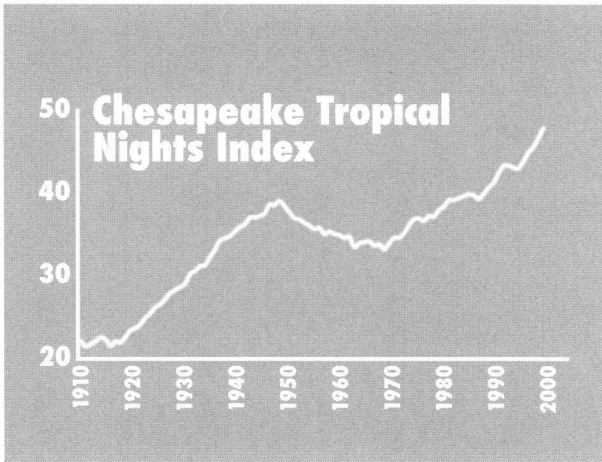

FIG. 16.5. The Chesapeake Bay experiences almost 30 more tropical nights a year than it did 100 years ago. Tropical nights are ones where the minimum temperature is higher than 68° F. From St. Laurent, Coles, and Hood, "Climate Extremes and Variability." Courtesy of Green Fin Studios.

TOTAL PRECIPITATION

The *total precipitation index* reflects the total amount of precipitation (measured in millimeters) that falls in a year. The measurement is made on *precipitation as water*, so precipitation falling in other forms (such as snow, sleet, or hail) is melted before it is measured. It appears that the amount of precipitation that the Chesapeake region receives each year is increasing (see figure 16.6). Since 1900, the Chesapeake has received 5.2 to 16.8 mm (0.2–0.7 inches) more precipitation each decade, about a 12% increase. The Chesapeake region receives about 4.5 more inches of rain per year than it did a century ago. Increased precipitation could potentially lead to increased nutrients and sediments washing into the Bay. More freshwater influxes to the Bay affects natural processes like spawning conditions for certain fish and germination thresholds for various species of aquatic vegetation. More rain can also mean better breeding conditions for mosquitoes, which can carry disease.

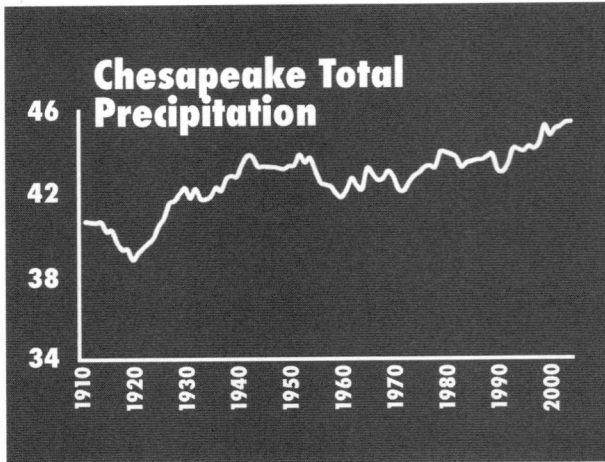

FIG. 16.6. This graph shows the total amount of precipitation, as water, in the Chesapeake Bay region each year. From St. Laurent, Coles, and Hood, "Climate Extremes and Variability." Courtesy of Green Fin Studios.

The results of this sensitivity study equip both scientists and planners with ways to consider how these trends may affect regional phenology. *Phenology* is how we study biological life cycles like migration, spawning, germination, and hibernation over time. These cycles are known to be triggered by environmental cues like sunlight, precipitation, and temperature. When the thresholds for these cues are disrupted with climate trends and extremes, it can affect everything from a species' first appearance, metamorphosis, breeding, overwintering or winter hibernation and torpor, and migration. For example, earlier warming in the spring might prompt plants to flower earlier, which may mean insects emerge too late to pollinate them. Or if insects emerge earlier than usual, migrating birds might arrive too late to feed on them. And on it goes: disrupted cycles disrupting other cycles. Most of us will notice it in our gardens, as hardiness zones change, and with recreational activities like hunting and fishing, whose regulations and open seasons are based on these cycles.

Other impacts not addressed in the Climate Sensitivity Study but considered among the greatest economic and human health threats in Maryland include the following:

SEA LEVEL RISE

The University of Maryland Center for Environmental Studies (UMCES) recently released its 2023 report on sea level rise. This report is issued every five years so that our civic leaders and private landowners can anticipate and plan for changes in flooding frequencies, marshes, shorelines, and salt-water intrusion. As modeling techniques improve, and confidence in the science increases (and whether or not we curb our rampant consumption of fossil fuels), the rates of change will adjust accordingly.

The report offers critical information for vulnerable coastal cities like Annapolis and Baltimore as they scramble to protect sensitive infrastructure. Smaller communities with working waterfronts, farms, and recreational economies are also heavily impacted. At the time of this writing, I am regularly receiving text messages from colleagues and friends who live and work in coastal counties, sharing their high tide flooding stories; from scientists sending me NOAA tide gauge data in awe of the extremes; and from parents whose kids' schools are closed for the next two days because buses can't safely access communities inundated with tidal flooding. Coastal flooding is not a new issue, but the increasing frequency of events is alarming.

The current rate of sea level change is between three and four millimeters per year, which amounts to about a foot over the last century. While this may not seem like much, we may see this again (and then some) by 2050. Another way of thinking about this: during the last century, sea levels rose a little more than one inch every decade. They are now rising by about one inch every five years. For the purposes of this guide, we are going to take a really broad interpretation of this increasingly impressive impact. This is a very complicated issue, and ideally an entire chapter could be devoted to it. What I find to be useful to the Master Naturalist course are the following points:

1. NOAA and USGS scientists measure sea level every few minutes via satellites, coastal buoys, and tide gauges. Even so, there

is no certainty about how fast land ice will melt or how rapidly our oceans will expand as they continue to absorb water released because of the excess heat trapped in our atmosphere.

2. By 2100, scientists anticipate that Maryland will experience a three-and-a-half-foot rise *even if we reach net zero* for greenhouse gas emissions by 2030. If we don't, and we continue with business as usual, that number will look more like five feet. It's important to remember that water encroachment spreads across the land. The greater the rise, the farther inland the impacts. Also worrying: almost one-third of our coastal communities have large populations of people who are socioeconomically disadvantaged, who (by extension) have fewer resources to address flood risk, recover from storm damage, or move to less vulnerable neighborhoods.

3. Maryland's rate of sea level rise is also faster than the global average. Why? Because the world's oceans are not actually "level." Sea surface height is influenced by gravity, ocean temperature, salinity differences, currents, and winds. These collectively cause variation in sea levels around the world. Additionally, Maryland (and the mid-Atlantic) is (slowly) sinking, a result of rebounding from the geologic bulge created from the weight of glacial activity from the last ice age. In parts of Maryland, tidal flooding events have already increased by 178% since 2000.

Remember: water goes where it wants to. Beyond our impact on climate, we have also altered our landscapes in ways that dramatically impact how water moves: into and out of our communities and homes, through reservoirs and pipes, and stormwater drainage networks. As seas continue to rise, and precipitation extremes build and become amplified by energy trapped in our atmosphere, these human systems (especially in older cities like Baltimore) are likely to fail.

Because sea level rise brings salt water inland, killing coastal forests, changing the vegetation of (and fragmenting) coastal marshes, as well as making coastal farm soils toxic and un-tillable (called saltwater intrusion), green infrastructure adaptation practices (like living shorelines, beneficial reuse of dredge material, and land acquisition for marsh migration corridors) will be needed rapidly to allow the water to move where it will without risk to our homes, farms, ecosystems, schools, and communities. These practices are already part of the suite of solutions Maryland is using to adapt to climate impacts.

HEAT WAVES

Unusually hot days and multiday heat waves are a natural part of weather. However, as the Earth continues to warm, these hotter-than-usual days and nights are becoming more common. Different than the growing season length and the warm tropical nights indices mentioned above, heat waves are a major threat to human health, disproportionately affect vulnerable communities, and are expected to become more frequent and intense with climate change.

The US Global Change Research Program has also found that the average *heat wave season* is 49 days longer than it was in the 1960s, and that 92% of their study area "experienced a statistically significant increase in heat wave frequency between the 1960s and 2020s."[10] Extreme heat threatens ecosystems, creating stressors that make it challenging for populations and species to handle "normal" threats. For the Chesapeake Bay, extreme heat can impact oxygen availability for aquatic organisms and exacerbate conditions that generate hypoxic dead zones.

As a Master Naturalist, the extreme heat will directly impact you and your audiences during outdoor programs and events in the summer months. It will be essential to care for yourself and your participants/volunteers and be prepared for the heat, knowing that it can impact individuals very differently.

So, given all this ominous news, how do we talk about climate change in a way that isn't terrifying but still activates a desire to do something about reducing fossil fuel use? How does a Master Naturalist discuss wetland plants, mammals, or the amazing adaptations of Maryland's reptiles, amphibians, insects, and birds and weave in messaging about the protection and responsible management of these resources from climate threats? Let's put these questions in a *frame*.

Strategic Framing as a Communication Tool

Even before you enrolled in the Master Naturalist course, people who knew you trusted what you had to say. Every single one of us is a trusted messenger. However, now that you are on track to become a Master Naturalist, you will have many additional tools to use to establish and build that trust with people you meet. We know that one tool, *storytelling*, is already wired into the human brain. But there is another useful tool, *strategic framing*, that involves organizing especially prominent considerations to shape the way an audience understands a particular problem and its potential solutions.

The way a story is told—the selective use of particular values, symbols, metaphors, and messengers—has the power to trigger shared *mental models* people already use to make sense of the world. We all rely on mental models—shortcuts to understanding based on our commonly held ideas shaped over long periods of time that have been informed and reinforced by the trusted messengers in our lives. Mental models play a major role in our decision-making and behavior.

Certain mental models are more helpful than others in how we see and support solutions. We can use *shared values* to tap into those shortcuts that activate more productive ways of thinking about a social issue like climate change. Combined, specific framing tools like metaphors, tone, connecting cause and effect, and openly talking about existing solutions can help us lead productive conversations and steer us away from debate.

So, what makes a carefully (or strategically) framed conversation? The following is excerpted from the National Network for Ocean and Climate

Change Interpretation (NNOCCI), a National Science Foundation–funded project of the Association of Zoos and Aquariums. In early training materials for the project, NNOCCI stressed their belief that, if we train enough voices in proven communication techniques, we can "change the national discourse around climate change to be productive, civic-minded, and solutions-focused."

With funding from the National Science Foundation, FrameWorks (a nonprofit research organization that helps mission-driven organizations build public will for progressive change by using rigorous social science methods to study how people understand social issues) researched more effective ways to communicate about—or frame—climate change. To do this, FrameWorks interviewed experts and members of the public about this issue and compared the differences in understanding between them. FrameWorks also tested how certain values (deeply held ideals in American culture) affected whether people care about climate change. . . . In a nutshell, strategic framing "changes the story."[11] Rather than gloomily repeating that "climate change is real and we're all going to die!" it seeks to leave people feeling a sense of hope, efficacy, and agency about climate change. The strategy uses three essential framing questions: Why does it matter? How does it work? And what can we do about it?

WHY DOES IT MATTER?

What we say matters. *How* we say it also makes a difference. Because our "information age" has produced a time and culture that portrays all sociopolitical and scientific issues as "debates between two sides," a reasonable *tone* indicates that climate conversations can be had by anyone, not just for those who already agree with the point of view being expressed. Avoiding a "crisis tone" allows for nuanced consideration of the problem instead of hopelessness.

Values connect us: Connecting to people's values right off the bat establishes why an issue matters and what's at stake. NNOCCI strategic framing uses two *big-picture values* that lead to productive consideration of climate change:

- *Responsible management:* This value emphasizes that "we are a nation of problem solvers. When we identify a problem, we figure out how to fix it. It is irresponsible for our country to ignore the problem of climate change when we know what needs to be done. We should do the right thing by putting in place a plan for reducing our country's fossil fuel use. Dealing with the problem now rather than letting it get worse is the responsible thing to do. We need sensible programs and policies that minimize current damage and prevent future damage to the marine ecosystem." (Quoted from early NNOCCI training materials)

- *Protection:* This value emphasizes the need to shield people from harm, "stepping in to ensure people's safety." It also underlines the importance of risk reduction, concern for the welfare of others, and preserving habitats. We must protect people and places from being harmed by the issues facing our environment. Jacques Cousteau said it best: "People protect what they love."

HOW DOES IT WORK?

Sticky Messages—Metaphors

Throughout our lives we hear short messages that direct our understanding. "Give a hoot! Don't pollute!" "Only YOU can prevent forest fires!" Do these sound familiar?

Sticky messages help to amplify your voice and cement ideas in people's minds. Generally, if what you say can be remembered and repeated easily, other trusted messengers can repeat your story. Metaphors can be sticky messages, helping bridge the gaps in our knowledge about an issue. They can explain how an abstract, unfamiliar, or misunderstood system or process works by making a carefully developed comparison to a concrete, familiar territory.

Within the NNOCCI strategic framing model, the metaphor that tested with the greatest impact is "the heat-trapping blanket." When we burn fossil fuels like coal, oil, and natural gas, we pump more and more carbon dioxide into the atmosphere, and this buildup creates a *blanket effect*, trapping in heat around the world. The ocean absorbs much of that excess heat, making it warmer too.

Connect the Dots for Context: Causal Chains

Metaphors are most powerful when links are made to something people care about. Clear, concise connections between the underlying problems and visible outcomes are called *causal links*. Making sure that the audience understands how an issue works is a cornerstone of strategic framing. Think of it as "connecting the dots." Clear *causal chains* fill the gaps between an expert and their audience, which can invite and empower the public to think through an issue and how to address it more productively. For example: "As we burn more fossil fuels for energy use, we release an excessive amount of carbon dioxide in the atmosphere. This creates a heat-trapping blanket effect on our planet and increases the global temperature. With higher temperatures, the oceans have warmed and expanded. Over the last century, the waters of the Chesapeake Bay have been rising over an inch a decade. By 2050, levels could rise two feet higher than they are today, affecting homes and businesses with more frequent flooding."

WHAT CAN WE DO ABOUT IT? SOLUTIONS

One thing critical to instilling hope, activism, and agency is that solutions *MUST* match the scale of the problem. While important (and collectively impactful), *individual actions* like changing to LCD light bulbs, biking to work, or recycling your newspaper are simply not enough. *Collective, community-level actions* have a far greater impact. Relating inspiring stories about communities coming together to *act as a group* is key. Exciting projects happening across neighborhoods, institutions, cities, or regions *inspire the "we."* (As in "we are all in this together.") So, ask your audiences: "What do you think *we* can do to fix this problem? What networks or groups are you a part of? What could they do together?"

For example, Project Drawdown is a leading resource for climate solutions whose mission is "to help the world stop climate change—as quickly, safely, and equitably as possible." The group is pursuing three key strategies: advancing effective, science-based climate solutions and strategies; fostering bold, new climate leadership; and promoting new narratives and new voices.[12]

Talking about solutions is actually *more* important and impactful than discussing impacts or climate science. This one key feature of communicating about climate change can help us see ourselves as heroes of the climate change story.

Art as a Solution

Storytelling and strategic framing aren't the only tools in our toolkit. We also have art. Art can also be effective at engaging our hearts and minds. Indeed, one of the best ways to make meaning of climate science data is to visually represent them in a way that tells a compelling story. Often, art succeeds where graphs and tables fail. "Art engenders empathy in a way that politics doesn't, and in a way that nothing else really does," says Lin-Manuel Miranda, the creator of the smash hit *Hamilton*. "Art creates change in people's hearts. But it happens slowly."[13]

Both art and science are born of wonder. When done well together, they can cause us to see the world more profoundly. They can even change our fundamental truths. If science creates changes in the understanding of *our minds*, and art activates empathy, creating a change in *our hearts*, then *science interpreted through art* has the power to affect not just what we know, but how it makes us feel. This connects what we know and how we feel about the issue to our *culture*.

There are many examples of science done well through art. Before photography there were scientific nature illustrations. Today, we have climate scientists like Ed Hawkins creating graphics like "Warming Stripes," and the watercolor work of researcher Jill Pelto, whose climate-inspired art made the cover of *Time* magazine in 2020. There are knitting projects like the Tempestry Project, which uses yarn colors in place of temperature data, and quilting fiber artists who tell data stories in color patterns. Again, climate science has known the connection between fossil fuel burning and the Earth's climate for decades, but that science has not been an accessible part of our culture. In 2005, the journalist and climate activist Bill McKibben argued (in "What the Warming World Needs Now Is Art, Sweet Art") that "although we know about it, we don't *know* about it."

"When people someday look back on our moment, the single most significant item will doubtless be the sudden spiking temperature," McKibben writes. "But they'll have a hell of a time figuring out what it meant to us."[14]

So, what *does* it mean to us? To a climate communicator, it's all about how we tell the story.

As a Maryland Master Naturalist, you will be in a strong position to elevate dialogue on climate action. You are already a trusted messenger and will now be seen as a subject matter expert in Maryland's natural history. Use this status to tell a *better* story about the climate crisis. Weave into it the wonder of Maryland's rich and diverse natural resources. Connect the dots using values and causal chains, and help the public see their role as heroes of the story by equipping them with easy-to-remember mechanisms for explaining just what is driving our current rapid global temperature intensification. Use metaphors to help people understand science. But most importantly, arm yourself with ready (and practical) community-level solutions that might serve as a seed for scaling up our efforts.

The climate crisis affects us socially, emotionally, economically, morally, and ecologically. The Maryland Master Naturalist community has the knowledge and trust to engage in real climate action support. Your training will advance and empower organizations and communities as they continue to plan for climate change adaptation. By actively including climate science (using metaphors like the heat-trapping blanket), values and solutions while connecting the dots between Maryland's abundant natural history for your audiences, Master Naturalists will generate (and elevate) productive conversations that help us all consider our role in Maryland's climate solutions stories.

Notes

1 Rachel Carson, *Silent Spring* (London: Penguin Modern Classics, 2000).

2 Yale Program on Climate Communication, "Global Warming's Six Americas," https://climatecommunication.yale.edu/about/projects/global-warmings-six -americas/.

3 *Global warming* is the term used in the climate opinion map survey questions. The maps and survey have been updated regularly since 2008, when "global warming" was the primary concern. In order for the survey to maintain consistency in data, the term *global warming* is still used instead of the term *climate change*.

4 Peter D. Howe, Matto Mildenberger, Jennifer R. Marlon, and Anthony Leiserowitz, "Geographic Variation in Opinions on Climate Change at State and Local Scales in the USA," *Nature Climate Change* 5 (2015), https://doi.org/10.1038/nclimate2583.

5 "What Is the Habitable Zone?," NASA, https://science.nasa.gov/resource/what-is -the-habitable-zone/.

6 "World of Change: Global Temperatures," NASA Earth Observatory, 2023, https://earthobservatory.nasa.gov/world-of-change/global-temperatures.

7 "World of Change: Global Temperatures."

8 Intergovernmental Panel on Climate Change, *Climate Change 2021: The Physical Science Basis*, Contribution of Working Group I to the Sixth Assessment Report of the Intergovernmental Panel on Climate Change (Cambridge, UK: Cambridge University Press, 2021), https://www.ipcc.ch/report/ar6/wg1/.

9 D. Chen, M. Rojas, B. H. Samset, K. Cobb, A. Diongue Niang, P. Edwards, S. Emori, et al., "Framing, Context and Methods," chapter 1 in IPCC, *Climate Change 2021*, https://www.ipcc.ch/report/ar6/wg1/chapter/chapter-1/.

10 "Heat Waves," US Global Change Research Program, https://www.globalchange .gov/indicators/heat-waves.

11 Susan Nall Bales, Julie Sweetland, and Andrew Volmert, *How to Talk about Oceans and Climate Change: A FrameWorks MessageMemo* (Washington, DC: FrameWorks Institute, 2015).

12 "About Project Drawdown," Project Drawdown, April 10, 2024, https:// drawdown.org/about#:~:text=Our%20mission%20is%20to%20help.

13 Goldie Chan, "10 Pieces of Storytelling Wisdom from Lin-Manuel Miranda," *Forbes*, May 25, 2021, https://www.forbes.com/sites/goldiechan/2021/05/25/10 -pieces-of-storytelling-wisdom-from-lin-manuel-miranda/.

14 Bill McKibben, "What the Warming World Needs Now Is Art, Sweet Art," Grist, April 22, 2005, https://grist.org/culture/mckibben-imagine/.

CHAPTER 17

Teaching and Interpretation

ANGELA YAU

In addition to learning countless ways to help restore Maryland's ecological systems, Maryland Master Naturalists can also harness the power of effective communication and presentation skills. An *environmental interpreter* needs to be an entertainer, a translator, a storyteller, and a revealer of secrets; to be the voice for the voiceless; to bring the past into the present; to connect hearts and minds; and to create a better future for our world.

Since your training includes a final project volunteering for at least one organization, you will almost certainly find yourself communicating with the public. Whether or not your volunteer project is "public facing," you may well engage in developing and offering public programs, displays, or other educational materials. And these skills are useful both within your volunteer work and in your daily life! The name of the training developed to give you these skills is *interpretation*. As with any profession, the foundations of interpretation can be both deep and wide. You could easily make interpretation a lifetime of study and practice!

The process of creating an *interpretive program*, in which you consider every aspect of a project's design, presentation, and evaluation, will broaden your skills and provide you with a fresh perspective on storytelling. Interpretation is a "living" practice, meaning it frequently changes and incorporates new research, social needs, and perspectives from the field. Resources provided at the end of the chapter will be helpful in staying up to date.

What Is Interpretation?

When you think of interpretation, you may immediately think of translation, sign language interpretation, or the interpretation of literature that you did in school. Here we will introduce *heritage* or *resource interpretation*, the kind you will share with audiences as you volunteer. The National Association for Interpretation (NAI) defines interpretation as "a purposeful approach to communication that facilitates meaningful, relevant, and inclusive experiences that deepen understanding, broaden perspectives, and inspire engagement with the world around us."[1]

In the most basic sense, interpretation is about sharing stories: about a place, an ecosystem, a person, a culture, an artifact, or even a creature—any of those things that make up our shared heritage on this planet. The practice is similar to some educational models like Montessori or the awareness to action model of environmental education. Although interpretation also focuses on the objective observation of data and facts, it differs by intentionally connecting hearts *and* minds. When we plan for an interpretive program or display, we both identify the *tangible* aspects of our object of study and use specific techniques to connect our visitors to a story to bring forth its *intangible* aspects.

History of Interpretation

Although the country's first "nature guides" were typically privately hired, by the late 19th century the practice was already inspiring people to join the conservation movement and being adopted by government agencies to enhance the public's sense of a place. John Muir is credited with being the first to use the term *interpret* for what we do. "I'll interpret the rocks, learn the language of the flood, storm and avalanche," he wrote in an 1896 journal entry. "I'll acquaint myself with the glaciers and wild gardens and get as near the heart of the world as I can."[2]

Two other people were instrumental in building a foundation for Natural History interpretation: Enos Mills, born in 1870, was a guide who led trips in the Rocky Mountains. Freeman Tilden, born in 1883, was a newspaper reporter, author, and commentator interested in effective techniques for reaching the public. An interpreter "is a naturalist who can guide others to the secrets of nature," Mills once said. "It is not necessary for a guide to be a walking encyclopedia. He arouses interest by dealing in big principles— not with detached and colorless information."[3]

Tilden was commissioned by the National Park Service in the 1950s to tour and study park communication practices. His *Interpreting Our Heritage*,[4] the first book dedicated solely to defining the profession of interpretation in 1957, remains the foundational work on the subject to this day. He defined six principles of interpretation, which I will elaborate on here with specific examples of how they are used as well as clarifying each with just one word to remember.

Principles of Interpretation

1. *Any interpretation that does not somehow relate what is being displayed or described to something within the personality or experience of the visitor will be sterile.*

    Interpretation draws a connection between the concepts or ideas you're trying to convey and your audience's own lived experiences. This means you, as the interpreter, need to understand your audience and consider ways to tap into their past experiences to help them relate to the content or resource you are sharing. You'll learn a variety of techniques to do this, but the keyword here is ***relate***.

2. *Information, as such, is not Interpretation. Interpretation is revelation based upon information. But they are entirely different things. However, all interpretation includes information.*

Telling a string of facts is not interpretation. We're not only conveying information when we interpret, we are revealing nature's secrets in a dramatic way. Think of it as those "aha" moments that occur when you "get it" or "see it." Effective interpretation gives just enough information for your audience to realize or deduce meaning relevant to their lives. It is not about serving your ego or showing off how many facts you've memorized. Rather, it is about leading audiences through the process of discovery in order to **reveal**.

3. *Interpretation is an art, which combines many arts, whether the materials presented are scientific, historical, or architectural. Any art is in some degree teachable.*

   This principle can be interpreted as: the art of crafting and delivering an excellent program; reading your audience and responding to the energy of the group; balancing the process of getting to revelation; and sharing the meanings of a resource. You can also consider how to integrate art within your program or how to engage all an audience's senses when designing an experience. In all of the ways you craft your program and become the voice of the resource, you can **create art**.

4. *The chief aim of Interpretation is not instruction, but provocation.*

   As Tilden's predecessor Enos Mills said, we don't need to be walking encyclopedias. We want our audiences to leave with an appreciation of nature and a desire to learn more. By inspiring our audiences to action or curiosity, our goal is to **provoke**.

5. *Interpretation should aim to present a whole rather than a part and must address itself to the whole man rather than any phase.*

   We want to both convey the "big ecological picture" and put our local setting into a broader context. And we also need to appeal to an audience's aesthetic, emotional, and intellectual levels. I liken this to our parallel commitments to equity and inclusion: creating a space of belonging where an audience can bring their authentic selves and see themselves reflected in and connected to the story. We aim to think of ecological and personal contexts in order to be **holistic**.

6. *Interpretation addressed to children (say up to the age of twelve) should not be a dilution of the presentation to adults but should follow a fundamentally different approach. To be at its best it will require a separate program.*

    Children are not miniature adults, and interpreting for kids requires a different mindset. If you know you will be working with children, prepare activities appropriate to their level of development. Parents love sharing interpretive experiences with their children, but interpretation for combined age groups can be tricky if you aren't well prepared. But remember: happy children equals happy parents. Also remember that elders may also require different interpretative approaches. We aim to keep programs **age appropriate**.

Tilden's principles were updated and expanded in 2002 by Larry Beck and Ted Cable in their book *Interpretation for the 21st Century: Fifteen Guiding Principles for Interpreting Nature and Culture*, which was again updated in 2011 as *The Gifts of Interpretation*.[5] In addition to updating Tilden's first six principles, Cable and Beck added another nine:

7. *Every place has a history. Interpreters can bring the past alive to make the present more enjoyable and the future more meaningful.*

8. *Technology can reveal the world in exciting new ways. However, incorporating this technology into the interpretive program must be done with foresight and thoughtful care.*

9. *Interpreters must concern themselves with the quantity and quality (selection and accuracy) of information presented. Focused, well researched interpretation will be more powerful than a longer discourse.*

10. *Before applying the arts and interpretation, the interpreter must be familiar with basic communication techniques. Quality interpretation depends on the interpreter's knowledge and skills, which must be continually developed over time.*

11. *Interpretive writing should address what readers would like to know, with the authority of wisdom and its accompanying humility and care.*

12. *The overall interpretive program must be capable of attracting support—financial, volunteer, political, administrative— whatever support is needed for the program to flourish.*

13. *Interpretation should instill in people the ability, and the desire, to sense the beauty in their surroundings—to provide spiritual uplift and to encourage resource preservation.*

14. *Interpreters can promote optimal experiences through intentional and thoughtful program and facility design.*

15. *Passion is the essential ingredient for powerful and effective interpretation—passion for the resource and for those people who come to be inspired by it.*

Key Terminology and Concepts

These key tools and terminology help create effective programs and interpretive media:

Resources: A resource could be as large as the planet or as small as a single tiny object. To "know your resource" is to research its background and history and to understand its cultural, historical, or ecological context. An interpreter aims to research accurate information through valid, objective, and current sources. This does not invalidate important subjective sources of information, such as oral histories, organizational memory, or long-term "site stories." But it is important to validate all information. You may find some "site stories" need to be modified (or left behind) in deference to documented facts.

Tangibles: The "touchable" features of an object or resource rely on senses like touch, color, size, texture, or identification. These features can be measured, analyzed, or quantified. For example, "This is a Northern bobwhite quail, *Colinus virginianus*, that inhabits blended meadow/shrub habitats and is identified by its call of bob-bob-whiiite."

Intangibles: Subjective "feeling" features—emotions; concepts of community and social structure; needs like survival, sustenance, or reproduction—may be inherent in the resource but are usually best revealed by the surrounding *story*. For example, "Bobwhite quail will pair off in the spring to build nests and raise chicks together. Both parents take an equal part in sitting on the nest, distracting predators, and rearing young. It is quite touching to see a quail family in the wild, with both parents wrangling their clutch of youngsters. By fall, territorial behavior and family groups merge with other quail in a covey to survive winter."

Universals: Those intangibles that are special because they transcend language, culture, and social structure to express a universal experience of life. These can be included to connect to everyone in your audience. From our example above, universals would include musing on a bird's sense of family, caring, and survival. These stories may be expressed differently, depending on whether they concern humans or the natural world.

Topic: The subject or informational area of your presentation. You might have a topic in mind, or be assigned a topic by the organization you are volunteering for, but you will need to consider both the tangible and intangible aspects of your topic to begin to reach for a *theme*—which will help you convert a bland "topic presentation" into *interpretation.*

Theme: Successful interpretation must have a theme, or a framework, for the story you will be telling. Once you determine the theme, it will be easy to articulate the message you want to convey and identify clear goals and objectives for your interpretive activity. There are three main steps to identifying your theme:

1. *Select your topic.* This should fit your setting or learning environment. For example, you might choose "Life in a meadow."

2. *Narrow your topic selection.* There are so many aspects to a meadow that it could be overwhelming to present (or for your audience to hear) so you focus on a key species, like the bobwhite quail.

3. *Write your theme statement.* A theme should be a complete sentence (which forces you to have a complete idea); make connections between the tangible and an intangible; and answer the "So what?" question. In museum studies this is often referred to as the "big idea."

In selecting and developing your theme, you should ask: Is the theme important to the site, and will it enrich the visitors' experience? Having an authentic immersion into a meadow will enrich most people's experience. They will see things they may not have noticed before, understand the value of a meadow in a new way, and have fun in the process. Will your audience be able to relate to this theme, and will they recognize the take-home message? If the visitor enjoys birding or hunting, will they understand the importance of the meadow to providing habitat for birds? If they experience firsthand the diversity of the meadow, they can relate to more complex ideas like the food web.

Is this a theme you care about? You need enthusiasm! Do you have the information you need to develop the program? If not, what resources do you need? Only after you have sketched out your theme can you assemble what you'll use to explain your theme. You don't need to shower your audience with facts; a well-developed theme will give the audience a clear, useful insight they can continue pondering long after your program.

> *Audience:* Who is the audience? Will your audience include families, children, people with disabilities, school groups, senior citizens, people who speak English as a second language? Will it be a mixture of these?

One way to engage a variety of audiences is to consider the eight *multiple intelligences*, otherwise known as "the ways people learn." These include *linguistic, logical, visual/spatial, musical, bodily/kinesthetic, intrapersonal, interpersonal,* and *naturalist.*[6] Some people find language naturally helps them understand new ideas; others rely on music or math. Most people draw on a variety of intelligences to make sense of a new situation or concept. If you design your programs to appeal to multiple intelligences, using a variety of methods to get your message across, you will satisfy more visitors because you will have addressed multiple styles of learning. For example, consider using props, sounds, touching, movement, pictures, or writing in your interpretive presentations. All audiences crave authentic experiences, and the sensory components bring these to life! How does the sun filter through the layers of vegetation in the meadow? Close your eyes and breathe deeply—smell the sweet drying grasses and the pungent soil, hear the insects and birds that make up the meadow.

There are great resources available to clarify some of the differences between the ways children and adults learn. Remember, if you aren't sure how to work with a particular audience, get advice from someone who is. Also, keep in mind that when participants arrive, they are often different than you anticipated. If that happens, don't despair! Be flexible and enjoy yourself. Remember, your audience wants you to succeed—because that's how they learn and have fun.

When thinking about your theme and your audience, keep in mind *the learning environment*, the physical setting where the instruction is taking place. Does the location contain any unusual features or hidden mysteries you should mention? What larger concepts, stories, or issues should you connect to your interpretation? Are there any dangers or major changes to alert your audience to? The analysis phase of interpretive planning is often overlooked, but it is the most important way to ensure your interpretive program or product will be successful. If you skip ahead to the design and development stages, you will often find yourself wondering if something will work or whether your activity will make sense. Starting your interpretive planning with a theme, audience, and site analysis is like eating breakfast: it starts you off right.

The POETRY of Interpretation

Perhaps you have seen a natural history program with your family and—like mine—they turn to ask you "was that good?" To answer, try using the acronym POETRY. An effective program should be:

Purposeful: Interpretation should support the mission of the organization or the site where it is being presented and have a clear set of goals and objectives. We can think of it like an *effective* (rather than a *fleeting*) New Year's resolution: your plans will only succeed if you break them into the basic pieces that will get you there step by step. An objective should be specific, measurable, achievable, relevant, and time-bound. Could you offer a brochure, ask a question, or have participants engage in an activity? Ask your audience who can recall one impact on bobwhite quail, or offer a sign-up for the next hiking series to explore other habitats.

Organized: Interpretation should follow a clear thought progression and avoid random tangents. We have all learned how to write a five-paragraph essay in school: we need the introduction ("tell them what you're going to tell them"); the body ("tell them"); and the conclusion ("tell them what you told them"). This model works for interpretation as well: your overall theme should be introduced (perhaps with a catchy "hook" to bring your audience into the program) and restated in the conclusion. Your "body" will be built of sub-themes that support the main theme, and three sub-themes is a great target number. When you plan the body, consider your goals and objectives, and make sure you cover them all. For example, you could have a visual presentation showing the changes to meadow ecosystems in the absence of fire and how these changes impact all aspects of the meadow. This presentation could take place on a walk, in a slide-illustrated talk, or both. In your conclusion, summarize what you have said with a restatement of your theme and perhaps give a call to action. For example, what could your audience do for their children to once again hear the call of the bobwhite?

Enjoyable: Interpretation should be enjoyable or engaging. Not every interpretive program will necessarily make your audience smile; some of the most powerful programs address difficult subjects, engage the audience in discussions, or challenge previously held beliefs. Although you might find this type of interpretation intimidating, don't shy away from exploring difficult topics or using specialized techniques like facilitated dialogue to broaden your audience's perspectives.

Thematic: A well-constructed theme answers a topic's "so what?" while connecting the tangible and intangible in a phrase that can be poetry in its own right. For example: "The call of the bobwhite is a song of survival in the sandhills."

Relevant: Interpretation needs to be relevant to both your audience and the site. In interpretation we put the audience first, wholistically considering their interests and their needs. Consider universal metaphors and how you will help your audience relate to your subject. For example: "A newly hatched bobwhite chick is the size of a golf ball, yet must meet countless challenges every day on its journey toward adulthood."

You: Freeman Tilden said great interpreters need *passion*, so remember: as the interpreter, you are a program's magic ingredient, and you can't fake enthusiasm.[7] Your audience will detect this within 30 seconds. Granted, as interpreters we are sometimes asked to take on projects for which we lack passion. If we are creative in our approach, and truly love the natural world (and the histories that we present), we can always find a way to care about a subject we've been asked to interpret. If you are struggling, find someone who *is* passionate and you will likely find yourself inspired by their enthusiasm.

You also need to do the continual personal development and self-care that it takes to keep your energy and focus with this work. There are many professional development opportunities offering new ideas in the field of interpretation. You can visit other sites and experience their programs (or interpretive media) and talk to staff or volunteers. It is important to spend time relaxing in nature, discovering new places, learning new things, and reflecting at sites of conscience. These practices will not only fuel your creativity, they will also energize you for the work of sharing stories.

Bringing It All Together

You may be wondering about the actual nuts and bolts of how to present a program. In some cases you might give a program only once, but many times you'll be able to repeat the program and hone it into a better version each time. Below is a list of things you should do to prepare for almost any interpretive activity.

- **Check equipment and props.** Check before the program to make sure everything is working properly. This includes items such as a projector, computer, props with moving parts, or any other device you may be using. Even with this step, you may still have problems, but a last-minute check saves frustration, stress, and embarrassment.

- **Arrive early.** You need time to have everything set up before your audience arrives. This gives you time to relax and take a few deep breaths, and also gives you time to greet participants, to answer questions, and to socialize. This sets the stage for a friendly and comfortable program.

- **Start on time.** It is unfair to make those who arrive on time for your presentation wait an extra 15 minutes for the laggards. Be respectful of those who show up on time, and try not to interrupt the flow of the presentation if people arrive late. On a hike, it is a good idea to ensure the first stop is within sight of the start for folks to catch up.

- **Dress appropriately.** Dress neatly but comfortably for the weather and location. Remember that you are trying to build respect for your message, site, and agency or organization. Also, think of your audience. You don't want to be dressed much more casually (or more formally) than they are. If you are tempted to wear historic dress, ensure that it is accurate for the time period, class, and person portrayed and that it supports the theme of the program (see figure 17.1).

- **Use notes appropriately.** It's perfectly fine to refer to notes during your presentation, but avoid reading them. Use them as a backup to help you keep on track or to remind you of facts and figures. Try minimizing notes to bullets or using a prop or a location to remind you of the points you are trying to make.

- **Speak clearly.** Enunciate well and use inflection in your voice. Make sure you face your audience, and use a microphone if you need to. Avoid repetitive and distracting words

FIG. 17.1. Demonstrations can blend nature and history while creating artifacts, as seen here with the author portraying subsistence farm life in 1870 Florida through basket weaving. Photo by Gary Paul.

(ah, umm, cool, OK, like, you know, etc.). The best way to become aware of these is to record yourself doing your presentation for review.

- Whenever possible, use pictures instead of words to make a point. When using projected images always make sure that they are high quality and large enough to see. Again: face the audience while you speak, and don't say "this is a . . ." as you move through each image. To maintain attention, project images for no longer than 10 to 15 seconds. Watch some TED talks to see how slides can be impactful.

- **Use props.** Props help to engage the audience. What you use will depend on your theme and could include natural items, field equipment, books, and even demonstrations. Allow your participants to touch the props if they're not too fragile. Pass objects around to your audience—but wait until the end of your presentation, since their attention will inevitably be drawn away from you when they look at the object.

A few additional things to keep in mind:

- **Safety first!** Safety is very important to consider when taking participants outside. If possible, inform participants before going on a walk about appropriate attire such as hats, sunglasses, sturdy shoes, raincoats, snow boots, or mittens. During tick season, or when encounters with poison ivy or stinging nettle may occur, advise your audience to wear long pants tucked into socks, and enclosed-toed shoes or boots, not sandals. This will help keep them safe from the elements and comfortable on the trail.

- **Be seen and heard.** When talking to a group, try to move to the middle, or to a high point, so everyone can see and hear you. If someone needs to face into the sun, it should be you—and not your audience. Do not wear sunglasses; people want to see your eyes.

- **Provide an overview.** Before the tour starts, tell your audience what to expect, how long the tour will be, if there are any special guidelines for them to follow, or if you will run into potential hazards along the way. Map out the time for them, but be flexible enough to allow for unexpected "teach-able moments." Discuss where (and if) bathrooms are available. Your audience will be more comfortable if they know what to expect.

- **Be a leader.** Remain in front of the group to help maintain control and to prevent losing people.

- **Plan specific stops.** Guided tours should have planned stops along the way that have a clear purpose and connect with the main theme. Most stops should be brief and give you time to gather your audience. Five stops on a guided tour that lasts about an hour is a good rule of thumb. The remainder of your time should be spent on self-directed discovery (giving the audience time to observe what interests them in addition to what interests you) and answering questions.

- **Plan to be spontaneous.** Keep your interpretation fluid enough so you can respond to teachable moments as they arise. If you are teaching about pond invertebrates and an eagle flies overhead, don't ignore the eagle; your participants certainly won't. Do your best to address the spontaneous moment, tie it to your theme if you can ("Eagles eat fish, which depend on invertebrates for survival"), and return to your regularly scheduled program.

Station and Roving Interpretation

Two common ways to interpret a site are by doing *roving interpretation* or by having one or more *information stations*. Roving interpreters move throughout a site or facility answering questions and interpreting a variety of features as visitor interest warrants. While it's more difficult to plan for these situations, there are things you can do to increase the value of the visitor's experience. Roving interpreters should also know where and how to find interesting things in the area, such as an active eagle nest. Be familiar with the site and its stories, and make yourself accessible. Position yourself strategically so visitors can find you.

At an *information station*, your main goal as an interpreter is to make the visitor feel welcome and comfortable and to encourage questions. It's a good idea to have touchable artifacts that relate to your site and that can be used to link the visitor with your location. You should be familiar with the site's notable features, plants, and wildlife so you can guide people and answer questions.

For all these situations, don't be afraid to say, "That's a great question. I don't know, but I can find out!" Ask for help and information from other participants. Offer to help find answers at the end of your program or give ideas on where the questioner could go for more information. Think of good questions as ways for you to grow in your own work that will also help you be prepared for similar questions down the road.

Evaluation

Your agency or organization may have specific evaluation requirements. However, even if it doesn't, program evaluations can be an important part of any interpretive process. You'll find out how people perceived your presentation; you'll probably receive kudos to inspire you to keep doing this kind of thing; and you'll also get feedback that will help you do an even better job the next time. Evaluations can also offer information on new, unanticipated insights the audience gained—which are sometimes the most exciting outcomes of a program!

There are many ways to evaluate your interpretive activities. You may simply decide to ask your audience a few questions verbally about what they learned or enjoyed during your presentation. In some cases, it may be appropriate to give them a few written questions. Commonly asked questions include: What did you most enjoy about this presentation? What was the most important or interesting thing you learned during this presentation? What do you wish had been covered today that wasn't? Do you have any suggestions for improving future presentations? Answers to these simple questions will alert you to the general perceptions of your presentation and will help with future program design. Another easy way to evaluate a presentation is to ask a friend or colleague, who isn't afraid of giving you constructive feedback, to join your audience. Sit down with this friendly evaluator right after the presentation and get their feedback on what went well and what didn't.

Developing Interpretive Products

In addition to communicating with audiences directly, you can provide interpretation through products such as signs, brochures, displays, websites, booklets, exhibits, games, stories, or songs. All of the steps for planning interpretation apply in written form as well.

Signs are connected to a particular location. They can range from a single large sign at a beautiful overlook to a series of smaller signs along a trail. They should be visually appealing, sturdy, appropriate to the setting, captivating, and eye catching. As with verbal interpretation, the best message on a sign is short and concise. A good graphic and a short title may be all that is needed. Often, excessive text detracts from the experience. A good rule of thumb is 50 words or less (a big block of text may make you feel that you are saying everything you want, but research shows that people won't stop to read it). Consider what is important to your site and what the visitor can physically see while standing in front of the sign. If there are many signs located along a trail, connect them by using a common content theme and by using the same visual design.

If you do create a sign with more information than just a label, be sure to use pictures or diagrams to illustrate your point. Include a theme-based title (or large heading) that ties the entire sign together. Visual artists tell us that odd numbers of things are more pleasing to the eye, so you should think about having three columns rather than two, or five pictures rather than four.

Brochures are inexpensive and effective tools for providing information. Just like other interpretive activities and objects, they should have a target audience and an educational objective or theme. Ask yourself who will be getting this brochure, where, and why? A few key tips can make your brochure more attractive, readable, and effective at getting your message across. *Trifold brochures,* which can fit in a standard business envelope, are constructed by folding an 8.5- by 11-inch sheet of paper twice to create three panels on each side. Professional brochure designers recommend using each panel to provide different parts of your educational message.

- The *front panel* should be visually appealing and provide enough content to invite the reader to pick up the piece, open it, and read more. This may simply include an image and title that describes the theme of the brochure.

- In most cases, the *back panel* should be limited to credits and additional contact information. It is also a good spot for a testimonial. This is the panel people are least likely to read, so avoid putting important messages here.

- The *inside three-panel spread* is the body of the brochure. When you open the trifold brochure fully, you have three panels available to develop your theme and accomplish your educational objective.

- The *inside front panel* is where information will be seen when the brochure is opened. It should provide a statement or testimonial that summarizes your message. Brochure designers recommend that you design this panel last. Writing the rest of the inside spread first will give you a better idea of what you want to summarize on the front inside panel.

You should be sure that the brochure is visually appealing and inviting to read. Avoid technical jargon and other forms of communication that alienate readers or obscure your message. Also avoid being long-winded. Short, sweet, and to the point is a good rule to follow. Use only high-quality photographs or images (at least 300 dots per inch) and include permissions and credits for borrowed material and logos.

Sample Program Outline

Your name: Angela Yau

Program title: *(Fill this in last)*   The Vanishing Call of the Bobwhite

Theme: Bobwhites need our help to survive as the sound of the sandhills.

Audience: Families coming to nature center

Goal(s): *(Why you are doing the program)*   Increase awareness of impacts on bobwhite quail and their role in the sandhills.

Motivate visitors to take action for bobwhite quail preservation.

Objective(s): *(What you want the audience to do)*

What is the objective?	How will you measure it?
>50% of visitors will identify at least 1 impact on quail	Ask question
>50% of visitors will identify at least 1 way to help quail	Ask question
>50% of visitors will seek more information	Take brochure *or* Go on hike

Resources/materials needed: Quail, sandhill pictures, fire pictures, quail call, egret and fire ant pictures

---

Program Outline

Introduction: Welcome, intro self and agency, housekeeping, state theme (call over pic of sandhills)

These beautiful longleaf sandhills are an endangered ecosystem with less than 2% of the original area remaining. The call of the bobwhite is now vanishing unless we all help these little birds with the big voice.

Body (subthemes):   Quail are unique little birds with a big voice that is known as the "sound of the sandhills." (picture with call)

Quail suffer from many natural and human impacts (egrets, fire ants, hunting, lack of fire).

Humans hold the key to quail recovery (support controlled burns, ground feeding, control fire ants).

Conclusion: Does anyone remember one of the things that impact quail? How about one thing you can do to help? What will you do to keep the call of the bobwhite ringing clear throughout the sandhills?

Offer brochures on quail conservation, invite them to follow me out on a hike to "hunt" for quail calls in the sandhills in old and recent burned areas . . . introduce attendees to a quail if we have one available.

Activity: Interpretive Program Outline

A *program outline* is a useful tool for developing your program. This tool comes from the National Association for Interpretation and is used with interpreters from all fields. Many agencies and organizations developed their own variation of this outline to include details needed for their particular missions and goals. This is an excellent resource to copy and use for any presentation that you will deliver or create on your own on the computer with added details or reminders as needed.[8]

Learning by Doing

Combining interpretation with community service projects that benefit the environment can help participants learn, build skills, and have a positive hands-on experience. These projects can give groups and individuals the satisfaction of fostering stewardship of the environment. As Bob Hirshon writes in his chapter on citizen science, there are countless kinds of service projects, and your work as an interpreter can add considerably to these projects. Interpretation plays a part in inspiring a group to take on a project as well as sharing the results—a school soliciting community support to install a pollinator garden needs to share the possibilities to gain that support and then to share the success stories of impacts to wildlife, schoolyard habitat, and learning outcomes. Once a project is completed, be sure to evaluate yourself. Reflect on the experience with the group. Record observations, ask how the project could be improved next time, and consider what worked well and what next steps might be. Lastly, put on your party hat! Celebrate your accomplishments. After all, you have made a difference for the environment!

Notes

1 "What Is Interpretation?," National Association for Interpretation, accessed January 10, 2023, https://nai-us.org/interp/nai/_About/what_is_interp.aspx.

2 John Muir, paraphrased in Linnie M. Wolfe, *Son of the Wilderness: The Life of John Muir* (Madison, WI: University of Wisconsin Press, 1978), 144.

3 Enos A. Mills, *Adventures of a Nature Guide* (Garden City, NY: Doubleday, 1920).

4 Freeman Tilden, *Interpreting Our Heritage* (Chapel Hill: University of North Carolina Press, 1957), 9.

5 Larry Beck and Ted Cable, *Interpretation for the 21st Century* (Urbana, IL: Sagamore Publishing, 2002), 8; Larry Beck and Ted Cable, *The Gifts of Interpretation* (Urbana, IL: Sagamore Publishing, 2011). Additional training and certifications are available through the National Association for Interpretation (NAI), www.interpnet.com; and Eppley Institute in partnership with the National Park Service, www.eppley.org. Some free tutorial courses are available online through Eppley Institute as well. Many organizations host regional and national meetings and workshops that are excellent places to learn more and network with fellow interpreters. You can also find regional opportunities through international discussion groups via social media.

6 Howard Gardner, *Frames of Mind: The Theory of Multiple Intelligences* (New York: Basic Books, 1983).

7 Tilden, *Interpreting Our Heritage*.

8 Program Outline Form from the National Association for Interpretation. Sample outline by author.

CONTRIBUTOR BIOGRAPHIES

MCKAY JENKINS is the Cornelius Tilghman Professor of Environmental Humanities at the University of Delaware. He is the author of many books about the natural world and social justice, including *The Delaware Naturalist Handbook*, *Food Fight: GMOs and the Future of the American Diet*, and *ContamiNation: My Quest to Survive in a Toxic World*. In addition to his research and teaching, he is active in a number of urban restoration ecology projects and helps oversee Baltimore's Rock Rose Food Justice Project. He has won the University of Delaware's Excellence in Teaching Award and both the Outstanding Teacher and Outstanding Scholarship Awards from the College of Arts and Sciences.

JOY SHINDLER RAFEY has administered the Maryland Master Naturalist Program since 2013 and is a Certified Maryland Master Naturalist herself. She holds a master's degree in recreation and parks management from Frostburg State University. Her research has focused on park-based environmental learning and volunteer stewardship as recreational pursuits, as well as assessing the impact these volunteers have on the facilities and organizations hosting the program. In 2024, she was presented with the Dean's Diversity,

Equity, Inclusion, and Respect Standard of Excellence Award to Faculty of UMD's College of Agriculture and Natural Resources for her efforts to make the Maryland Master Naturalist Program more accessible and build community engagement throughout the state. As a longtime board member and 2024 president of the Alliance of Natural Resource Outreach and Service Programs, she supports efforts to establish and sustain Master Naturalist–type programs throughout the country.

RAYMOND V. BOSMANS is professor emeritus at the University of Maryland, a past president of the Mid-Atlantic Turtle and Tortoise Society, and a graduate of the University of Maryland College of Agriculture in horticulture with a major in landscape design and master's degree in extension education with a minor in entomology. Since 1973, he has served as a county extension agent in two counties, a regional extension specialist for horticulture, and a regional director (administration) for the University of Maryland Extension. Since retirement, he has served as an adjunct professor at Montgomery College and Howard Community College, where he teaches courses in plant sciences, woody plants, landscape design, and aquatic gardening.

GWENDA BREWER is the science program manager for the Natural Heritage Program, Wildlife and Heritage Service, of the Maryland Department of Natural Resources. She leads conservation planning projects, coordinates revisions of Maryland's threatened and endangered species list, regularly carries out bird surveys, and has led projects to inventory old growth forest and to develop a rapid ecological assessment for wetlands in riparian areas. She also serves on the steering committee for the Maryland & DC Breeding Bird Atlas 3 project, represents the state in several regional bird conservation partnerships, and received the Bird Conservation Award in 2008 from the North American Bird Conservation Initiative and Association of Joint Venture Management Boards. She has taught in the Maryland Master Naturalist Program since its inception. A past member and chair of the Ornithological Council, she currently serves on the Rainforest Trust Advisory Council and the Women for Conservation Board.

LETHA GRIMES has worked with the Maryland Department of Natural Resources, Fishing and Boating Services, for 37 years. Most of those years were working in the Rivers and Reservoirs Program, conducting surveys and monitoring fish populations in the Potomac River from Cumberland to DC and all major impoundment in the state. Currently, she works on freshwater regulations and sport fish restoration grants. She also conducts outreach and youth angler education, administering programs that include the Backyard Fishing Initiative, the Envirothon (an environmental competition for high school students), and Maryland Becoming an Outdoors-Woman. She teaches in the Master Naturalist Program.

J. MORGAN GROVE joined the USDA Forest Service in 1996 and is a social scientist and team leader for the Forest Service's Baltimore Urban Field Station. A co-principal investigator of the National Science Foundation's Baltimore Ecosystem Study since its beginning in 1997, he is a lecturer at Yale University. He is the lead author for *The Baltimore School of Urban Ecology: Space, Scale, and Time for the Study of Cities*, which advances a new school of urban ecology for the 21st century, and coedited a companion book, *Science for the Sustainable City: Empirical Insights from the Baltimore School of Urban Ecology*. He has a BA from Yale College with a dual degree in architecture and environmental studies, an MFS in community forestry from Yale University, and a PhD in social ecology from Yale University.

BOB HIRSHON is a Maryland Master Naturalist, lifelong natural history fan, and founder of Springtail Media. He hosts and produces a podcast, *Citizen Science: Stories of Science We Can Do Together*, for SciStarter, a National Science Foundation–supported hub for participatory science projects. He also produces Science Update, a multimedia science news and information platform, answers science questions from kids on the Sirius XM show *Absolutely Mindy*, and created Kinetic City, a multimedia science channel for children. He reports on science for *Science News*, *Scientific American*, *Discover*, and other magazines.

LUKE MACAULAY conducts applied research and extension (outreach and education) in wildlife management and conservation, with a focus on practical approaches to enhance wildlife habitat and mitigate wildlife problems in agricultural, forested, and grassland/shrubland systems. He worked for the US Department of Justice from 2002 to 2007, first in the Antitrust Division and advancing to the position of spokesperson for the US Attorney's Office in San Francisco. He returned to graduate school to earn a master's degree in range management (grazing lands) from the University of California, Berkeley, and continued on to earn his PhD in rangeland and wildlife management. Upon earning his PhD, he worked for several years as an extension specialist in rangeland planning and policy at the University of California, Berkeley, until starting his current position with the University of Maryland Extension. In December 2023, the Natural Resources Extension Education Foundation elected him as a trustee to its board.

STEWARD T. A. PICKETT is a distinguished senior scientist at Cary Institute of Ecosystem Studies. His research interests include urban ecology, ecology of racialized segregation, landscape ecology, and ecology of disturbance and disaster.

DAVID RUPPERT earned BS and MS degrees in physics from the University of Dallas and Dartmouth College, respectively. He worked as a high school teacher for the Archdiocese of Washington before earning his PhD in soil science from the University of Maryland. Since that time, he has been a faculty member at Texas A&M University–Kingsville, teaching and researching soils, and a faculty member at the University of Maryland in the Department of Environmental Science and Technology. In 2023, he served as president of the Mid-Atlantic Association of Professional Soil Scientists. He currently works as an in-house soils consultant for a branch of the Department of Energy.

MARTIN F. SCHMIDT JR. has been teaching physics, chemistry, and geosciences to grades 9–12 near Baltimore since 1978. He is author of the book *Maryland's Geology*, which is designed to give a basic geologic background and then apply it to the state for students, teachers, and laypeople. He has given frequent talks to Maryland Master Naturalist classes. Through his teaching he hopes to help people develop understanding of and connections to the amazing Earth systems around us.

DOUGLAS TALLAMY is the T. A. Baker Professor of Agriculture in the Department of Entomology and Wildlife Ecology at the University of Delaware, where he has authored 112 research publications and has taught insect-related courses for 42 years. Chief among his research goals is to better understand the many ways insects interact with plants and how such interactions determine the diversity of animal communities. His books include *Bringing Nature Home*; *The Living Landscape*, coauthored with Rick Darke; *Nature's Best Hope*, a New York Times bestseller; and *The Nature of Oaks*, winner of the American Horticultural Society's 2022 book award. In 2021, he cofounded Homegrown National Park with Michelle Alfandari. His awards include recognition from the Garden Writer's Association, Audubon, the National Wildlife Federation, Allegheny College, Ecoforesters, the Garden Club of America, and the American Horticultural Association.

COREEN WEILMINSTER works for the Maryland Department of Natural Resources Chesapeake and Coastal Services Unit, where she is the education coordinator for the Chesapeake Bay National Estuarine Research Reserve in Maryland. She has served in many active roles regionally, including support of the Maryland Master Naturalist training program, the National Park Service's Storytellers of the Chesapeake training program, the Mid-Atlantic Climate Change Education Collaborative, and as a strategic framing facilitator for the National Network of Ocean and Climate Change Interpretation. She is committed to building awareness of climate change science and resiliency within local communities utilizing the best available data and focusing on productive communication surrounding solutions. She has a BS from Penn State University in political science and an MA in environmental studies with a concentration in environmental education from Montclair State University, New Jersey School of Conservation. Coreen is also an artist and published poet and lives with her husband, two daughters, and two Bedlington terriers in Arnold, Maryland.

TIMOTHY WHEELER is associate editor and senior writer for the *Bay Journal*. He has covered the Chesapeake Bay and other environmental issues for most of his career, including during nearly 32 years with the *Baltimore Sun* and *Evening Sun*. He is a former president of the Society of Environmental Journalists and has won many awards, including the 2010 Excellence in Journalism Award from the Renewable Natural Resources Foundation. A native of West Virginia, he grew up eating oysters and followed his taste buds to the Bay. He has worked for the *Richmond Times-Dispatch*; a regional news service in Washington, DC; and the Norfolk *Virginian-Pilot* and *Ledger-Star*.

He holds a BA from the University of Virginia and a master's degree in journalism from Columbia University.

KERRY WIXTED is the amphibian, reptile, and invasive species program manager for the Association of Fish and Wildlife Agencies. She has worked in many different roles, including as a regional biologist and education specialist for the Maryland Department of Natural Resources, Wildlife and Heritage Service, and as a guest lecturer for Nature Forward's Natural History Field Studies program. She has a BS in wildlife and fisheries management from Frostburg State University, an MS in environmental and evolutionary biology from West Virginia University, and a certificate in teaching adult learners from Anne Arundel Community College. She enjoys connecting people to nature through different mediums. In her spare time, she is an amateur photographer, volunteer naturalist, and a guinea pig foster.

ANGELA YAU is the nature program and facility manager at Locust Grove Nature Center with Montgomery Parks and a certified interpretive trainer with the National Association for Interpretation. She earned a master's degree in wildlife ecology and conservation from the University of Florida, where she specialized in human dimensions and interpretive and ecotourism planning. She has worked in interpretation of cultural and natural resources and training staff and volunteers for over 25 years and loves helping people be the voice of the resource. Angela has a passion for sharing the stories that help people to connect, care for, and conserve these resources.

INDEX